Oppenheimer

Oppenheimer

THE TRAGIC INTELLECT

Charles Thorpe

The University of Chicago Press   CHICAGO & LONDON

The University of Chicago Press, Chicago 60637
The University of Chicago Press, Ltd., London
© 2006 by The University of Chicago
All rights reserved. Published 2006
Paperback Edition 2008
Printed in the United States of America

17 16 15 14 13 12 11 10 09 08   2 3 4 5 6

ISBN-13: 978-0-226-79845-5 (cloth)
ISBN-13: 978-0-226-79846-2 (paper)
ISBN-10: 0-226-79845-3 (cloth)
ISBN-10: 0-226-79846-1 (paper)

Library of Congress Cataloging-in-Publication Data
Thorpe, Charles, 1973–
     Oppenheimer : the tragic intellect / Charles Thorpe.
        p.   cm.
     Includes bibliographical references and index.
     ISBN-13: 978-0-226-79845-5 (acid-free paper)
     ISBN-10: 0-226-79845-3 (acid-free paper)
     1. Oppenheimer, J. Robert, 1904–1967.   2. Physicists—United States—
Biography.   3. Scientists—Intellectual life—20th century.   4. Science—Moral
and ethical aspects.   5. Science and state—United States.   6. Atomic bomb—
United States—History.   I. Title.
QC16.062T56   2006
530.092—dc22
[B]   2006015223

FOR ANGELA

The individual event, the act, goes far beyond the general law. It is a sort of intersection of many generalities, harmonizing them in one instance as they cannot be harmonized in general. And we as men are not only the ingredients of our communities; we are their intersection, making a harmony which does not exist between the communities except as we, the individual men, may create it and reveal it.

J. Robert Oppenheimer, "The Sciences and Man's Community" (1953)

CONTENTS

This book traces the life and career of physicist J. Robert Oppenheimer. But it aims, through looking at his life, to analyze more general themes: the shaping of self; vocation; the cultural and political authority of science; charisma; and individual moral responsibility. Framing all of this is the way in which science became, in the twentieth century, a central instrument of violence, transforming the capacity and scope of violence and, in so doing, becoming a vital resource of state power.

Problems of power and violence, in light of the atomic bomb, were central to Oppenheimer's reflections after World War II on the meaning of science. In his 1948 lecture "The Open Mind," Oppenheimer pointed to the paradox that this activity, held in modern culture to be at the polar opposite to coercion, has become perhaps the primary medium of technological violence. A central faith of modernity, and perhaps the core idea of the Enlightenment, was that science and reason offer a solution to the problem of violence. In the middle of the twentieth century, such a view of the social order of science as antithetical to coercion took on particular significance as part of the liberal response to Fascism and Communism. Science, it was said, flourished in, and helped to preserve, a peaceful and free society. The founder of the academic history of science in America, George Sarton, articulated this faith most clearly when he wrote, "Science makes for peace more than anything else in the world; it is the cement that holds together the highest and the most comprehensive minds of all countries, of all races, of all creeds." It was a view that strongly informed the statement by his student, sociologist Robert K. Merton, of the universalistic values that, Merton argued, constituted the normative structure of science.[1]

Yet contemporaneous developments of the twentieth century began to make such ideals of science sound increasingly hollow. Twentieth-century history attests to the intimacy of the modern relationship between science and violence, which has cast a shadow over visions of scientific progress. Modern violence has taken on an increasingly "scientific" character: impersonal, institutionalized, and rationally organized. And science has become integral to the technological sophistication and power of modern warfare. If the nineteenth century saw the mechanization and industrialization of warfare, the twentieth century has been shaped by the scientization of war—a development indicated by the characterization, albeit caricatured, of World War I as the "chemist's war" and World War II as the "physicist's war." America's chief wartime science administrator, Vannevar Bush, called for a "science" of total war.[2] And the scientization of warfare is today reflected in the language with terms such as *smart bombs* and *surgical strikes*. Yet despite the pervasiveness of the modern integration of science and violence, the atomic bombings of Hiroshima and Nagasaki stand out as having particular significance. This has to do not only with the degree of destruction unleashed, but also with the way in which the release of the power of the atom was cast as the high point of scientific modernity.[3]

These interconnections between science and violence raise the problem of the responsibility and role of the scientist. Perhaps the most articulate and complete twentieth-century formulation of this problem was sociologist Max Weber's 1918 lecture "Science as a Vocation." For Weber, the essence of science as a vocation was acceptance of the divide between fact and value and, therefore, eschewing of professional concern for ends or ultimate values. Science, Weber said, serves "self-clarification and knowledge of interrelated facts"—in other words, awareness.[4] Weber insisted on the separation of fact from value so as to preserve the autonomy of science from politics—and to protect science against political violence (which he saw in, for example, nationalist students' disrupting lectures of those they identified as political opponents).

Herbert Marcuse later pointed out the paradox that Weber's insistence on the separation of science from any substantive values makes science more vulnerable to being subordinated to external forces: "Your 'neutrality' is as *compulsory* as it is *illusory*. For neutrality is only real where you have the power to repel interference: if you do not, you become the victim and assistant of any power that chooses to use you."[5] Marcuse's point is particularly significant when one considers problems of technology and the contemporary situation in which sophisticated research and development organizations are in place to rapidly convert scientific findings into military applications. When in modern technological warfare the scientist becomes a servant of state power, that role does not contradict, but is arguably a fulfilment of, the requirements of the Weberian vocation. Weber's ethos of science as a vocation, while defending the life of science against the irrational violence of the political campaign, provides no ethical safeguards against modern scientized violence. Instead, such an ethos of value-neutrality, entailing a discipline not altogether different from that of bureaucratic and military organizations, facilitates the mobilization of science in the rationalized violence of the modern state.[6]

This book examines how Oppenheimer, as wartime leader of the Los Alamos atomic weapons laboratory and as senior postwar scientific adviser to the U.S. government, sought to construct his role and attempted to handle his responsibilities in relation to science, politics, and the problems of warfare and violence. Oppenheimer formulated a vocational ethic close to the one outlined by Weber. Yet the physicist's struggles in his role as atomic bomb scientist revealed ethical problems that ultimately could not be adequately handled in terms of the compartmentalized ethics of vocation. Oppenheimer's moral conflicts and struggles demonstrate tensions in, and the limitations of, the ethics of vocation.

Oppenheimer's struggle with vocation also attests to the difficulty of formulating and maintaining an ethical stance that goes beyond this compartmentalizing ethos. A limited and fragmented ethical orientation is powerfully fostered and maintained by modern, specialized technobureaucratic culture and institutions.[7] A key message of the Personnel Security Board finding against Oppenheimer in 1954 was that scientists overstep the boundaries of their authority when they concern themselves ethically with the consequences of their work. In a democratic society, the board insisted, questions of ends should be left to elected representatives. That bureaucratic and instrumental conception of the scientist's role seems to be the dominant one in modern Western societies. And as science journalist Daniel Greenberg has recently argued, scientists, overridingly concerned with protecting their sources of funding, have become increasingly unwilling to rock the boat by challenging this narrow role.[8]

Adopting a broader conception of the scientific or intellectual responsibility of the scientist means challenging the assimilation of science into the state and its corresponding divorce from civil society. It means questioning the way in which the conception of science as a resource or instrument has overwhelmed and pushed out the place of science within a public conversation constitutive of civil society. In this regard, it is worth mentioning a communication

I recently received from the National Archives. It was in response to an inquiry about the famous report of the Atomic Energy Commission's General Advisory Committee (GAC), which, under Oppenheimer's chairmanship, took a position against the development of the hydrogen bomb. The archivist told me:

> AEC historical document no. 349, the report of the General Advisory Committee to the Atomic Energy Commission, Oct. 30, 1949, re: their 17th meeting, whose topic was the "Super," i.e., development of a hydrogen bomb, had its classification cancelled on March 15, 1954, by the Atomic Energy Commission. On February 17, 1994, it was stamped "classification still retained" by Department of Energy reviewers and was withdrawn from the open AEC records. It currently remains classified in spite of the fact that anyone can read it in its entirety in the Appendix of Herbert F. York's 1976 book, "The Advisors: Oppenheimer, Teller, and the Superbomb."[9]

It is paradoxical that the document was declassified in the run-up to the 1954 security hearing. In order to produce a publicly "successful degradation ceremony,"[10] the AEC was forced to make public much information that was previously hidden behind the security curtain of the nuclear state. The publicly released transcript of the hearing was a revelation of the workings of science and policy within the Cold War state. I have no information about the reasons for the reclassification of the GAC report in 1994, nor an understanding of why this action would be seen as useful or appropriate when the document is now, in practice, irrevocably in the public domain. But the symbolic maintenance of this dissenting document as a state secret speaks eloquently to what sociologist Chandra Mukerji has called the state's appropriation of the "voice of science."[11] It makes clear the difference between science as an instrument or resource of its funders and patrons (whether the state or business) and science as a constituent of the broader culture of civil society and free public discourse. This difference was an important part of what was at stake in the Oppenheimer security hearing.

There is currently an intense interest in Oppenheimer among historians of science and scholars of American history. This book follows the publication of very fine studies by historians Silvan S. Schweber and, very recently, Kai Bird and Martin J. Sherwin, David C. Cassidy, and Priscilla McMillan.

For Bird and Sherwin, Oppenheimer was an authentic voice of American scientific, intellectual, and political liberalism. For McMillan, he was a defeated moderating voice in American foreign policy.[12] In contrast to these books,

I emphasize what I see as Oppenheimer's failure to develop a critical political perspective as his liberalism was shaped by the culture of the Cold War. I argue that (paradoxically, in light of the security hearing) Oppenheimer in significant ways accommodated himself to and internalized the culture and mentality of the national-security state.

Schweber is critical of Oppenheimer's personal inability to live up to the model of responsibility that he put forward in his writings and reflections on science. In particular, Schweber has criticized Oppenheimer's inconsistent response to McCarthyism, particularly in relation to the security problems of his graduate students and their ordeals with the House Un-American Activities Committee. Schweber suggests that Oppenheimer was "too fractured an individual" to handle the ethical and political dilemmas presented by Hiroshima and the Cold War, and he instead presents physicist Hans Bethe as the more consistent embodiment of an ethic of scientific responsibility. Schweber celebrates Bethe as a model of the working craftsman in science, whose research was his "anchor in integrity." This is an ethic of duty in a calling, or vocation, and Schweber notes that Bethe "responded to the intellectual and social world around him by adopting a Weberian stance: he would deal with the world rationally, to the utmost limits that rationality would allow."[13] In contrast, I take Oppenheimer's dilemmas not so much as indicative of a purely personal failure to live up to an ethic, but as indicative of both a broader ethical uncertainty and inherent problems with the ethics of vocation in relation to problems of war and state power. Cassidy has admirably contextualized Oppenheimer's struggles within the framework of America's rise to global power and the centrality to this of the alliance among science (physics in particular), industry, and the military. Cassidy suggests that Oppenheimer's career reflects how this alliance, while strengthening American science financially and in some ways politically, led to a sacrifice of the independent cultural authority of science.[14] My account is broadly in agreement with this analysis. But at the same time, I aim to connect these contextual themes with issues of self-shaping, the idea of vocation, the ethics of responsibility, and the changing cultural identity of the scientist.

The recent biographies all, in different ways, place Oppenheimer's life in the context of the transformations of science and American society and politics during the Cold War. My aim in this book has been to provide a biography that draws together individual character structure and social structure, looking at the social processes and collective work though which individual identity is constituted. It is a sociological biography, which looks at the collaborative and interactional shaping of the individual in a web of relationships. In that sense, it aims to break down the division between individual and context, treating both in terms of social process. This is a difficult task. Sociologist Norbert Elias has written, "Wherever one looks, one comes across the same antinomies: we have a certain traditional idea of what we are as individuals. And we have a more or less distinct idea of what we mean when we say 'society.' But these two ideas, the consciousness we have of ourselves as society on the one hand and as individuals on the other, never entirely coalesce . . . What we lack, let us be clear about it, are conceptual models and, beyond them, a total vision with the aid of which our ideas of human beings as individuals and as societies can be better harmonized."[15] This study attempts to use the narrative form of a sociologically conceptualized biography to weave together the threads of the "individual" and the "social."

This book can also be read as a study of themes that emerge from the work of Max Weber: vocation, responsibility, cultivation and expertise, charisma, bureaucracy, instrumental reason, fact and value, means and ends. While I did not consciously begin thinking of the research as Weberian, these concepts and themes seemed to quite naturally emerge from and fit seamlessly with the historical material. Many of the concepts employed by Weber—charisma, problems of specialization, fact and value—also occur in the discourse of the World War II generation of atomic scientists and are in that sense actors' categories for this study. That may have something to do with the character of Weber's sociology—he did not seek to replace history with an abstract sociological model, but rather to define a series of concepts that would facilitate interpretive historical understanding. But of course, in relation to Weber's own lifetime, the events I am describing are not history, but the future. It is also possible, therefore, that the fit may have something to do with the cultural impact of Weber's own work. For example, it seems that the widespread modern use of the term *charisma* to describe secular leadership, even if not always faithful to Weber's analysis, owes something to his formulation. But the correspondence is most likely due to Weber's picking out and codifying problems and themes that were emerging in his time from a variety of sources and that would—again, through many influences and sources—become central to thinking about problems of science and modernity in the twentieth century. The "high modern" bureaucratic world of the Manhattan Project and the early Cold War was arguably more neatly "Weberian" than our contemporary postmodern society.[16] Nevertheless, the issue of whether and to what extent these Weberian categories are an analytic *resource* or a cultural and historical *topic* is indicative of what Anthony Giddens has called the "double hermeneutic" of the social sciences. Giddens reminds us that sociological concepts and understandings of the world are constitutive elements of the modern cultures we seek to interpret.[17]

Finally, it is worth mentioning a paradox inherent in the writing of this history. In the book, I have given considerable attention to others' impressions of Oppenheimer and also to the way in which he was described and his life interpreted within his own lifetime (as well as after his death). This is because I want to question the idea of a discrete "real" Oppenheimer separate from these impressions, descriptions, and interpretations (which is not to say that there is nothing but representation). Representations and expectations of him, whether explicit or implicit, were a key part of Oppenheimer's social context and social existence. In his own lifetime, it is possible to see Oppenheimer responding to, resisting, and shaping but also enacting, playing to, and being shaped by these various representations and associated expectations. The narrative structure of this book ends with Oppenheimer's death. But it is interesting to ask to what extent this is the natural ending point of a biography. There is a sense in which death *is* such an ending point—there is no longer an Oppenheimer to play "Oppenheimer." But there is also a sense in which it doesn't have to be. Sociologist Charles Horton Cooley wrote that "there is no separation between real and imaginary persons; indeed, to be imagined is to become real, in a social sense."[18] Oppenheimer *as a social item,* "real" therefore "in a social sense," is continuing to be constructed by the act of writing and representation. As communications scholar Bryan C. Taylor has put it, Oppenheimer is "an enduring discursive form through which audiences discover and contest the ideologies of modern science and the national-security state." Since Oppenheimer's death, "history and popular culture have 'saved' the *sign* of Oppenheimer as an opportunity to explore the formidable social problems associated with nuclear weapons and the possibility of their solution."[19] Biographies "imagine" their subjects, and also participate in creating them as socially real.

ACKNOWLEDGMENTS

There are many people to thank for their help with this project. The book grew out of my doctoral dissertation, and my greatest debt is to my doctoral supervisor, Steven Shapin, for years of patient advice and encouragement and for his intellectual example. I also want to thank the other members of my doctoral committee at the University of California, San Diego—Gerald Doppelt, Harvey Goldman, Naomi Oreskes, and Andrew Scull—for their advice, critical comments, interest, and encouragement. I am, in addition, very grateful to Herbert York at UCSD for sharing his knowledge and experience.

I learned a great deal from fellow participants in the 1998–99 research group on Scientific Personae, organized by Lorraine Daston at the Max Planck Institute for the History of Science; this work benefited greatly from that experience. It has also benefited from discussion with colleagues in the School of Social Sciences at Cardiff University, in particular Harry Collins, Robert Evans, Ian Welsh, and Barbara Adam. I would also like to thank colleagues in the Department of Science and Technology Studies at University College London for the enthusiasm and encouragement that they communicated as I was completing the writing. For conversation, correspondence, the provision of documents, and their published and unpublished work that I have read, I am grateful to Michael Day, Shawn Mullet, James Hijiya, John Rigden, Albert Christman, and Silvan S. Schweber. I am very grateful to Cathryn Carson and David Hollinger for inviting me to participate in the Oppenheimer Centennial Conference at Berkeley. There I had also the opportunity to meet historians Kai Bird, Martin Sherwin, and David Cassidy, whose scholarly work on Oppenheimer I very much admire. David Kaiser read and commented on a number of drafts of the full manuscript, and I have benefited a great deal from his insights, criticisms, and advice.

This work also owes much to the archivists who helped me—in particular Roger Meade at Los Alamos National Laboratory Archives and Hedy Dunn, Patricia Goulding, and Shelley Morris at the Los Alamos Historical Museum, but also staff at the Library of Congress, the American Institute of Physics, and all the other archives I visited (a list of which appears in the bibliography). I am also extremely grateful to the Manhattan Project veterans and scientists who allowed me to interview them.

The research and writing were made possible by a Science Studies Dissertation Fellowship at UCSD, a dissertation writing fellowship from the Charlotte Newcombe Foundation, and a pre-doctoral fellowship at the Max Planck Institute for the History of Science. Subsequent research was funded by a British Academy Small Research Grant.

I am grateful to Katie Crawford, Philip and Diane Olsson, and Susan and Michael Dimock for their hospitality while I was carrying out research. I would also like to thank my family for their support, and I especially want to thank my wife, Angela, for her loving support and understanding during this work.

Any errors, omissions, or shortcomings in the work are, of course, entirely my own fault and responsibility.

CHAPTER ONE

Introduction: Charisma, Self, and Sociological Biography

Physicist J. Robert Oppenheimer (1904–1967) occupied a nodal position in the emergence of late modern technoscientific culture and in the compact between science and the state that developed from World War II. To trace the constitution of Oppenheimer's wartime and postwar scientific identity is to trace the key struggles over the role of the scientist in relation to nuclear weapons, the state, and culture. This is a study in biography, but it is one that reveals the individual—Oppenheimer—as a point of intersection of social forces and interests and that describes the collaborative, social, and interactional fashioning of his identity, his scientific role, and his intellectual, political, and cultural authority. It examines how he negotiated the opportunities created and the constraints imposed by the institutional positions he occupied and by the relationships and networks in which he was embedded. It traces the social and interactional constitution of a unique individual scientific identity and role. In so doing, it provides a history of the making of broader forms of power and authority entwining science and the late modern state.

Between 1943 and 1945, Oppenheimer was director of the Los Alamos Laboratory—the remote site in northern New Mexico where the atomic bombs that destroyed Hiroshima and Nagasaki were designed and built. It was the key installation of the Manhattan Project, a vast military-industrial-scientific endeavor organized under the Army Corps of Engineers. Employing at its peak nearly 129,000 workers and costing $2 billion, the Manhattan Project was the largest technoscientific project to that time. It was a hybrid organizational network incorporating not only scientists and engineers, but also a long list of America's major industrial corporations, including DuPont, Monsanto, Tennessee Eastman, Westinghouse, Chrysler, Union Carbide, Bell Labs, and other large chemical, electrical, and construction firms. At Hanford, Washington, and Oak Ridge, Tennessee, sprawling factories and industrial towns were erected to produce plutonium and to separate out the fissionable uranium-235 isotope. The project linked these industrial sites with university laboratories at Chicago, Columbia, Berkeley, and elsewhere. Los Alamos was the culminating point of the work of these disparate sites. It brought together mathematicians, theoretical and experimental physicists, chemists, metallurgists, high-explosives experts, and engineers, combining this expertise to produce a novel form of technoscientific power and a new method of total war.[1]

The bomb project catapulted scientists into a position within America's political and administrative elites, and Oppenheimer emerged from the war as the chief representative of this new power of the scientist. In 1947, he was appointed to the country's top science advisory position: chairman of the Atomic Energy Commission's General Advisory Committee (GAC). However, Oppenheimer's power was beset by tensions and contradictions. Since his earliest involvement in the bomb project, he had been under investigation by military intelligence and the Federal Bureau of Investigation (FBI) for his Communist associations and political involvements of the late 1930s and early 1940s. In 1949, when the GAC advised against the development of the hydrogen bomb, Oppenheimer was widely suspected of spearheading opposition to the new weapon. During the early 1950s, H-bomb proponents (including physicist Edward Teller, AEC chairman Admiral Lewis L. Strauss, and powerful figures in the military) began a behind-the-scenes campaign to remove Oppenheimer from any governmental role. This struggle culminated in the security hearings of 1954, when an AEC Personnel Security Board declared Oppenheimer a security risk. The withdrawal of Oppenheimer's security clearance suddenly severed his connection with government, consigning him to the political wilderness. He was only partially rehabilitated when, in 1963, he received the AEC's prestigious Fermi Award, given the previous year to Teller. Though his past work for the government was now officially recognized and rewarded, his security clearance was not renewed.

This, in outline, is a well-known story. Even during his lifetime, Oppenheimer was a focal point for reflection on the place of science and scientists in the modern world. That remains the case today: in academia and in popular culture, the narrative of Oppenheimer as tragic hero has become a parable neatly encapsulating the moral and political dilemmas of the nuclear age. It is a tale that has been the subject of many biographies, historical studies, novels, plays, and movies.[2] Commonly, the Oppenheimer story relies on tropes of purity and danger: Oppenheimer represents the corruption of the pure scientist, overwhelmed both by encroaching militarism and by his own desire for power. Oppenheimer's role in building the atomic bomb represents a fall from grace, the scientist's original sin. The security hearings are often portrayed as a kind of martyrdom or crucifixion, and Oppenheimer's subsequent exile from power as a retreat from a corrupt world, a chance for purity and salvation. Oppenheimer appears sometimes as a saint, sometimes as Faust, with the atomic bomb as a diabolic device.[3]

This narrative has found a central place in our understanding of the scientifically modern. Sociologists, philosophers, historians, and other social commentators examining the role of the scientific intellectual have all attempted to come to terms with the figure of Oppenheimer. In *Brighter Than a Thousand Suns,* the journalist Robert Jungk's celebrated study of the atomic scientists, Oppenheimer appears in a field of struggle between pure science and the will to power. He is presented as embodying a unity between science and humanistic culture, a unity that is shattered by the one-sided technical-instrumental orientation that led to the atomic bomb. For Jungk, Oppenheimer was the tragic representative of the scientists' Faustian bargain with military technology. Jungk wrote in 1958, nine years before Oppenheimer's death: "Oppenheimer . . . reveals . . . why the twentieth century Faust allows himself, in his obsession with success and despite occasional twinges of conscience, to be persuaded into signing the pact with the Devil that confronts him: What is 'technically sweet' he finds nothing less than irresistible."[4] Oppenheimer's former friend, Haakon Chevalier (their connection was to be the key subject of interrogation in the 1954 hearings), concluded that Oppenheimer was "a Faust of the twentieth century, he had sold his soul to the bomb."[5]

For sociologist Lewis Feuer, Oppenheimer represented the rise of managerialism, technocratic power, and militarism in science. "During our generation," he wrote, "science has become the bearer of a death wish," and he quoted Oppenheimer's famous reaction to the first atomic bomb test: "I am become death—the shatterer of worlds."[6] Lewis Coser was also interested in Oppenheimer as a leading representative of the scientists' new public role in confronting the problems of atomic weapons. Like Feuer, Coser was worried that scientists were becoming "the domesticated retainers of their bureaucratic masters." But in contrast to Feuer, he saw Oppenheimer as exemplary of scientists who "have cultivated uncommon sensitivity to the values of our culture and the fate of our society." In Coser's view, Oppenheimer was a "true scientific intellectual."[7]

Philip Rieff similarly dwelled on Oppenheimer's "charismatic" and symbolic role: "His thin handsome face and figure replaced Einstein's as the public image of genius . . . He had actually become the priest-scientist of Comtean vision, transforming history as well as nature." But Rieff argued that the scope for such a charismatic role for scientists in modern America was limited. Without a vibrant humanistic public culture to support them, the scientists' engagement with politics was doomed to failure. For Rieff, Oppenheimer's denunciation by the AEC signified the reduction of the scientific elite to the merely technical function of a "service class."[8]

The security hearings have frequently been taken to instantiate a deep-rooted, or even inevitable, conflict between the intellectual and the powers. Historian Giorgio de Santillana was directly inspired by the Oppenheimer case in writing *The Crime of Galileo,* published in 1958. In both cases, he argued, the free "scientific mind" was at odds with "Reasons of State."[9] Political scientist Sanford Lakoff compared the Oppenheimer hearings with the Athenians' persecution of Socrates and argued that "the trial of Dr. Oppenheimer was also the trial of liberal democracy in America." But above all, Lakoff argued, the "tragedy in Dr. Oppenheimer's predicament . . . stemmed . . . from his internal struggle with the scientific vocation." For Oppenheimer, unlike Socrates, "the center of his life is not the city but his vocation." Oppenheimer symbolized for Lakoff the "alienation" of the modern intellectual and the severance of specialized knowledge from a moral and political engagement with the world.[10]

NUCLEAR PHYSICS, RESPONSIBILITY, AND VOCATION

Oppenheimer has been a focus for reflection on the relationship between truth and worldly power: between the intellectual and the polis, "pure science" and technology, charisma and bureaucracy. Oppenheimer's personal trajectory represents a key moment in a larger story of social changes impacting the organization of science and intellectual life: bureaucratization, professionalization, the rise of science as a career, the routinization of career patterns, and, above all, the ever closer integration of science into the affairs of state.

Max Weber linked the rise of modern rational bureaucracy to a particular character structure, that of the "personally detached and strictly 'objective' *expert.*" This figure of the expert stood in conflict with, and in Western societies has gradually replaced, the older type of humanistic cultivated man. The education of the cultivated man aimed at producing a particular kind of "bearing in life" rather than expert knowledge per se. Weber wrote, "Behind all the present discussions of the foundations of the educational system, the struggle of the 'specialist type of man' against the older type of 'cultivated man' is hidden at some decisive point . . . This fight intrudes into all intimate cultural questions."[11] The decline of the cultivated man and the rise of the specialist reflected the increasing cultural dominance of science, expertise, and rationality and their separation from other frameworks of value. In the disenchanted world of modernity, science has had to stand independently from religion, art, or humanistic moral values. All "former illusions" such as science as the "way to true God" or the "way to true happiness" have been dispelled. Weber agreed with Tolstoy that science could give "no answer to . . . the only question important for us: 'What shall we do and how shall we live?' " Instead, the value of the scientific enterprise in a rationalized and disenchanted world was limited to the service of factual knowledge: "Science today is a 'vocation' organized in special disciplines in the service of self-clarification and knowledge of interrelated facts. It is not the gift of grace of seers and prophets dispensing sacred values and revelations." Weber's conception of the ethos of science was set in tension between the twin connotations of both the German *Beruf* and the English *vocation:* on the one hand, the more archaic and spiritual value of the calling; on the other, the modern secular occupation. Weber's concern was whether it was possible to sustain a sense of the meaning and value of science while it was becoming a secular, routinized profession.[12]

Michel Foucault has also centrally grappled with the implications of the specialization and disenchantment of the intellectual role during the twentieth century, and he has pointed to Oppenheimer as a pivotal figure in these transformations. Like Weber, he emphasized the modern divorce of knowledge from sacred religious and moral values: "Truth is a thing of this world." Instead of speaking for transcendent values or universal truths, the modern intellectual-as-expert provides techniques of power: the intellectual "is no longer the rhapsodist of the eternal, but the strategist of life and death." And Foucault wrote, "It seems to me that this figure of the 'specific' intellectual has emerged since the Second World War. Perhaps it was the atomic scientist (in a word, or rather a name: Oppenheimer) who acted as the point of transition between the universal and the specific intellectual."[13] Foucault suggested that Oppenheimer and the atomic scientists were able to combine the narrowly focused expertise of the specific intellectual with the claim to speak for all people that had been the mark of the universal intellectual. The global scope of the atomic threat enabled the scientists to be understood as speaking for humanity when they addressed the problems of the nuclear age. This universality, however, was rooted not in claims to universal truth or transcendent moral law, but rather in a new kind of global technological power.

Foucault's account points to the way in which the Manhattan Project drew together and intensified those processes identified by Weber, which in more dispersed ways were already changing the nature of the scientific vocation. Foucault, however, did not adequately address the ethical tensions and ambiguities in the new scientific role that emerged. The claim to "universality" of specialized expertise remains contested, and the Tolstoyan problem of meaning, emphasized by Weber, has not disappeared. The threat of atomic warfare gave rise to moral problems that could not be addressed by specialized expertise alone. The atomic bomb was the culmination of the rise of technical expertise, but it also called into question the nature of expert authority and its adequacy to deal with the crises of the modern world. The bomb project put scientists in a new situation, in which they had to either claim some sort of moral authority or publicly divest themselves of it entirely. Weber's problem of vocation was at the heart of struggles over the nature and scope of scientific authority in the wake of World War II.[14]

This book tells a particular story, about how these tensions played out in Oppenheimer's life and career. It aims to capture the particularity of his situation and of his interventions, while at the same time drawing attention to the broader institutional and cultural context that he was negotiating. It was a particular social and institutional trajectory that shaped Oppenheimer's personal identity and his historical significance. Of course, there are other individuals whose trajectories offer similarities and who responded in interestingly similar and different ways to the challenge of atomic weapons. But more than any other figure, Oppenheimer had the potential to combine the emerging technocratic power of the scientist within the state with a humanistic and critical perspective on the development of nuclear weapons. He therefore stood in notable contrast with such scientists as Albert Einstein and Niels Bohr, and others who criticized the national-security state from positions outside it. He equally stood in contrast with institutional insiders, such as Edward Teller, who defined their role as scientists strictly in instrumental terms, exclusive of any obligation to consider questions of ultimate ends.

Einstein was the most important representative of the view that scientists have a moral obligation to address the ends to which research is applied. His only direct involvement with the atomic bomb project was in signing a letter to Roosevelt urging that the U.S. government take seriously the possibility of developing an atomic weapon. This step was motivated by his fear of the Nazis. But after World War II, Einstein became a vigorous advocate of arms control and world government. For example, the (Bertrand) Russell–Einstein manifesto of 1955 highlighted the threat of nuclear holocaust and called on scientists to work toward the goal of ending war. It led to the institution of the Pugwash conferences, aiming to promote scientific internationalism as a vehicle for peaceful international cooperation. Einstein was never included in, nor did he seek inclusion in, formal government advisory bodies. His political engagement was always as an outsider, drawing on moral authority rather than political power.[15]

Others more embroiled than Einstein in the atomic bomb project could nevertheless foresee an arms race and tried to take steps to prevent one. Bohr spent the war trying to convince the British and U.S. governments to support a plan for international control of atomic energy. Although he was an important influence on other scientists, including Oppenheimer, his own direct interventions met with little success. For example, Bohr's meeting with Winston Churchill in May 1944 was a disaster. Bohr, characteristically, mumbled in a barely audible voice, and Churchill understood only that he was advocating telling the Soviets about the atomic bomb. Churchill thought Bohr dangerously naive, and Bohr later said, "We did not speak the same language."[16]

A group of scientists on the Manhattan Project at Chicago, including James Franck, Leo Szilard, and Eugene Rabinowitch, tried to prevent the military use of the atomic bomb by arguing for a technical demonstration instead. Szilard circulated a petition urging restraint against the military use of the bomb. Versions of this petition were signed by more than a hundred scientists in the Chicago and Oak Ridge laboratories. But in the face of the powerful institutional and bureaucratic momentum toward use of the bomb, such efforts proved of little avail. Szilard told Oppenheimer at the time that although the petition was unlikely to influence wartime decisions, nevertheless, "from a point of view of the standing of the scientists in the eyes of the general public one or two years from now it is a good thing that a minority of scientists should have gone on record in favor of giving greater weight to moral arguments."[17]

The connection of science to political activity has often been associated with a belief in the possibility of a rational solution to political problems. Rabinowitch and other figures associated with the Federation of American Scientists and the *Bulletin of the Atomic Scientists* believed that, as sociologist Edward Shils summarized it, "the scientific method that led to the monstrosity [of the atomic bomb] could also lead the way to the solution." Rabinowitch "thought that the scientific method could replace the vagaries of political passion, ideology, and self-righteousness."[18] The same spirit was expressed by Linus Pauling, the 1962 Nobel Peace Prize winner, co-founder of the Emergency Committee of the Atomic Scientists in 1946, and an important force behind the 1963 Partial Test Ban Treaty. Joseph Rotblat was working along the same lines in Britain after the war. The only scientist to leave Los Alamos on moral grounds, he later became a founder of Pugwash and won the Nobel Peace Prize in 1995 for his work in promoting arms control. These scientists explicitly sought to connect their professional identity with a moral and political agenda, and they have done so largely outside governmental channels, pursuing their campaign within civil society.[19]

Others, while harboring moral objections to the buildup of nuclear weapons, have been more ambivalent about the place these objections should occupy in relation to their professional scientific life. For example, historian Silvan S. Schweber suggests that the theoretical physicist Hans Bethe wrestled with moral problems but ultimately decided that only through working on weapons could he have any real influence on policy. While he articulated technical arguments for a test ban and, later, against the Strategic Defense Initiative, he did not—unlike, say, Einstein—consistently oppose nuclear weapons on moral grounds.[20]

Teller, in contrast, showed no ambivalence whatever. For him, scientists in the relevant fields had a clear-cut, positive obligation to push the boundaries of what is technically possible; in the field of weapons research, this meant developing ever more powerful and sophisticated means of destruction. Teller argued that "scientists have responsibilities that are real and great. The scientist must try to understand nature and to extend man's use of that understanding. When a scientist has learned what he can and built what he can build, his work is not yet done. He must also explain in clear and simple terms what he has found and what he has constructed. And there his responsibility as a scientist ends."[21]

Teller's argument is premised on the radical incommensurability between ultimate values. In an argument closely paralleling Weber's, he suggested that science as a vocation is one value-sphere among many and that the responsibility of the scientist is strictly limited to that particular value-sphere:

> There are three things of great importance which philosophers like Plato said must be answered together in a positive way: What is good? What is true? What is beautiful? I disagree with Plato. I think that these are three entirely different questions. What is true is up to the scientist. What is good is up to the politician and maybe the religious leaders. What is beautiful is up to the artist. These are three very important questions. And neither of them should be handled by what the other two answers are.[22]

The modern condition of the separation between value-spheres and the fragmentation of authority implied for Teller the limitation of the weapons scientist to the production of weapons—he should not concern himself with their consequences.

Teller's position is the official one of modern Western society. It is institutionalized in the official culture of bureaucratic organizations that divide professional from personal life and that restrict responsibility to narrow institutional roles. While Teller himself may be a controversial figure, his conception of the scientific role and its responsibilities is not. It is one that is tacitly subscribed to by the many thousands of scientists across the industrialized world who work in weapons laboratories and for whom perfecting the means of mass killing is a bureaucratic job requirement.[23] Assuming a responsibility beyond the instrumental role advocated by Teller has generally meant becoming an outsider in relation to the bureaucratic apparatus of the national-security state.

The predicament in which I situate Oppenheimer is therefore general across Western societies. It involves, first, the fragmentation of cultural authority endemic to bureaucratic and industrial modernity and to the scientific vocation in the twentieth century; and, second, the crisis posed by the atomic bomb and forms of technoscientific power that break down the institutional boundaries of science as a distinct sphere. Specialized modern intellectual authority appears increasingly inadequate to deal with the global dimensions of the problems of late modern technoscience. As Einstein put it, "By painful experience we have learned that rational thinking does not suffice to solve the problems of our social life. Penetrating research and keen scientific work have often had tragic implications for mankind, . . . creating the means for his own mass destruction."[24]

The integration of science with state power and violence cut to the heart of postwar liberal political culture. Science was a key motif in the reconstruction of Western liberalism after the end of World War II. Sociologist Shiv Visvanathan has argued that in the face of the twin crises of the Great Depression and the plunge into world war, "science took over as the sustaining force of the liberal imagination . . . The scientific method was substituted for the invisible hand and [Karl] Popper and [Michael] Polanyi became the Adam Smiths of this new regime."[25] Popper's *The Open Society and Its Enemies* (1945) and Polanyi's *Science, Faith and Society* (1946), though they expressed different epistemologies, converged on the notion that science instantiated and expressed the core values of liberal democracy and that the professional values of science made it incompatible with totalitarianism.[26]

Yet this vision of science as a beacon of liberal humanist values was disturbed by the nightmares of Hiroshima and Nagasaki. Even as the atomic bomb made science a sword and shield against totalitarianism, it also rendered the gap between high culture and barbarism uncomfortably narrow, even nonexistent. As Oppenheimer saw it, the scientific endeavor, "fostered throughout the centuries, in which the role of coercion was perhaps reduced more completely than in any other human activity," had culminated in the construction of "a secret, and an unparalleled instrument of coercion."[27]

Oppenheimer occupied a pivotal position in these moral, political, and cultural dilemmas and conflicts. He emerged from the war as a nodal figure in the new relationship between science and the American state and as the embodiment of the new cultural significance of science. Until 1954, he played a key mediating role between government and the scientific community. Oppenheimer attempted to hold together the competing scientific roles of the humanistic critic, the technocrat, and the weaponeer. He was seen as uniquely able to combine a technocratic advisory role within the state with more archaic forms of cultural authority. To explore Oppenheimer's individual trajectory is therefore to examine tensions between humanism and technological expertise that lie at the core of modern science and society.

CULTIVATION, CHARISMA, AND SCIENTIFIC AUTHORITY

Oppenheimer's persona seems to pitch directly against dominant sociological ideas about the nature of self and authority in modern science, large-scale organizations, and government. He was not only a nuclear expert but also the kind of "cultivated man" whom cultural commentators since Weber have repeatedly pronounced extinct. In contrast to the narrowly specialized focus of modern bureaucratized "big science," Oppenheimer came to be celebrated for his general or "humanistic" intellect. He once described himself as "a properly educated esthete." Chemist Glenn Seaborg wrote of "the scope of his knowledge and interest—in languages, literature, the arts, music, and the social and political problems of the world" and of "his fervent desire to see and relate an order and purpose in the entire spectrum of human existence and experience." William L. Laurence, the only journalist to witness the Manhattan Project's first atomic bomb test, described Oppenheimer as not merely a scientist but "a poet and a dreamer." At one of his last public appearances, to receive an honorary degree at Princeton, Oppenheimer was introduced as a modern Renaissance man: a "physicist and sailor, philosopher and horseman, linguist and cook, lover of fine wine and better poetry."[28]

Aesthetic values were central to Oppenheimer's personal appeal and public image. The wife of physicist P. A. M. Dirac wrote to Oppenheimer in 1964 commending him on his impeccable taste: "Many people have a lot of money, but few people have taste." The Oppenheimer house was a relief from this general philistinism; "to wake up in surroundings so pleasant to the eye, wherever one turns to, is to me a great feast indeed." Oppenheimer was "a darling host." The Oppenheimers' home in Princeton (a house dating from 1696), with its collection of French literature and paintings by van Gogh, Vuillard, and Derain, was described by a journalist as "the perfect mirror of their cultivation." It was characteristic that in 1958, *Look* magazine photographed Oppenheimer in the Princeton house standing in front of his inherited van Gogh.[29]

Because he was financially independent, Oppenheimer could not be perceived as a mere bureaucratic hireling. While science was increasingly professionalized and bureaucratized, he retained some remnants of the qualities of the gentleman-amateur. Physicist Isidor I. Rabi, for example, had been struck by the way in which Oppenheimer, as a postdoctoral student in Zurich, would talk about literature rather than his studies in physics, and also by the apparent effortlessness and unconcern with which Oppenheimer approached his work. Even at Los Alamos, according to physicist Charles Critchfield, Oppenheimer "didn't talk about weapons or physics. He talked about the mystery of life . . . He [would] walk around the room . . . He would rub his palms together and look to the side . . . He kept quoting the Bhagavad Gita."[30]

Oppenheimer's cultivation was often invoked in relation to social virtues. In his novel *The Man Who Would Be God,* Chevalier's character of Sebastian Bloch—clearly recognizable as Oppenheimer—is celebrated for his general intellect: "He's said to have what they call a universal mind. He doesn't limit himself to one speciality like most scientists, but seems to be completely at home in the whole realm of science and the arts as well." And this general cultivation is crucially linked in the description to gentlemanly qualities of disinterestedness: "He seems to have little or no personal ambition . . . Most of the important contributions he has made have been ideas passed on to colleagues and students, or work done in collaboration, for which he has never claimed credit."[31]

Chevalier wrote of Oppenheimer, "His presence . . . seemed to bring out in each of the assembled what was most genuine and most expressive in his true nature." This mirrors the refrain of the Los Alamos scientists, who noted that Oppenheimer tended to draw out and clarify the key points of any technical discussion, his gentlemanly disinterestedness ensuring a genuinely civil, rather than autocratic, role. According to Victor Weisskopf, "Oppenheimer gave us an example of how large scientific enterprises can be more than the sum of the collaborative effort of their groups. They can be imbued with a creative spirit based upon a common heritage and a common aim." Physicist James Tuck, part of the British mission to Los Alamos, similarly said, "His function here [at Los Alamos] was not to do penetrating original research but to inspire it. It required a surpassing knowledge of science and of scientists . . . A lesser man could not have done it. Scientists are not necessarily cultured, especially in America. Oppenheimer had to be." And it was because they had "a great gentleman to serve under," said Tuck, that the Los Alamos scientists invariably "remember that golden time with enormous emotion."[32]

Oppenheimer's breadth and cultivation were not only humanistic and aesthetic, but also technical and scientific. Physicist Robert R. Wilson, for example, highlighted the aesthetic in Oppenheimer's influence on and leadership of American physics. When Oppenheimer received the Fermi Award in 1963, Wilson wrote to him that "American physics is beautiful to behold," and this beauty "has to do with a style, and an intensity, and a depth that we can relate to your profound example." Rabi wrote, "Oppenheimer understood the whole structure of physics with extraordinary clarity, and not only the structure, but the interactions between the different elements." In addition to his expertise in theoretical physics, he "would continually amaze experimenters by his great knowledge of their own subject." In an age of specialization, Oppenheimer stood out as a general scientific philosopher, and this was a key part of what was described as his "intellectual sex appeal."[33]

Oppenheimer is remembered as the man who insisted that the first atomic bomb test be called Trinity, invoking John Donne's sonnet "Batter my heart, three-person'd God"; who summed up the new place of the scientist in the atomic age by quoting a line from the Bhagavad Gita, "I am become Death, destroyer of worlds"; who told Truman, "Mr. President, I have blood on my hands"; and who recognized that "the physicists have known sin."[34] Oppenheimer was seen to transcend the role of the scientific specialist even as he gave voice to the predicament of this figure and to the meaning of science in the postwar nuclear age. This was a key aspect of what was, and still is, repeatedly referred to as Oppenheimer's "charisma."

The term *charisma* was introduced into sociological vocabulary by Weber to pick out a form of authority that was being lost from the modern scientific and bureaucratic world. In Weber's usage, charismatic authority derived from an attributed "gift of grace" or "specific gifts of body and spirit" that set an individual apart as a "natural leader." Weber described charisma as a revolutionary force, fundamentally opposed to any form of institutional routine and especially to bureaucratic organization. "The charismatic structure," he said, "knows no regulated 'career,' 'advancement,' 'salary,' or regulated and expert training." The modern "disenchanted" world, devoid of belief in magical or supernatural forces and qualities and increasingly pervaded by bureaucratic structures, is also a world emptied of charisma. Since science is the source of this "disenchantment of the world," and since Weber observed the increasingly close affiliation of science with the bureaucratic state, he suggested that scientists would be one of the most unlikely groups in which to find a charismatic leader.[35]

For sociologist Daniel Bell, it was just this contrast and paradox that made Oppenheimer's persona such an appealing, but also elusive, one. He described Oppenheimer as a "gnostic figure" who "seemed to have stepped more from the world of thaumaturgy than of science." To Bell, Oppenheimer exemplified the "messianic role of the scientist" and the "charismatic dimension" of modern science; and, he added, it is "the tension between those charismatic elements and the realities of large-scale organization that will frame the political realities of the post-industrial society."[36]

Oppenheimer's scientific colleagues have recurrently drawn on the notion of charisma to describe what they saw as Oppenheimer's special personal qualities of leadership. Rabi wrote that Oppenheimer succeeded Einstein "as the great charismatic figure of the scientific world," linking Oppenheimer's charisma to his "spiritual quality, this refinement as expressed in speech and manner." Wilson reported how he was "caught up by the Oppenheimer charisma" at Los Alamos. Seaborg dwelled on Oppenheimer's "magnetic, really electric, personality, [and] his charismatic presence." And Teller spoke of the "brilliance, enthusiasm and charisma" with which Oppenheimer led Los Alamos.[37]

As descriptions of Oppenheimer as "charismatic" are elaborated in terms of his intellectual powers and qualities of leadership, these, in turn, are pervasively linked with descriptions of his physiognomy. Almost no commentary on the force of Oppenheimer's personal presence failed to remark upon his tall, thin frame and especially upon the color of his eyes and the intensity of his gaze. Before the war, his Berkeley students were used to what they called the "blue glare treatment": "when aroused," journalist Peter Goodchild reports, "Oppenheimer's eyes seemed to turn from a grey-blue to a vivid blue." Los Alamos resident Bernice Brode also wrote about his blue eyes, which "had that special intensity, peculiar to him." Bell commented on Oppenheimer's "bony features and translucent eyes . . . set in a face that seemed to have been etched by inner anguish." French social and political thinker Raymond Aron, who associated with Oppenheimer after the hearings, recalled that "he created a strong impression by the contrast between the purity of his blue eyes and the nervous tension in all his gestures and statements." Roger Robb, the AEC's counsel in the security hearings, expressed his dislike of Oppenheimer in physiognomic terms: "he had the iciest pair of blue eyes I ever saw."[38] And Chevalier described Oppenheimer as

> tall, nervous and intent . . . But it was the head that was the most striking: the halo of wispy black curly hair, the fine, sharp nose, and especially the eyes, surprisingly blue, having a strange depth and intensity, and yet expressive of a candor that was altogether disarming. He looked like a young Einstein, and at the same time like an over-grown choir-boy.

It was a face that made Chevalier think about glorified faces he had seen in paintings: "I associated it with the faces of apostles . . . A kind of light shone from it, which illuminated the scene around him." One wartime Los Alamos resident thought that "something about his eyes gave him a certain aura."[39]

It was a face and a body that summoned up similar images in Robert Wilson's wife, Jane; when, toward the end of the Manhattan Project, Oppenheimer was ill with chicken pox and reduced to about 115 pounds, she thought that "our thin, ascetic Director looked like a 15th-century portrait of a saint." In describing Oppenheimer, Rabi was reminded of a friend he had known as a student at Cornell: "Physically and perhaps intellectually and emotionally, he was very like Oppenheimer. One day he announced: 'I give the lie to the materialist. I am a disembodied spirit.' In Oppenheimer," Rabi concluded, "the element of earthiness was feeble." The physicist Leona Libby contrasted Enrico Fermi's earthiness with the "the poetic, disembodied, spiritual emanations that were the basis for Oppenheimer's charisma." And journalist John Mason Brown thought that "the power of [Oppenheimer's] personality" was reinforced by "the fragility of his person. When he speaks he seems to grow, since the largeness of his mind so affirms itself that the smallness of his body is forgotten."[40] So Oppenheimer was identified, in part, as an ascetic, with the moral authority that has been associated with the ascetic way of life over a great span of Western history.[41]

These sensibilities, archaic in tone and texture although recently expressed, clash with our understanding of ourselves as rational, egalitarian moderns, and their overtones of Carlylean "great man" history may raise the hackles of sociologists and social historians. There might therefore be a natural tendency to dismiss this talk as mere "legend-making" or "mythmaking."[42] To disregard the language of "charisma" and "cultivation," however, and consider it merely epiphenomenal would be a mistake. Such imagery was constitutive, in very real ways, of Oppenheimer's cultural authority. These were images and responses that Oppenheimer himself, as we will see, acted so as to encourage and foster. The "real" Oppenheimer can in no sense be separated from this symbolism. Rather than attempt merely to debunk the "Oppenheimer legend," it is, in my view, more interesting to seriously engage with the issue and topic of his embodied cultural authority and with its place, however tension-ridden and contradictory, within modern technoscience. Yet one cannot, of course, stop with this imagery and these sensibilities; we must ask *how* Oppenheimer fashioned his self, identity, and authority, for what purposes, and with what consequences.[43]

BIOGRAPHY AS SOCIOLOGY

The cover photograph for the first issue of *Physics Today*, in May 1948, showed Oppenheimer's trademark porkpie hat resting on the piping of a cyclotron accelerator. Awkward against the symmetric architecture of the machinery, the hat sits as it would be worn—pulled down to one side. It was a touch of personal style; yet underneath the hat, instead of a face, is still more piping, merging into shadow. Oppenheimer himself is physically absent from the photograph, lurking as an invisible presence. The very iconography through which Oppenheimer is remembered and celebrated has often seemed to hide the man rather than illuminate him. There was, and still is, an elusive quality to him. An obituary aptly described him as "the equivocal hero of science." He was a man who was at once "loved and hated, trusted and mistrusted, admired and reviled."[44] Often, colleagues, friends, and even enemies found these conflicting responses intermingled, almost inseparable in their opinion of the man. The physicist David Bohm, one of his former students, thought that the "infinite sadness" of Oppenheimer's face on the cover of *Time* magazine made him look like Jesus Christ. But, Bohm reflected, "I think a better image would be a linear combination of J[esus] C[hrist] and Judas, or of Judas trying to look like J[esus] C[hrist]. An interesting case of mistaken identification." This sentiment was echoed by Chevalier: in portraits, he said, Oppenheimer's expression "could be Christ-like and Mephistophelean by turns."[45]

"Who was he, what was he? Who knew him?" These questions, asked in an obituary of Oppenheimer, had been the subject of intense, ongoing reflection during Oppenheimer's lifetime. The questions remained unresolved at the time of his death, and they remain so today. As one commentator put it, "If what he did is familiar, exactly how he did it is not so fully understood; and who he was, and what he became in the process of doing it, is probably beyond full knowing."[46] In Oppenheimer, there was a convergence of the roles of the scientific specialist and the broadly cultivated humanist, the ascetic and the high-living socialite, the intellectual bohemian and the governmental insider, the pure scientist and the atomic bomb builder. His public identity was a confusing union of opposites. Nor can coherence easily be found in his private self.

The task of finding a more basic private self behind Oppenheimer's public personas proves frustrating. Even his closest colleagues felt that they did not know him intimately. Particularly after the war, if there was a private man behind the public facade, he was so well hidden as to be, for all intents and purposes, nonexistent. There was just the public face.[47] When interviewed by the FBI in 1947, physicist Enrico Fermi said that as far as he knew, "Oppenheimer has no really close personal friends . . . The friends which he does have are largely associated with him because of scientific matters." Fermi did not recall that Oppenheimer "had any real intimates on the project at Los Alamos." A journalist wrote that "again and again, as I talked with his friends, I heard the refrain, 'I'm devoted to Robert. And yet, when I think about it, I can't say that I really know him well.'"[48]

Others expressed a different sense in which Oppenheimer was an enigma. Physicist Felix Bloch once asked Rabi, "You know, I wonder of all the Oppenheimers I have seen, which is the real one?" Rabi replied, "I'll bet you Robert doesn't know himself." Rabi, perhaps Oppenheimer's closest confidant in the war years, could find no underlying unity to Oppenheimer. He was, Rabi said, "a man who was put together of many bright shining splinters." Abraham Pais, a physicist who worked with Oppenheimer at the Institute for Advanced Study at Princeton, wrote, "In all my life I have never known a personality more complex than Robert Oppenheimer." This "explains, I think, why different people reacted to him in such extremely varied ways." Pais summarized his own attitude toward the man as one of "ambivalence."[49] Journalists and historians have also expressed these sentiments. Jungk referred to Oppenheimer's "complex and inconsistent" character, and Richard Rhodes described him as "a man of disturbing contradictions." More recently, Schweber has critically contrasted Oppenheimer's fragmentation with what he sees as Bethe's stronger sense of self rooted in his vocation.[50]

This complexity was given particular significance in Teller's testimony during the 1954 loyalty-security hearings that Oppenheimer's "confused and complicated" behavior made him "exceedingly hard to understand." Teller said that he would rather the security of the country be "in hands which I understand better, and therefore trust more."[51] In the hearings, conflicting understandings of the role of the scientist in the polity and of the legitimate scope of scientific authority were played out as a struggle over the person of Oppenheimer. At stake in this examination of, and contestation over, Oppenheimer's biography and character was the entire history of the relationship between science and the state as it had developed during and following World War II.

The making of Oppenheimer's life into public drama did not happen only after his death. His life *was* public drama, and it was recognized as such by his friends, colleagues, and enemies. Describing Oppenheimer's reaction to the stagnation of efforts at international arms control in 1946, AEC chairman David Lilienthal wrote in his diary, "He is really a tragic figure . . . Here is the making of great drama; indeed, this *is* great drama." Journalist Alfred Friendly described the transcript of the security hearings as "Aristotelian drama," "Shakespearian in richness and variety," with a "plot more intricate than *Gone With The Wind.*" It was almost inevitable that a stage play would be based on the transcript of the hearings. Heinar Kipphardt's production of *In the Matter of J. Robert Oppenheimer* opened in 1964, three years before Oppenheimer's death.[52]

Libby described Oppenheimer as "an accomplished actor." Rabi said that he "lived a charade." These perceptions are partly connected with the fact that Oppenheimer was, after the war, an archetypal scientist of not only the nuclear age but also the age of mass media, representing science for *Time, Life,* and *Look* magazines and for radio and television audiences. He followed Einstein in being a scientific celebrity. But he was ambivalent about this role, both resisting and playing to his various public scripts. And this ambivalence in turn exacerbated his fragmentation and disunity of self. Jeremy Bernstein witnessed a television newsman's interview of Oppenheimer:

> When the cameras were rolling he spoke in a soft voice, his eyes lowered; he wore an almost martyred look. I think [they] were discussing what he used to refer to contemptuously as his "case," the security trial. When the cameras stopped, there was a different Oppenheimer—the "smiling public man"—discussing with Smith the "in" French restaurants in New York and the Caribbean, where Oppie had a home. It was an eerie form of show business. At some point he caught sight of me and said I could watch so long as he couldn't see me—an odd request to which, of course, I agreed.[53]

Such testimony by colleagues and associates would suggest that to understand Oppenheimer is just to appreciate a series of performances and roles enacted in different settings.

Oppenheimer fashioned different identities, in different places and at different times—hero, martyr, saint, philosopher, technocrat, detached expert, engaged intellectual—and he was helped to do so by friends and colleagues, allies, and sometimes opponents, all of whom were willing participants or were drawn into the charade. His identity was the outcome of collective work, in his interaction with others. It was also shaped by the various institutional contexts in which he operated and through which he had a particular trajectory. Through collective interactional and representational work, Oppenheimer's unique identity became inscribed as a central reference point of late modern technoscientific culture. To write the biography of Oppenheimer is therefore to write simultaneously both the account of an individual life and the history of the making of social, institutional, and cultural forms; it is in that sense *sociological* biography.[54]

Oppenheimer occupied a nodal point at which competing cultural tendencies converged and intersected.[55] He emblematized, in particular, the tension between a broad political and cultural role for the scientist as intellectual, and the instrumental role for science as a means and source of power. Oppenheimer came to embody cultural tensions rooted in the development of the secular, modern conception of truth as "a thing of this world." As bomb builder, public spokesman, and symbol of esoteric pure science, Oppenheimer came to exemplify an emerging understanding of truth *as* power as well as cultural tensions *between* truth and power. As both expert and cultivated man, he personified tensions between, on the one hand, the narrowing and fragmentation of cognitive and cultural authority, due to specialization; and, on the other hand, the increasingly far-reaching impact of science on politics and culture.

At Los Alamos, Oppenheimer became the focal point for conflicts of scientific collegiality and openness versus security, secrecy, and compartmentalization. After the war, he was centrally implicated in tensions in the relationship of science to the state and to international affairs. Could the atomic bomb—an unprecedented instrument of military and political power—provide a basis for transcending worldly conflicts and divisions, and for the creation of a newly universal role for intellectuals? Did the incorporation of scientists into technological, military, and advisory roles in the state mean the politicization and instrumentalization of science—or did it, as some hoped, augur a new rationality in politics and international relations? Could science be treated as a model of democratic civil society even while it was being mobilized as a technological and military resource of the state? What sorts of individual and collective responsibility, and what kinds of intellectual authority, were demanded by the new atomic technoscience and the new formations of knowledge and power?

If Oppenheimer is a figure whose historical significance is chronically unresolved, it is because the cultural struggles in which he was implicated and in which his identity was forged remain unresolved and ongoing.[56] This book is a study of the making of individual self and identity, but also of the making of cultural forms and historical meanings.

Chapter 2 will trace Oppenheimer's youthful struggles for self in relation to Jewish cultural identity, education and self-cultivation, and the specialized scientific vocation. Becoming a physicist meant negotiating a set of cultural and moral repertoires for the disciplining and fashioning of the self. The chapter examines in particular how Oppenheimer dealt with tensions between *Bildung* and science, between humanistic breadth and technical specialization, and the ways in which these antinomies were interwoven with more personal struggles for acceptance and identity.

Chapter 3 examines how Oppenheimer's understanding of self and vocation and their relationship to social responsibility, politics, and power changed during his period as a professor of physics at Berkeley, from the early 1930s to the beginning of his involvement in the war and Los Alamos. The chapter traces Oppenheimer's shift from inwardness to involvement in the political world. I suggest that this shift expressed changing personal, social, and institutional relationships. In particular, I argue that Oppenheimer's surprising emergence as a scientific organizer and leader of Los Alamos was the result of collective work by colleagues as well as countervailing pressures from military authorities with an interest in shaping his role and function to their ends.

Chapter 4 describes how the construction of Los Alamos as an institution involved the collaborative fashioning of Oppenheimer as its charismatic leader. I develop a sociological analysis of how the complex organization of Los Alamos was constructed and defined and the role that Oppenheimer's charismatic authority played in this process. In order to examine the social constitution of Oppenheimer's charisma, I provide an analysis of his location in the organizational and political hierarchy of the Manhattan Project and an account of the meaning and functions of his leadership in quotidian life at Los Alamos. I therefore aim to situate Oppenheimer within the history of Los Alamos, from above *and* below, and to describe his role in relation to the moral texture of social and organizational life.

The Los Alamos scientists deeply internalized the institutional goals of the Manhattan Project. In order to understand this collective dedication to building the "gadget," and to apprehend the organizational functions of Oppenheimer's charismatic leadership, it is necessary also to grasp how everyday life at Los Alamos was subtly but powerfully structured around the project's military goal. Chapter 5 examines the more coercive side of Oppenheimer's leadership in enforcing the schedule and the instrumental mission of the laboratory. It introduces dilemmas about the relationship between moral responsibility, technology, and bureaucracy that would become central to the problems of the scientific role in the postwar period.

During the war, Oppenheimer mediated between rank-and-file Los Alamos scientists and the project's military and civilian administrative elites. His wartime position evolved into a postwar role as a key representative of the broader scientific community in its relations with government. In chapter 6, I analyze how he negotiated his scientific role in relation to the politics of the Atomic Energy Commission and the power of the military, and I examine the relationship between these political struggles and Oppenheimer's changing formulations of the scientific vocation.

Chapter 7 examines the 1954 loyalty-security hearings as a key site of conflict over the role of the scientist in relation to the national-security state and over the legitimate content and scope of scientific authority. Oppenheimer's personal identity became a focal point of contestation over the definition of the social and political role of the scientist.

Chapter 8 is a concluding examination of how Oppenheimer tried to reconstitute his public role after the security hearings had cut his ties to government. Oppenheimer survived this defeat by embracing it, becoming the very embodiment of the despair of the humanist intellectual in a technological world. For this performance, he fell back on personal resources of cultivation and general culture. As he made speeches lamenting the death of the humanist intellectual and the impossibility of cultivation in an age of specialization and technology, Oppenheimer presented himself precisely as the embodiment of that humanistic ideal—as the last intellectual.

CHAPTER TWO

Struggling for Self

THE OPPENHEIMER FAMILY:
IMMIGRATION, ASSIMILATION, AND CULTURE

J. Robert Oppenheimer was born into an upper-middle-class German Jewish family in Manhattan. His father, Julius, was born in Hanau (near Frankfurt) and immigrated to the United States in 1888, at the age of seventeen.[1] Julius joined two of his relatives who had immigrated around 1870 and set up an enterprise called Rothfeld, Stern and Company, which imported cloth for the lining of men's suits. He later became the manager of the firm, and by the time Robert was born, in 1904, the business was very successful and the family quite wealthy. Julius's wife, Ella Friedman, was an artist, also of German Jewish background, whose family had come to the United States in the 1840s. Robert and his brother, Frank, younger by eight years, grew up in a nine-room apartment on Riverside Drive in the Upper West Side, overlooking the Hudson.[2]

The Oppenheimers were secular and culturally assimilated. Oppenheimer's father achieved his "principal ambition" when he became an American citizen. The family minimized any identification with Germany. German was not spoken at home; Robert later said, "My mother didn't talk it well, my father didn't believe in talking it." Before World War I, and then again in 1921 and 1924, the family accompanied Julius on business trips to Germany, which were extended into family vacations.[3] Most of Julius's family had remained in Germany. Oppenheimer last saw his grandfather at the age of seven; he described him as "an unsuccessful business man, born himself in a hovel, really, in an almost medieval German village, with a taste for scholarship."[4]

According to Isidor Rabi, Robert's parents were "delightful persons of great cultivation and taste." Julius was described by his employees as a "proper gentleman." Oppenheimer's youthhood friend Paul Horgan thought of Julius as "desperately amiable, anxious to be agreeable." Remembering Mrs. Oppenheimer, Horgan said, "I always think of her in tones of pearl gray, because she kept wearing chiffon of that color, and she wore a string of pearls. She had an artificial hand, which she supported with the other hand, always [wearing] a gray silk glove." She was, he said, "very handsome . . . a very delicate person, I think probably highly neurotic, highly attenuated emotionally, and she always presided with a great delicacy and grace at the table and other events." Like her husband, she was anxiously polite and attentive to others, and from her Oppenheimer inherited his decorous manners.[5]

Before her marriage, Ella had been active as an artist, studying and painting in Paris. On her return to the United States, she had set up a fashionable studio in New York, and she taught art at college level. She did not paint very much after marriage; instead, she and Julius became collectors, and they were early buyers of van Gogh.[6]

As a child, Robert was surrounded by wealth and attended by nurses and servants. But his home was also strict and claustrophobic—offering, he later said, "no normal healthy way to be a bastard." According to Horgan, the household was "very formal" and had "a sadness, there was a melancholy tone." Frank remembered the family home as having "a feeling of friendliness and warmth and gentleness and quite a lot of conversation." But at the same time, "it was reasonably formal." A college friend who visited the family's holiday home in Bay Shore, Long Island, later described the atmosphere of the family as "European sort of . . . not free and easy like Americans, though they wouldn't have thought of themselves as foreigners I'm sure."[7]

It was a highly sheltered environment, even cloistered. "Everything was done in the home," said Frank, and there was a strong sense, which the parents communicated to their sons, of this domestic space being protected against a threatening outside world. Frank recalled, "One couldn't buy apples off the street because the vendors would spit on them and contaminate them, and the barber came to the house and cut one's hair, and I had my tonsils out in my own room. Just a general distrust of the pollution of the outside world." This cosseting was a reaction to the death, soon after birth, of the Oppenheimers' second child, when Robert was four years old. The emotional repressiveness of these upper-class surroundings is clear from Oppenheimer's later description of himself as having been "an unctuous repulsively good little boy." "My life as a child," he said, "did not prepare me in any way for the fact that there are cruel and bitter things."[8]

At the same time that the parents' fears protected the boys from the outside world, Robert was aware of his mother's worry at the result, which was his isolation from other children. When interviewed later in life by the philosopher and historian of science Thomas Kuhn, Oppenheimer said, "I think my mother especially was dissatisfied with the limited interest I had in people of my own age, and . . . I know she kept trying to get me to be more like other boys, but with indifferent success." When he was fourteen years old, his parents sent him to a summer camp. But his physical delicacy and his naïveté about the world made him an obvious target for bullying. Throughout his teenage years, Oppenheimer was awkward in his body: a former classmate said that "there was something strangely childish about him . . . There was a sort of *déséquilibre.*"[9]

Between 1911 and 1921, Oppenheimer attended the Ethical Culture School on Central Park West near Sixty-third Street. He immersed himself in school-work and study. His hobby, from the age of five, was mineralogy; at age eleven he was a member of the New York Mineralogical Club, and at age twelve he presented a paper there. His mother and father "were pleased that I was a good student, were pleased that I was highbrow, were perhaps somewhat mockingly proud of my vigor in collecting and learning about minerals."[10] The family's interests were, in general, aesthetic rather than scholarly; they were "literate" but not "bookish." They placed great emphasis on education, however. Julius was on the board of the Ethical Culture School between 1907 and 1915 and was a great admirer of its founder, Felix Adler (1851–1933). And beyond this educational interest, Julius was interested in Ethical Culture as a philosophy and a worldview. The adolescent Oppenheimer quipped that his father "swallowed Dr. Adler like morality compressed."[11]

Adler, the son of a Reform rabbi, was born in Germany but was raised in the United States, where he founded the Ethical Culture movement. In developing his educational philosophy, Adler argued that the universal moral message of Judaism should be stripped of the trappings of religious doctrine and ritual. Ethical Culture was an attempt to move beyond religion per se to an emphasis on worldly goals of social reform and, in particular, education. Education was deemed by Adler to be in itself a moral process.[12] Fundamental to this philosophy was the German Enlightenment idea of *Bildung.* It is not a concept that translates easily, but at its core was the idea of self-cultivation through education. Education was broadly conceived as moral self-development, aimed at drawing forth the individual's unique personality. The inward search for the "true self" meant attempting to transcend particularistic forms of identity. The understanding of self as being formed through an active process of disciplined work and study held the meritocratic implication that distinction and status could be achieved through education, rather than being given by birth. In such egalitarian and universalistic implications lay the appeal of *Bildung* for emancipated German Jews. *Bildung* promised Jews a new, secular identity, which would allow assimilation and inclusion without requiring an outright rejection of their Jewishness. According to historian George Mosse, it was "an ideal ready-made for Jewish assimilation, because it transcended all differences of nationality and religion through the unfolding of the individual personality."[13]

Yet these ideals never had broad appeal across class lines, even within Germany. *Bildung* was embraced initially by parts of the German upper class and aristocracy, then by civil servants and sections of the middle classes. Among the middle classes, *Bildung* tended to be a decorative ideal, maintained for prestige value only. And over the course of the nineteenth century, it became increasingly detached from the progressive humanism of the Enlightenment and associated instead with conservative nationalist ideals of *Volk* culture. The Jews maintained allegiance to the original Enlightenment conception of *Bildung* even while the majority of the German bourgeoisie and intellectuals were turning away from it. By the turn of the century (and into the twentieth century), the German middle classes were remaking *Bildung* as an idea of national *Gemeinschaft,* or the particular virtues of German culture. Paradoxically, then, as Jews clung to the original, pure conception of *Bildung,* they found themselves increasingly culturally isolated from the German middle classes. At the same time, their assimilationist orientation and attachment to German high-cultural ideals separated German Jews from eastern European Jews. The German Jews shared the deep prejudices of German speakers against all eastern Europeans. The world of the Jews was highly stratified by country of origin, language, urban versus rural background, social class, and strength of religious identification.[14]

These divisions were also present in the New World, in tensions between the different immigrant Jewish communities of New York City. There was a sharp divide between the more established and wealthy German Jews, who had immigrated after 1848, and the much larger numbers of eastern European Jews, who had immigrated after 1880, fleeing pogroms and persecution.[15] In addition to marked economic differences between the communities, there were divergent cultural and religious orientations. The eastern European immigrants tended to be devoutly religious and Orthodox, in contrast to the secular or Reformist German Jews. These differences led to strains between the Jewish communities, heightened by the fact that the new wave of immigration was met with an anti-Semitic backlash among the American public and press. The German Jews themselves carried their German prejudices against Yiddish-speaking eastern Jews, and many were anxious not to be associated with the newcomers. What is more, the poverty of the new immigrants confronted the emancipated German Jews with an image of the ghetto, which they had long ago left behind. Their reactions were complex. Social-reform zeal aimed at "Americanizing" the newcomers, and philanthropy aimed at improving conditions in immigrant slums; but these efforts coexisted with forms of social distancing and restrictionism that, as historian Howard Sachar wrote, "swiftly emulated Gentile snobbery in every respect."[16]

The appeal of Ethical Culture was framed and restricted by these divisions. Drawing heavily on German intellectual and cultural traditions, Adler's philosophy attracted mainly relatively well-to-do families of German Jewish background. Even among German Jews, Adler's program enjoyed only limited appeal in comparison with Reform Judaism. As a result, the Ethical Culture society had, as one historian noted, only "marginal status, if even that, in Jewish communal life." The assimilationist outlook of Ethical Culture was partly a response to the increasing anti-Semitism of late nineteenth-century America. Adler himself promoted assimilation as the solution to anti-Semitism.[17]

Perhaps the greatest impact of the Ethical Culture School on Oppenheimer came from the two teachers who took a strong personal interest in his education and his life as a young man. The first was Augustus Klock, who taught chemistry and physics. Early on, Klock regarded Oppenheimer as having great potential and gave him private tutoring outside school hours. During two winters and one summer vacation, they spent about five days a week together, working on chemistry and physics experiments and on Oppenheimer's hobby of mineral collecting. Oppenheimer had a microscope and a polarizer at home, which he would use to test minerals. He recalled that he "loved" laboratory work and that, for him, the "whole bifurcation" between theory and experiment came later.[18]

In addition to Klock, Oppenheimer was strongly influenced, in an altogether different intellectual direction, by his English teacher Herbert Smith. Students would go to Smith's home and discuss poetry and literature. Oppenheimer continued to send Smith examples of his poetry from university. Oppenheimer's great affection for him is clear in the vigorous and flamboyant correspondence that they kept up, particularly while Oppenheimer was at

Harvard and later Cambridge (England). Oppenheimer's interests ranged from science to literature, poetry, and art. His close school friend Francis Fergusson expected him to become a writer.[19] One teacher at the Ethical Culture School, in view of Oppenheimer's twin enthusiasms, suggested to him that "your vocation is to be a science writer," but this did not appeal to him. A career as a mining engineer seemed more romantic. While at Harvard, Oppenheimer wrote a short story for Smith in which the protagonist was a mining engineer. But, revealingly, Oppenheimer described his fictional hero as "a sophisticated and introspective person, and the filth, the phosphorescent manager, and the miserable, indifferent miners only make him laugh and look smugly at the sunset."[20]

The summer of 1922, before he went to Harvard, had a profound impact on Oppenheimer's sense of self. The previous summer, while in Europe with his family, Oppenheimer had ventured off to prospect for minerals in the old mines of Bohemia. He contracted a very serious case of "trench dysentery"; in its wake came a persistent colitis, which forced him to postpone his entrance to Harvard for a year. A trip to the high desert of New Mexico, something often prescribed for American consumptives, was suggested by his parents, and Smith offered to take him. They stayed with Fergusson's family in Albuquerque. Francis's older brother, Harvey, was a novelist, and his mother would later publish classic works on the cultures of the Southwest. It was on this trip that Oppenheimer met Paul Horgan, who in adulthood would become a major American novelist.[21]

In a direct reaction against the claustrophobic propriety of his upbringing, Oppenheimer took with alacrity to New Mexico's hardy outdoor life and became proficient at horseback riding. He and Horgan would travel on horseback over trails across the Sangre de Cristo Mountains and stay at the guest ranch of Katherine Chaves Page, a place Oppenheimer found magical.[22] Page—a beautiful, aristocratic woman from an old hidalgo family, living in reduced circumstances—took Oppenheimer under her wing. Horgan remembers that "she was a very beautiful woman and very intelligent, sweet-natured," and that Oppenheimer "was deeply devoted" to her. When Oppenheimer read Willa Cather's *A Lost Lady* (published in 1923), he was reminded of Mrs. Page.[23]

The New Mexico experience was particularly significant for Oppenheimer, because of the continuing exclusion that Jews faced in American society. At the start of the trip, Oppenheimer asked Smith if he could travel as his brother; this request was sharply refused. But even without such overt disguise of his ethnic background, Oppenheimer was able to adeptly traverse social boundaries. That the trip to New Mexico was bound up with questions of identity and selfhood is suggested by Fergusson, who said that Robert felt "his Jewishness and his wealth, and his eastern connections, and his going to New Mexico was partly to escape from that."[24]

Oppenheimer's discomfort about his background was evident when, in early 1923, Katherine Page was in New York City teaching Spanish at Finch Junior College and attended a party at the Oppenheimer home. Robert wrote to Smith that "Mrs. Page started bravely enough, but soon grew silent under the weight of paternal banalities and Ethical gossip." Robert found the evening very trying, a "dismal gathering." His anxiety about the occasion suggests that he was straddling separate social worlds that appeared to him to be incompatible. Another occasion on which the world of his New Mexico escape collided with the world of his upbringing was when his family drove Paul Horgan to Buffalo en route to Quebec. Robert thought the trip was a disaster. The mixture of Horgan with his parents was, he said, an "explosive agglomeration." Robert believed that his parents were "a little jealous of Paul, and a little irritated at the ease with which he disregarded obstacles whose conquests formed the central jewels in the Oppenheimer crown." He referred specifically to what he perceived to be the source of their insecurity—a "complex" due to which, he felt, they "tried to apologize for being Jews."[25]

Robert's relationship with his father was a tense one, and it is probable that his insistence throughout his life that his first initial, J., stood for nothing (rather than "Julius," as recorded on his birth certificate) reflected these strains.[26] In the late 1940s, Smith told a journalist that Robert's father had a touch of "business vulgarity which acutely embarrassed Robert, although he would never mention it." The importance of this observation is clear in relation to the negative stereotypes of Jews that were prevalent in American society when Julius immigrated, in which the notion of "vulgarity" was a central trope. The tragedy suggested by Smith's statement was that the Oppenheimers were haunted by these stereotypes, and that their insecure and self-conscious son was particularly attuned and sensitive to such images. Smith thought that Robert's problems, including the "psychosomatic" persistence of his colitis, were due to a "pronounced oedipal attitude" toward his father. Whatever the psychological roots may have been, there was an important sociocultural dimension to the tension in Robert's relationship with his father. At one point, on the New Mexico trip, Smith asked Robert to fold his jacket for him. Robert snapped back, "Oh yes. The tailor's son would know how to do that, wouldn't he?"[27] Allusions to Shylock and references to "skinflint Jews" in his letters suggest, in the context of Oppenheimer's difficult and complex relationship with his family, some internalization of anti-Semitic attitudes.[28] This was part of what was behind Oppenheimer's almost neurotic self-consciousness, his constant monitoring of himself for any slip or faux pas. He later said, "My feeling about myself was always one of extreme discontent."[29]

Rabi thought that Oppenheimer actively distanced himself from his Jewish background, and he saw this as absolutely key to understanding this complex man: "Oppenheimer was Jewish, but he wished he weren't and tried to pretend that he wasn't." Felix Bloch agreed: Oppenheimer "tried to act as if he were not a Jew and succeeded well because he was a good actor." Rabi, having no such problems with his own Jewishness, believed that this was why Oppenheimer "never got to be an integrated personality."[30]

Rabi very strongly disapproved of what he saw as Oppenheimer's confusion about Jewish identity. Rabi had been raised in poverty in an Orthodox, Yiddish-speaking home in Manhattan's Lower East Side and later in Brooklyn. The differences between the German and eastern European Jewish communities in New York could not have been presented more starkly than in the contrast between Oppenheimer's and Rabi's childhoods. And although Rabi rebelled against Orthodox Judaism, his religious roots always remained central to his self-identity.[31] Rabi's stake in these issues does not lessen the validity of his observation. The question of Jewishness was at the heart of Oppenheimer's youthful problem of identity, and Oppenheimer's project of active self-creation followed from this rejection of, and escape from, Jewish identification.

BECOMING A SCIENTIST: SELF-SHAPING AND IDENTITY CRISIS

Oppenheimer entered Harvard University in August 1922. Going to Harvard was a fulfillment of the hopes of advancement through education that were expressed in his upbringing. It was a period of intellectual exploration and self-development. But in the background of his years there were the realities of institutionalized discrimination and exclusion. At Harvard, Oppenheimer was faced with the problem of navigating his way through an institution shot through with anti-Semitism.

At the time he entered the university, it was embroiled in controversy over a proposed quota designed to reduce the number of Jewish students. The university's establishment felt that the rising number of Jewish students threatened the "social standards" of the school—specifically, its character as an Anglo-Saxon Protestant institution.[32] The quota, proposed by university president Lawrence Lowell, led to a massive public controversy and was temporarily defeated. However, Lowell partially achieved his goal in January

1926, when Harvard adopted a plan whose ostensible purpose was to create a "student body . . . properly representative of all groups in our national life," by increasing the weight given to personal information other than examination results in the selection of candidates.[33]

Oppenheimer's quip that his arrival at Harvard was "like the Goths coming into Rome" not only suggests his excitement at being able to "ransack" the libraries and the courses, but also strongly evokes a sense of being an outsider, breaking his way into a place that had erected barriers to try to prevent his entry. The barriers Oppenheimer faced are suggested by the experience of Rabi a few years earlier at Cornell. Rabi reported that because of the "discrimination and anti-Jewish sentiment" that he encountered at Cornell, "I never had enough confidence to get to know any faculty member. I went through all the years not knowing a single faculty member." Oppenheimer's later friend and colleague from Berkeley and Los Alamos, David Hawkins, has said that it was "not a negligible fact in Robert's background that he had been a victim of considerable anti-Semitism at Harvard and elsewhere."[34]

Oppenheimer's friend Jeffries Wyman remembered, "He found social adjustment very difficult, and I think he was often very unhappy. I suppose he was lonely and he didn't fit in well with the human environment." But Oppenheimer did cultivate a small group of close friends, most of whom felt themselves to be at odds with the mainstream of student culture at Harvard. A couple of months into his first semester, Oppenheimer reassured Smith that "I have not suffered from loneliness. There are plenty of amusing fellows with whom to read, talk, play tennis and make expeditions into the hills and toward the water."[35]

Oppenheimer kept up his correspondence with Horgan, who went to study at the Eastman School of Music in Rochester, New York, and he also maintained a lively correspondence with Smith. Fergusson was with Oppenheimer at Harvard for one year, before going to Oxford on a Rhodes scholarship. Perhaps feeling responsible for his insecure and less mature friend, Fergusson initiated Oppenheimer into a circle of intellectually serious seniors and graduate students. Fergusson was concentrating in biology at Harvard, turning to literature and the arts when he went to Oxford. He introduced Oppenheimer to a science discussion group of about twelve people that met on Mondays, sometimes inviting a professor to give a talk. Wyman, also a biology student, remembered Fergusson's insistence that he meet Oppenheimer: "Francis was full of talk about Bob Oppenheimer."[36]

Other members of Oppenheimer's circle at Harvard were John Edsall, a senior premedical student; William Clouser Boyd (whom Oppenheimer called "Clowser"), a classmate in his freshman chemistry course; and Frederick Bernheim, who had been in the class behind Oppenheimer at the Ethical Culture School, although they did not meet until college. On arrival at Harvard, Bernheim had opted to live with a missionary from Andover: "I wanted not to be involved in a sort of Jewish enclave." Bernheim's background, like Oppenheimer's, was German Jewish. He said, "I had come from a school where . . . eighty percent [were Jewish]—and at that time there was a good deal of anti-Semitism, and . . . [I wanted to] be able to go around with the non-Jewish students, which I proceeded to do for the first year." But once he met Oppenheimer, "there was no question about whom I would want to room with next year, because he was then quite obviously much more congenial to me than my missionary roommate. So despite my prejudices, I roomed with him for the next two years."[37]

Oppenheimer shared with other graduates of the Ethical Culture School an attitude of intellectual seriousness. They viewed their studies as important above all else, and they felt that this set them apart from the mainstream of undergraduate culture. One of Oppenheimer's Ethical Culture contemporaries, who went on to the all-female Vassar College, wrote to Smith complaining about the debutante culture among the undergraduates: "The manners and costumes and ancestors of the girls here seem to count more than Phi Beta Kappas. In other words the atmosphere is refined while at school it was rather tough." She was told by one of the faculty, "O, you know a lot more than these people here if you come from the Ethical Culture School."[38] Fergusson felt similarly about Harvard:

> Instead of five thousand keen, intellectually awake, well-read young men . . . I find five thousand tawdry yokels, yanked from fat farms and smug small towns to bellow at football games . . . I did not come here to be made a 100% American; I am not going to be a "bizzness" man. I came here to acquire an education, and I hope to be a person of intelligence some day.

Nevertheless, Fergusson concluded that "if one is willing to embrace misanthropy, the education is here for the taking."[39]

Misanthropy was the effective antidote to the enforced sociability and Babbittry of undergraduate culture. A degree of alienation and isolation was fitting for a young scholar. Like Fergusson, Oppenheimer was disengaged from collective undergraduate activities; his friendships were "wholly on an intellectual basis." He recalled that "I had a few very close friends, but we really all of us were working almost to the limit of our capacity and, I would have thought, not talking too much."[40] Such aloofness was, in part, a response to the exclusiveness of Massachusetts society, in which social standing was closely linked to ancestral lineage. In response to a remark made by Boyd, who was of Scots-Irish and German descent, Oppenheimer said, "Well, neither of us came over on the Mayflower, you know." To Boyd, "that meant, I think, that he didn't try to cultivate the socially prominent, just as I didn't. We wanted to know people who had some brains."[41]

Oppenheimer may not have come over on the Mayflower, but he had what Horgan called an "innately aristocratic" quality. This was due partly to his "exquisite manners"; Horgan said, "I've rarely known anyone with more beautiful manners." In addition, he had a generous allowance from his father and would splurge on evenings with his friends. Oppenheimer had, according to Horgan, a "nice feeling for good restaurants." For example, he took Horgan to the "very grand" Voisin restaurant in New York, a favorite of his father's, and after dinner they saw actress Jeanne Eagels's celebrated performance in *Rain*.[42] Boyd recalls, "He was a very nice person to know. He was the soul of generosity. That was practically a religion with him."[43]

Oppenheimer does not seem to have had any interest in politics during this period. His friends do not remember him ever discussing political matters. So it is curious that he joined the Liberal Club, which was a meeting point for liberal and internationalist students disillusioned with the world situation after the war, in particular the collapse of American support for the League of Nations. But, as historian David Cassidy points out, many of the student clubs and societies excluded Jews, and Oppenheimer was not predisposed to identify with specifically Jewish student organizations. The Liberal Club stood out from the other clubs in its more open membership policy and atmosphere, so Oppenheimer may have joined simply in order to belong somewhere. Also, the liberal-left political attitudes were perhaps comfortably familiar in relation to the social-reformist ethos of Ethical Culture, and one of Oppenheimer's former schoolmates, Algernon Black, was a leading member of the club. Through Black, Oppenheimer met Edsall, who persuaded him to help edit the club's new magazine. Oppenheimer suggested its name—*The Gad-Fly,* after Socrates, who was the "gadfly" of the Athenian people. Oppenheimer helped edit the first two issues and may have contributed some articles, though none bore his name. Still, he remembered feeling "very much like a fish out of water" in the club. In a letter to Smith, he lambasted the members' "assinine [*sic*] pomposity" and "the methodical expletives of our drunken patriots."[44]

Oppenheimer shared Fergusson's disdain for team sports and heavy drinking. Even upon graduation, when Bernheim and Boyd got "plastered" on alcohol they'd smuggled out of the labs, Oppenheimer "only took one drink and retired to his room." Bernheim remembered, "I don't think he ever enjoyed alcohol." He was also a bit prudish. Fergusson observed that a friend of theirs had earned "the condemnation of Bob [Oppenheimer] by the ruthlessness and pride with which he pursues low women."[45]

Oppenheimer was ambivalent about women. He was happy to contemplate the opposite sex from a distance or to construct romantic literary fantasies, but women in the flesh made him uneasy. He wrote to Smith, for example, "I have come, lugubriously, to the conclusion that the two people at Wellesley and the dozen or so here that even pretend to pursue me are a sorry and worthless lot." He had come across "several New England women trying to live down a congenital primness and psychic sterility."[46] On the other hand, he referred to a former Ethical Culture schoolmate, whom he bumped into in Cambridge, as "a spiteful and unprofitable slut."[47] There were a couple of women in whom he showed mild interest. At one point he found himself diverted from his studies by "the contemplation of a most beautiful and lovely lady who is writing a thesis on Spinoza—charmingly ironic, at that, don't you think?" But it seems that, like the protagonist of one of his favorite novels—Aldous Huxley's *Crome Yellow*—Oppenheimer was too intensely self-conscious and introspective to do more than contemplate.[48]

Oppenheimer's friends from that period do not remember him spending time with women. Wyman said, "We were all too much in love with problems in philosophy and science and the arts and intellectual life to be thinking about girls at that time . . . We were young people falling in love with ideas right and left and interested in people who gave us ideas . . . But perhaps we lacked some of the mundane forms of love affairs that make life easier—perhaps Bob Oppenheimer did."[49] Partly as a result, Oppenheimer's relationships with his small circle of male friends were particularly intense. Boyd remembered a classmate who said, "Who is this guy Oppenheimer who keeps coming to you? I think he's a pest." "Well," said Boyd, "I didn't think he was a pest."[50] But Oppenheimer's roommate, Bernheim, did find his attachment sometimes overly intense.

In fact, he was a little bit possessive . . . I think, first of all, he felt he was very inadequate with girls, and he would resent it very much if I went out with a girl. And . . . when I wanted to invite somebody to go to dinner with us and I invited him too often [Oppenheimer] would object. There was not homosexuality at all. I don't think he was homosexual in any way. I had no sexual feelings for him or he for me, as far as I know, but he had . . . a sort of feeling that we should make a unit.[51]

Oppenheimer's letters suggest an almost romantic quality to his relationships with his male friends. For example, in a letter to Smith, he wrote: "I begin to believe in eternal passions, now, when I see that each note from you still sends me into a violent schoolgirl flutter of excitement." He also wrote florid, and fairly excruciating, poetry, which he would describe apologetically as "exhibitionism," "debauches," and "masturbation" but which he nevertheless circulated within his literary clique.[52]

In such letters and poetry, Oppenheimer was carrying on a performance: his greatest fear was that his correspondents would get bored with him.[53] Smith was an enthusiast for Freud. Through his displays of neuroses and psychosexual ambiguity, Oppenheimer attempted to hold Smith's interest by making himself into a "case" for analysis. Sending him one composition, Oppenheimer apparently hoped that Smith would accuse him of "morbidity and neuroticism."[54] And although he frowned upon drinking and womanizing, he also liked to think of himself as morally degenerate.[55] Above all, he refused to accept the mundane. In Oppenheimer's hands, descriptions of day-to-day life at Harvard quickly gave way to melodrama: "I make stenches in three different labs, . . . serve tea and talk learnedly to a few lost souls, go off for the weekend to distill the low grade energy into laughter and exhaustion, read Greek, commit faux pas, search my desk for letters, and wish I were dead. Voilà."[56]

Oppenheimer did not have a sturdy constitution. As well as enduring a persistent colitis, he was frequently ill with minor complaints. Bernheim thought that his roommate "was somewhat of a hypochondriac." Oppenheimer took an electric pad to bed with him every night, which one night overheated, almost causing a serious fire. Sickness became part of his self-fashioning. For example, apologizing to Smith for the "surliness" of one of his letters, he explained, "I must assure you that it was written in one of those spells of colitic insolence and misanthropy from which you have already suffered so." On another occasion, he asked Smith's forgiveness for his "colitic execrabilities." In February of his freshman year, Oppenheimer wrote to Smith from Harvard's Stillman Infirmary, "I have had a terrific fever, and have read another Conrad. So this may sound a bit strange."[57] Fergusson submitted an independent report on the patient: "At present he is in the infirmary recovering from an attack of grippe. He is cheerful, well-fed, and not very sick. He frugally consumes a *Hamlet* or so while under the weather."[58] Together with his advertised neuroses, Oppenheimer cultivated physical frailty as an aspect of the intellectual and literary self.

It was, after all, his colitis that Oppenheimer had to thank for his discovery of New Mexico and the literary and artistic circle he met there. In his letters to Smith, New Mexico appeared as an intellectual and artistic utopia.[59] The charm of the Pecos Valley even provided Oppenheimer with a new adjective: "Harvard," he wrote, "has a serene and ridiculous appearance . . . [which is] as amusing as *Crome Yellow*, and . . . at least as delightful in a somewhat Pecosian way." In his search for identity, Oppenheimer cast himself, Smith, and his own small circle of highbrow teenage friends as an elite avant-garde. He could thus rationalize his sense of isolation and could see Smith as his soul mate in an imaginary higher form of intellectual community: "For me, and, I suspect, for you, it was never the opinion of the multitude that counted so much; it was the opinion and the conduct of the great—I don't mean absolutely, but to us."[60]

Oppenheimer was still very much influenced by Smith and his milieu in modeling his own identity. Oppenheimer called Fergusson "my revered master and tutor in the science of Smithology." In a letter to Smith, Fergusson described Oppenheimer's "conversation" as "a caricature of yours, ornamented with some of Paul [Horgan]'s and my more elaborate affectations." Literature became for Oppenheimer, Horgan, and Fergusson their own language that set them apart as a clique. Horgan remembered, "[We were enchanted] with the notion of great enrichment of language so that we often conversed in a baroque lingo."[61] These interests, however, tended to demarcate his friendship with Horgan and Fergusson from that with Bernheim and Boyd. Bernheim did not share Oppenheimer's literary bent, finding Oppenheimer "a little bit precious in the way he quoted French poetry, Verlaine, Baudelaire, and so on. And I tended to resent it." Boyd found Oppenheimer's compositions "very *avant garde* for my taste."[62]

Oppenheimer maintained a literary and cultivated posture even while he was increasingly fixing his intellectual course toward a specialized scientific training. Despite idolizing Smith and adopting many of his tastes, Oppenheimer was to make a vocational choice for professional science, in which his literary and poetic interests were widely regarded as unusual. Whether or not Oppenheimer was aware of it, academic anti-Semitism was particularly strong in the humanistic disciplines of history and English literature. Historian Leonard Dinnerstein wrote that in the first decades of the twentieth century, it was widely regarded as "inconceivable for a Jew to transmit or comprehend the culture and traditions of an American Christian society." Max Lerner, who graduated from Yale in 1923, was told that as a Jew, he could not hope to teach literature at a college. It was not until 1939 that Lionel Trilling became Columbia University's first Jewish tenure-track professor in English literature. The University of Michigan was exceptional in that at the end of the 1920s, it had Jewish faculty chairing the departments of English, Romance languages, and economics. Scientific disciplines were more open to Jews than were the humanities.[63]

Thinking that he would be a mining engineer, Oppenheimer began his college career as a chemistry major. But in the spring semester of his second year, he took advanced thermodynamics from the famous experimental physicist Percy Bridgman; this was a small seminar group that met for an hour and a half twice a week.[64]

In his third and final year at Harvard, Oppenheimer began "spending a great deal of time in Bridgman's Laboratory." This was, he remembered, "as far as science goes . . . the really great part of my time at Harvard." Oppenheimer felt "enormous joy and admiration" for Bridgman, regarding him as "a man to whom one wanted to be an apprentice."[65] Oppenheimer was learning from Bridgman not only techniques and theorems, but also what it was to be a scientist. Bridgman's pragmatic philosophy of science was that "scientific method is what working scientists do, not what other people or even they themselves may say about it." Oppenheimer told Thomas Kuhn that Bridgman "didn't articulate a philosophic point of view, but he lived it, both in the way he worked in the laboratory, which, as you know, was very special, and in the way he taught."[66]

However, in order to gain the patronage of men like Bridgman, Oppenheimer had to carefully manage his presentations of self, particularly in relation to his Jewishness. In his last year at Harvard, he applied to Cambridge, hoping to study at the Cavendish Laboratory with its director, the Nobel Prize–winning physicist Ernest Rutherford. Bridgman supported Oppenheimer's application, telling Rutherford that Oppenheimer was a good "betting proposition." At the end of the letter, he added, "As appears from his name, Oppenheimer is a Jew, but entirely without the usual qualifications of his race. He is a tall, well set-up young man, with a rather engaging diffidence of manner, and I think you need have no hesitation whatever for any reason of this sort in considering his application."[67] The Harvard mathematician George D. Birkhoff, with whom Oppenheimer also studied, recommended Oppenheimer in a similar fashion a few years later: "He is Jewish but I should consider him a very fine type of young man."[68]

Oppenheimer's cultural capital allowed him to cross social boundaries. Horgan said that in his student years, Oppenheimer had "this lovely social quality that permitted him to enter into the moment very strongly, wherever it was and whenever it was." This quality was a mixture of "a great superiority" and "great charm." But others noticed Oppenheimer's underlying awkwardness and self-consciousness. One incident that stuck in Bridgman's mind was when Oppenheimer was admiring a picture of a Greek temple at Bridgman's house. When Bridgman mentioned the date and style of the architecture, Oppenheimer responded, "Oh, that's interesting, because from the style of the capitals I would have put it at fifty, a hundred years earlier than that." Oppenheimer's performance was rough around the edges; his intended graceful displays of erudition could come out as pedantic or clumsily patronizing. He lacked the insider's self-assurance. Birkhoff also reported, "I have been told that he is highly strung."[69]

Oppenheimer worked tremendously hard at Harvard, taking a six-course load each semester and graduating in three years. He occasionally revealed the intensity of his academic work in letters to Smith. Nevertheless, he appeared to his friends never to study—or, as Boyd put it, if he did study, "he was pretty careful not to let you catch him at it." And he would treat with studied casualness what to others were difficult intellectual accomplishments. Boyd thought that "he just seemed to know everything without trying . . . He never seemed to have to study."[70] This self-presentation by Oppenheimer was significant in another respect: through his display of effortlessness, Oppenheimer distinguished himself from the derogatory image of Jews as ungentlemanly "overachievers."[71]

Oppenheimer emphasized to Kuhn the unevenness and lack of structure in his education: he "never had an elementary course in physics . . . There's nothing there—just a skin over a hole." Because of "never having the same underpinning in physics or in collateral preparatory courses," Oppenheimer felt that among his physics and chemistry student contemporaries, "I had no 'co-moving coordinates;' there wasn't anyone else who was having the same lack of preoccupations and the same preoccupations."[72]

Oppenheimer also told Kuhn that his education did not channel him in a rigidly scientific direction.

At school and at Harvard I learned a lot of things that had no immediate connection with chemistry or physics. I learned Greek at school which was even then somewhat exotic . . . and I continued doing things like that at Harvard on a quite massive scale so that the notion that I was travelling down a clear track would be wrong. I determined to get a mastery of French and its literature which I knew very poorly; I had a very exciting time reading the *Principia* with Whitehead.[73]

He wrote to Smith that he was going to have to debate with Whitehead in one of the seminars and was "already trembling" at the prospect.[74] But the courses Oppenheimer took focused on chemistry and physics.[75] The addition of, for example, "A General View of French Literature" and "Theory of Knowledge" does not represent a remarkable eclecticism, particularly in the context of a liberal education in the United States.

Oppenheimer's search for an intellectual identity or vocation became particularly pressing upon his graduation from Harvard in 1925, when he went to study at Christ's College, Cambridge, and the Cavendish Laboratory. He told Kuhn, "I don't know why I picked Cambridge, but I wanted to go to Rutherford's laboratory. I know I talked to Bridgman about it."[76]

It was especially peculiar that Oppenheimer chose this center of experimental physics, because two months before his graduation he had written to Fergusson that his work with Bridgman "convinces me that my genre, whatever it is, is not experimental physics."[77] While praising Oppenheimer's intellect and academic record, Bridgman told Rutherford that Oppenheimer's mind was "analytical, rather than physical, and he is not at home in the manipulations of the laboratory."[78] But Oppenheimer told Kuhn that as far as theoretical physics was concerned, at that time "I didn't know you could earn your living that way [nor] . . . whether it was a way of life." Probably for that reason, "it didn't occur to me to go to Germany. It didn't occur to me to go to Copenhagen."[79]

At Harvard, Oppenheimer was removed from, and did not appreciate, the revolutionary significance of the developments in theoretical physics in Europe, particularly quantum mechanics: "If I'd been asked if this was a 'hot time' [in physics] I'd probably have said no more than what was normal." He was also still a novice. "It wasn't yet my job . . . I was still, in the bad sense of the word, a student." Oppenheimer joined the large number of American physics students who went to Europe for graduate and postdoctoral work and, on their return, transformed the state of American physics. But at this stage, Oppenheimer had no idea of the things he would be bringing back. He later said, "I don't even know why I left Harvard, but I somehow felt that [Cambridge] was more near the center."[80]

Rutherford did not take Oppenheimer on, but he made arrangements for him to work in the laboratory of J. J. Thomson, who had discovered the electron and was now almost seventy years old. Oppenheimer spent his time at Cambridge in a corner of Thomson's basement laboratory engaged in what he later called the "miseries" of making thin films of beryllium, to use in experiments on metallic conductivity. He found the laboratory work dull, and his lack of manual skill contrasted with the ability of the English physicists, who blew their own glass and seemed "uncommonly skillful." Oppenheimer's own "inability to solder two copper wires together" was, he said, "probably succeeding in getting me crazy."[81] This, together with lack of success with women, resulted in his feeling "about as manly as a tadpole or a cauliflower."[82]

Cambridge was a great disappointment to Oppenheimer on a number of fronts. Fergusson's horror stories about Oxford made Cambridge seem slightly better. "But," Oppenheimer said, "its excellences are just as fantastically inaccessible, and there are vast, sloppy strata where there is nothing, absolutely nothing, to be found." He described the lectures as "abominable" and "vile."[83]

At Cambridge, Oppenheimer became increasingly isolated. Although friends from Harvard—Bernheim and Wyman—had also gone on to Cambridge, and Fergusson was at Oxford, the intimacy of their life at Harvard did not survive in England. Bernheim, studying biochemistry, was at a different college, and they roomed separately. Bernheim met his future wife at his laboratory in Cambridge and spent most of his free time with her. He found it a relief to be separated from his former roommate; Oppenheimer's "intensity and his drive," said Bernheim, "always made me feel slightly uncomfortable."[84]

Wyman was at St. John's College, but he did not see much of Oppenheimer, which he puts down to their both being "fairly busy and occupied." He was, however, aware that Oppenheimer was "very unhappy." During a holiday reunion with Fergusson in Paris, Oppenheimer suddenly, in the middle of a conversation, leapt on his friend as if to strangle him. Fergusson easily fended off the attack, and Oppenheimer wrote to him with abject apologies when back in England. But it was another sign of his inner troubles.[85]

In March of 1926, he took a trip to Corsica with Wyman and Edsall. On this trip, Oppenheimer's psychological troubles apparently came to a head. Toward the end of the vacation, the three were having dinner when a waiter interrupted in order to give Oppenheimer the time of the boat back to the mainland. Oppenheimer told his companions that he had left a poisoned apple on the desk of the Cavendish physicist P. M. S. Blackett and that he had to go back to "see what happened." Oppenheimer did return to Cambridge, but neither Wyman nor Edsall knew whether his story was true. Blackett was unharmed. Nevertheless, whether the apple was fact or—as is more likely—fantasy, both his companions saw the episode as an expression of jealousy.

Oppenheimer's work with Thomson was going poorly, and he resented Blackett's success and burgeoning reputation as an experimental physicist. This jealousy was made harder to bear by the fact that Blackett was so well and widely liked and seemingly immune from the sorts of self-doubt suffered by Oppenheimer. Edsall, who was first introduced to Blackett by Oppenheimer, thought that Oppenheimer's "feeling toward Blackett was one of tremendous admiration, combined perhaps with an intense jealousy—jealousy because of his feeling that Blackett was brilliant and handsome and a man of great social charm, and combining all this with great brilliance as a scientist—and I think he had a sense of his own comparative awkwardness and perhaps a personal sense of being physically unattractive compared to Blackett and so on."[86]

The seriousness of Oppenheimer's psychological and emotional crisis was indicated by the fact that he had been to see a Harley Street psychiatrist. He told Edsall that June (1926) that he had been diagnosed with dementia praecox, for which, the psychiatrist said, treatment would do more harm than good. Horgan thought the dementia praecox was merely a "typical" example of Oppenheimer's "baroque tendency to exaggerate." But those with him at Cambridge sensed that there was more to it. Oppenheimer told Wyman that he would sometimes just lie on the floor of his room and roll from side to side.[87]

THE DISCIPLINING OF SELF AND THE CONSTRUCTION OF A PROFESSIONAL IDENTITY

Oppenheimer's inner struggles of identity were inextricably interwoven with the broader question of finding his place in the social world. The search for vocation was a matter both of finding an institutional position and formal career and of finding a disciplining framework that would structure his identity and provide a sense of meaning and direction. The process of becoming a mature scientist was a struggle involving the active transformation of the self.[88] It also required a monomaniacal focus on a specialized task. For Oppenheimer, this focus was the most problematic aspect of a scientific career. At the end of his year at Cambridge, his direction was still uncertain.

Yet the beginnings of a new direction were in fact appearing during Oppenheimer's time at Cambridge. Under the tutelage of physicist Ralph Fowler, Oppenheimer started to develop an interest in quantum mechanics. He began to attend meetings of the Kapitza Club, an informal physics discussion group, and read extensively in the published literature on quantum mechanics. Visitors to the Cavendish were also crucial in his developing sense of direction. Meeting Niels Bohr was formative: "At that point I forgot about beryllium and films and decided to try to learn the trade of being a theoretical physicist." He also met and became friends with Paul Dirac, although Oppenheimer took a while to warm to him. When Wyman told Oppenheimer how impressed he was by a seminar that Dirac had given, Oppenheimer's competitiveness showed through. He said that he "didn't think Dirac really amounted to anything." But it was not long before Oppenheimer changed his mind and came to see Dirac as "absolutely grand." Blackett introduced Oppenheimer to Paul Ehrenfest, who was "extraordinarily warm and friendly" the two men would often go "out on the river" together and talk about "collision problems, Coulomb's law." Oppenheimer also met and struck up a warm relationship with Ehrenfest's assistant, George Uhlenbeck, during a week with other American physicists at the University of Leiden (which Oppenheimer welcomed as a break from England). Oppenheimer thus found himself already gaining acceptance by this impressive group. Max Born visited the Cavendish in spring 1926 and invited Oppenheimer to continue his work on the problem of continuous spectra at his Institute of Theoretical Physics at Göttingen. Oppenheimer accepted the offer. Upon leaving Cambridge, he felt himself to be embarking on a new course as a theoretician: "I felt completely relieved of the responsibility to go back into the laboratory. I hadn't been good, I hadn't done anybody any good, and I hadn't had any fun whatever; and here was something I felt just driven to try."[89] He earned his PhD at Göttingen and stayed on another year as a postdoctoral student.

Göttingen provided for Oppenheimer the intellectual "center" he had been seeking. On Christmas Eve 1926, soon after he arrived, his article "On the Quantum Theory of Continuous Spectra" was accepted by the journal *Zeitschrift für Physik*. In the following year he wrote four more articles, including one that would become a classic—"The Born-Oppenheimer Approximation," co-authored with his mentor, Max Born. At Göttingen, Oppenheimer found himself accepted as an equal and "part of a little community of people" with "common [intellectual] interests and tastes."[90] Here he met figures such as Richard Courant, James Franck, Werner Heisenberg, Gregor Wentzel, Wolfgang Pauli, Pascual Jordan, Maria Göppert Mayer, and John von Neumann.[91] Another young American physicist, Earle H. Kennard, wrote that "there are three young geniuses in theory here"—Pascual Jordan, Paul Dirac, and Oppenheimer.[92]

Oppenheimer lodged with a physician, Dr. Cario, who was facing ruin due to inflation and was taking in guests in order to make ends meet. Oppenheimer was joined there by Dirac—with whom he took long Sunday walks—and by Karl T. Compton, who brought his wife and daughter. Oppenheimer tended to keep his distance from the other American physicists, whom he patronizingly described as "pretty good at physics, but completely uneducated and unspoiled." He added, "They envy the Germans their intellectual adroitness & organization, & want physics to come to America." Oppenheimer was fascinated by Dirac, who was notorious for being either silent or terse and obscure. Oppenheimer later told Kuhn that Dirac's creativity was "never initially verbal but initially algebraic" and that he regarded words as "a way to make himself intelligible to other people, which he hardly needs." He was thrilled when Dirac, on arriving in Göttingen, gave him the proofs of his paper on the quantum theory of radiation.[93]

Oppenheimer was at Göttingen during what he later described as the "heroic time" of the quantum revolution in physics, of which his new mentor, Born, was a key proponent. It was, Oppenheimer remembered, "a period . . . of many false starts and many untenable conjectures . . . a time of earnest correspondence and hurried conferences, of debate, criticism, and brilliant mathematical improvisation." Even more than the seminars he attended, it was friendships and informal discussion that, Oppenheimer said, "gave me some sense and perhaps more gradually some taste in physics." The American theoretical physicist Edwin Kemble, during his visit to Göttingen in June 1927, reported back to Harvard, "Oppenheimer is turning out to be even more brilliant than we thought when we had him at Harvard. He is turning out new work very rapidly and is able to hold his own with any of the galaxy of young mathematical physicists here." He added, however, "Unfortunately Born tells me that he has the same difficulty about expressing himself in writing which we observed at Harvard."[94] Historians Kai Bird and Martin Sherwin have suggested that Kemble had in mind the contrast between the expressiveness of Oppenheimer's literary writing and conversation and the overly terse character of his scientific writing at that point. Kemble thought that Oppenheimer seemed to be "two different people" when discussing science on the one hand and general, cultural topics on the other.[95]

Born found Oppenheimer to be arrogant and difficult to teach: "He was a man of great talent, and he was conscious of his superiority in a way which was embarrassing and led to trouble." In particular, Born was frustrated about frequent interruptions of the seminar on quantum mechanics, when Oppenheimer would take the chalk and announce, "This can be done much better in the following manner . . ." Born said, "I was a little afraid of Oppenheimer, and my half-hearted attempts to stop him were unsuccessful." Eventually, the students wrote a petition in which they threatened to boycott the seminar unless Born cracked down on Oppenheimer's interruptions. Born dealt with the matter by placing the petition in a conspicuous position on his desk and then leaving the room during a meeting with his troublesome American student. When he reentered the room, Born found Oppenheimer "rather pale and not so voluble as usual. And the interruptions in the seminar ceased altogether." On another occasion, Born gave Oppenheimer a copy of a paper he had just finished on the collision of electrons and hydrogen atoms. Born wanted Oppenheimer to check over the complex calculations. He remembered how Oppenheimer "brought it back and said: 'I couldn't find any mistake—did you really do this alone?' The astonishment expressed by these words and visible on his face was rather excusable, for I was never good at long calculations and always made silly mistakes. All my pupils knew that, but Robert Oppenheimer was the only one frank and rude enough to say it without joking. I was not offended; it actually increased my esteem for his remarkable personality."[96]

Oppenheimer was something of an anomaly at Göttingen, and as such he was a subject of gossip. His "highly-developed table manners" made the students lodging with him in Göttingen "feel like barbarians." He was much wealthier than his fellow students and was sometimes resented as being ostentatious. For example, it was one of his "well known peculiarities that, if anyone admired something of his, he would find some pretext of making a present of it." He did this when a graduate student in physics whom he was courting, Charlotte Riefenstahl, admired his fine pigskin luggage. When he left Göttingen, he gave Born a valuable first-edition copy of Lagrange's *Mécanique Analytique*. But his peers often felt that he was "showing off, putting on a performance."[97]

The hostility Oppenheimer faced from fellow students was part of an increasingly nationalist, anti-Semitic climate. Göttingen University was a hotbed of nationalist activity during the 1920s. For example, in 1923, three visiting French theology students had been driven out of town by a right-wing student mob. The university took no action against the nationalist students but instead censured the professors who had hosted the French visitors. Both anti-Semitism and hostility to the Weimar Republic were rife among the student population. The students, who were predominantly middle and upper class and affiliated with fraternities, bitterly resented their newly pinched economic circumstances, caused by inflation. This resentment often focused on foreign students, whose currency stretched further, and it was fueled by xenophobic and anti-Semitic articles in the newspaper *Göttinger Tageblatt*. For his part, Oppenheimer began to find Germany austere and depressing. He later told Kuhn that a friend whose family owned a publishing house and who therefore could afford a car always parked the car in a barn outside

Göttingen, "because he thought it was dangerous to be seen driving it" in town. In November 1926, Oppenheimer wrote to Fergusson about the social climate: "Neuroticism is severely frowned upon. So are Jews, Prussians and French." Looking back later in life, Oppenheimer reflected, "[I] got out of Gottingen when I could. It was a bitter time. It was before the Nazis but it was . . . at a time when the beginnings of this bitterness were extremely clear. Inflation had occurred—people who were ruined by it were—you could see the hostility and harshness in them." He got out of the university early by petitioning to take his doctoral examination in the spring of 1927, rather than wait until the end of the summer.[98]

Oppenheimer returned to the United States for a year in July 1927. That summer, Riefenstahl visited America, and Oppenheimer took her and some friends, including physicist Samuel Goudsmit, on a tour of New York. He gave them what Goudsmit called "the real Oppenheimer treatment . . . He met us in this great chauffeur-driven limousine, and took us downtown to a hotel he had selected in Greenwich Village. Then he took us to dinner at the Prince George Hotel . . . And there we sat, in the kind of restaurant I've hardly ever been to before or since . . . Very memorable."[99]

Oppenheimer worked at Harvard until Christmas and then moved to the California Institute of Technology in Pasadena, where he observed experimental work going on under Robert Millikan. While in America, he kept up correspondence with European physicists. He turned down a job offer from Harvard, choosing instead a joint appointment at the University of California, Berkeley and Caltech. Berkeley was an intellectual "desert" with "no theoretical physics." But Oppenheimer thought "it would be nice to try to start something." Worried about being too far out of touch, he decided to maintain his connection with Caltech.[100] And before he started the appointment, he was able to take another postdoctoral year (1928–29) in Europe. He visited Leiden to work with Ehrenfest and then went to Utrecht to work for one month with the Dutch theoretical physicist Hendrik A. Kramers. He had planned, following this, to join Bohr's Institute for Theoretical Physics in Copenhagen but was dissuaded by Ehrenfest, who arranged for him to work under Pauli in Zurich instead.

Ehrenfest thought that Bohr's "largeness and vagueness" would not be helpful to Oppenheimer. Rather, Oppenheimer said, Ehrenfest believed "that I needed someone who was a professional calculating physicist and that Pauli would be right for me . . . He thought in other words that I needed more discipline and more schooling . . . It was clear that he was sending me there to be fixed up." Ehrenfest wrote to Pauli, "For the development of his great scientific talents Oppenheimer needs right now to be morally and intellectually spanked into shape!" Victor Weisskopf, who was Pauli's assistant between 1934 and 1936, said that Oppenheimer did not stay long enough in Zurich to get the full "Pauli treatment"—but "those of us who stayed longer were all spanked into shape by Pauli, and we loved it." Nevertheless, Oppenheimer said, "I got to be not only extremely respectful but also extremely fond of Pauli and I learned a lot from him." Oppenheimer's chronic colitis still dogged him, however. He had succumbed to occasional illness while at Göttingen. Ehrenfest added health grounds to support Oppenheimer's application to the International Education Board to study in Zürich: "So long as he keeps that obstinate cough ... it is better that he perhaps remains where the climate is more favourable than it is here." Despite the milder climate in Zurich, Oppenheimer had to take a six-week break from working with Pauli, with doctor's orders not to do any physics.[101]

Rabi, another American studying at Zurich, was struck by the fact that although "Oppenheimer worked very hard that spring [he] had a gift of concealing his assiduous application with an air of easy nonchalance." Pauli remarked to him that Oppenheimer "seemed to treat physics as an avocation and psychoanalysis as a vocation." It was characteristic of Oppenheimer, Rabi wrote, that even while working on complex problems of astrophysics—such as a calculation of the opacity of surfaces of stars to their internal radiation— "he spoke little of these problems and seemed to be much more interested in literature, especially the Hindu classics and the more esoteric Western writers." At the time, Oppenheimer was learning Italian in order to read Dante in the original.[102]

His breadth of interests clashed with the disciplined focus required of the professional physicist. Göttingen, Oppenheimer had remarked, was "almost exclusively scientific, & such philosophers as are here are pretty largely interested in epistemological paradoxes and tricks." Dirac had reproached him: "I hear that you write poetry as well as working at physics. How on earth can you do two such things at once? In science one tries to tell people, in such a way as to be understood by everyone, something that no one ever knew before. But in the case of poetry it's the exact opposite!"[103] German scientific education in the 1920s was just as utilitarian as the American system, if not more so. Oppenheimer told Fergusson that at Göttingen, "they have an enormous respect for America, for Ford and Compton and the bogus realism of Sinclair Lewis"—and that "they are working very hard here, and combining a fantastically impregnable metaphysical disingenuousness with the gogetting habits of a wall paper manufacturer . . . Everyone else seems to be concerned about trying to make Germany a practically successful and sane country."[104] In this situation, Oppenheimer's general culture was sometimes interpreted as suggesting amateurism or dilettantism.

Between 1925 and 1929, then, Oppenheimer was immersed in and trained in what he called "the more critical, more disciplined, more professional science of Europe."[105] It was on the basis of this training that he made his key early contributions—in 1927, the Born-Oppenheimer approximation; in 1928, an article on the quantum mechanical idea of "tunneling," published in the *Proceedings of the National Academy of Sciences.* Under Pauli's guidance, Oppenheimer began work on the fundamental problems of quantum mechanics and on analyzing remaining difficulties with quantum electrodynamics.[106] By 1930, he was a member of a small, specialized elite circle, working at the esoteric forefront of the new physics. At the same time, however, he maintained a self-presentation as a generally cultivated aesthete and leisured gentleman— identities frequently regarded as incongruous with the role of the scientific specialist, which Oppenheimer was pursuing in his education and career.

This chapter has suggested that in Oppenheimer's youthful self-formation, the interplay and tensions between general aesthetic cultivation and specialized expertise were closely bound up with questions of Jewish identity and assimilation. The German Jewish attachment to *Bildung* was expressed in the habitus of Oppenheimer's parents and was reinforced and refracted through his Ethical Culture education. This attachment was an aspect of the German Jewish search for a secular, universalistic conception of self and identity, an ideal developed in an American context by Adler. Paradoxically, however, this model of universalistic identity became identified in a particularistic way as (German) Jewish. Just as Bernheim, on entering Harvard, hoped to break out of the "Jewish enclave" of the Ethical Culture School, so Oppenheimer's rebellion against Adler's "morality compressed" was a further escape from Jewish identification. His attraction to a flamboyant literary and artistic persona, made available by Smith and his circle of young disciples, was in part a reaction against the sober moralism of his school and his father. New Mexico, too, seemed to offer a space where he could create himself in an unfettered way. This active self-creation, at once rebellious and unconfident, would in turn help Oppenheimer negotiate the institutionally anti-Semitic environment of Harvard. Through his cultivation of an elite habitus, displaying effortless superiority, Oppenheimer, whether consciously or not, constructed himself in opposition to stereotypes of the earnestly achieving, upwardly mobile Jewish student. This was his "diffidence of manner" that Bridgman found both "engaging" and un-Jewish. This mode of self-construction, however, became more problematic as Oppenheimer entered the world of theoretical physics, a professional community within which Jewishness was not an obstacle (though during his time at Göttingen, Oppenheimer was aware of the climate of German nationalist anti-Semitism that was soon to impinge on this world). Membership in this scholarly community meant adjusting to its own countervailing pressures toward disciplined intellectual focus and the professional self. Oppenheimer had to choose between making physics a vocation or, as Pauli put it, an avocation.

Yet even as Oppenheimer pursued the vocation of theoretical physics, he combined it with his earlier literary-creative habitus. This characteristic combination became part of his appeal as a flamboyant, cultured teacher of physics, with a tinge of the outsider, at Berkeley in the 1930s.

CHAPTER THREE

Confronting the World

BERKELEY: THE TEACHER

Oppenheimer returned to California in the summer of 1929 as a recognized member of the European community of theoretical physicists and a participant in the revolutionary new field of quantum mechanics. His emerging professional identity combined this expertise with his general aesthetic cultivation. The chair of the Berkeley physics department, Raymond T. Birge, remarked on the scope of Oppenheimer's knowledge and understanding in physics and his cultural interests outside science. Oppenheimer's cultural capacities and presentation of self were particularly important in his role as a teacher. Oppenheimer most powerfully transmitted the culture of quantum theory not through the impersonal mechanism of publication, but rather through face-to-face teaching and the establishment of a school of theoretical physics at Berkeley. Indeed, his role in the development of quantum theory in America was most importantly in transmission, rather than extension. His main influence was in building up a first-class school of theoretical physics.[1]

Oppenheimer was brought to California as a representative and carrier of the largely European culture of the new physics. In founding the first American school of theoretical physics at Berkeley in the 1930s, he formed a bridge between the "great tradition" of the European theoretical schools and the emerging vitality of an American program in both theoretical and experimental physics.[2]

Oppenheimer was also able to bridge the divide between the worlds of experimental and theoretical physics. At Berkeley, his research was centered around Ernest Lawrence's cyclotron and the experimental program of the Radiation Laboratory (Rad Lab). Birge was enthusiastic about Oppenheimer's ability to provide theoretical interpretations of experimental findings: "He has a more extensive knowledge of experimental results than any theoretical physicist that I know . . . [and] is a perfect team-mate for Professor Lawrence."[3]

Experimentalists and theorists met regularly, at weekly colloquiums and the weekly Journal Club. Robert Wilson, as a young experimentalist working in the Rad Lab, was urged by Lawrence to attend the theory seminar. "We won't understand a word," Lawrence told him, "but it's good for you to hear what the theorists are talking about." Wilson continued to attend these seminars and found them useful for acquiring "a flavor of what was going on . . . [to] get a feeling for what's exciting, and maybe think about a theoretical problem myself for a bit. Or, sometimes, it was like going to Quaker Church, where you just sit and think."[4]

Oppenheimer's influence was, however, strongest within his small circle of students, for whom he took on a charismatic aura. Birge noted the role of Oppenheimer's personality in his teaching: he was, "through his inspiration, interest, and personality, attracting graduate students to him." He worried about Oppenheimer's tendency to "go over the heads" of his students and about his impatience with those less intellectually agile than himself. Nevertheless, Birge saw improvement on this front over time, and Oppenheimer developed quickly as a teacher. Indeed, part of what attracted students to Oppenheimer was his oracular quality. According to Isidor Rabi, he "never expressed himself completely. He always left a feeling that there were depths of sensibility and insight not yet revealed." Oppenheimer's group of eight to ten graduate students and about half a dozen postdoctoral fellows met every day in Oppenheimer's office. They would discuss each other's work in progress, ranging from electrodynamics to cosmic rays, astrophysics, and nuclear physics. Oppenheimer would advise each student on how to proceed. After he left, most of the students would stay behind so that Oppenheimer's close intellectual collaborator, postdoc Robert Serber, could interpret, explaining to each student "what Oppie had told him to do." "They were," Serber recalled, "much more willing to display lack of understanding to me than they were to Oppie." Such close, daily interaction instilled a sense of collective endeavor, a group identity centered upon a common apprenticeship to a master—an apostolic model of education, so to speak.[5]

Oppenheimer's powerful personal influence was compounded by the fact that the physics group at Berkeley had relatively little direct contact with the wider theoretical physics community. Although there was the odd distinguished visitor, notably including Niels Bohr in 1933 and Enrico Fermi in 1935, the general isolation was an important factor in Oppenheimer's becoming, for the young scientists, the personal embodiment of the role of theoretical physicist. A close friend and admirer of Oppenheimer's, the philosopher David Hawkins, recalled that "he had devoted students who would stay up all night studying something in order to be able to say something rather casually that he would approve of." They were famous for imitating his mannerisms. Pauli referred to the theoretical physics group at Berkeley as Oppenheimer's "nim nim nim boys," because they adopted their professor's tendency to mumble incomprehensibly while in thought. They copied him down to the last detail, including his flat-footed walk and his habit of jumping up to light others' cigarettes or pipes. They also adopted his habit of signaling assent or thought with a Germanic "ja, ja." According to Wilson, Oppenheimer "affected a German mode." The connection with Europe was thus not only intellectual but was also inscribed in the cultural orientation and habits of Oppenheimer's theoretical physics group.[6]

Following Oppenheimer's example of what it meant to be a theoretical physicist and an intellectual also involved emulating his cultural and aesthetic tastes. So, according to Wilson, Oppenheimer's students would listen to "Bach. Always Bach. The B Minor [Mass] . . . must have been a favorite of Oppie." They cultivated a style that marked them as a set. "They were," Wilson remembered, "a separate group from the students in the Rad Lab. I think they somewhat disdained those of us who were working in the lab." Oppenheimer's students went everywhere with him, following him to Caltech, where he taught in the summers, and to his holiday ranch in New Mexico. Wilson reflected, "I was jealous because I was not one of the theory students. After you had been a student of Oppie's for a certain time, then you became a member of his circle. They would go off to fancy dinners with Oppie, to his home, and they did social things . . . They lived a high kind of life that we would hear about. I looked at them with great envy and hoped to become a member of that clique somehow, but I never made it."[7]

Oppenheimer regarded dining at expensive Bay Area restaurants and sampling fine wines as part of a "total education" for his students. Theoretical physicist Rudolf Peierls wrote that Oppenheimer "had strong views on questions of style in food and drink. Martinis had to be strong. Coffee had to be black." Peierls recalled that when Oppenheimer took the members of a committee to a steak house after a meeting, nearly everyone took the cue from Oppenheimer that the steak should be ordered rare. Only one person ordered his beef well-done. "Robert looked at him and said, 'Why don't you have fish?'" Wilson remembered that an evening at the Oppenheimer house in Berkeley typically involved "the driest of martinis mixed by the hand of the master, sophisticated guests, gourmet food (but on the scant side), and amorphous buzz of conversation, smoke, alcohol." One speciality was Oppenheimer's version of the spicy Indonesian dish *nasi goreng,* which among his guests earned the nickname "nasty gory." Edward Teller, the first time he met Oppenheimer (in 1937), was taken to a Mexican restaurant in San Francisco. Teller found the combination of the spicy food and Oppenheimer's personality so "overpowering" that he lost his voice.[8]

Oppenheimer's ability to be the center of the group, to treat his students to dinner, to throw lavish parties, was due to his inherited wealth.[9] This material capital was a necessary support for his social capital. Others may have aspired to his munificence but could not carry it off. Oppenheimer regularly paid for dinner for large groups of students and colleagues at expensive restaurants in San Francisco. Serber remembered that on one such occasion, after a joint seminar with Stanford, "[Felix] Bloch grew expansive, and leaned over and picked up the check. He looked at it, blinked, leaned over again and put it back down."[10] In the midst of the Depression, Oppenheimer's wealth and extravagance certainly added to his personal aura. His money enabled him, to some extent, to adopt the attitude of the gentleman-amateur, in contrast with that of the harried professional. His lavish lifestyle and personal sophistication were a source of authority that contributed to his ability to lead and bind together a new school of theoretical physics.

A friend and colleague from Berkeley remembered that Oppenheimer "found it very difficult to form proper relationships with equals" and tended to surround himself with "adoring disciples." Outside this small circle, such adulation was looked at with some bemusement. The Italian physicist Emilio Segrè was always skeptical about Oppenheimer. When Segrè was working at the Rad Lab, he found that Oppenheimer "was considered a demigod by himself and others at Berkeley, and as such he spake in learned and obscure fashions." To Segrè, Oppenheimer's physical research hardly merited worship. Oppenheimer and his students at Berkeley tended to attack problems prematurely, "resulting in indifferent success." Oppenheimer's work on cosmic rays and atomic and nuclear physics "embodied many good and prescient ideas, but was often inconclusive." His best work, in astrophysics, proved "truly prophetic" only after his death. Moreover, Segrè was not impressed by the high culture of Oppenheimer and his "acolytes": "Oppenheimer and his group did not inspire in me the awe that they perhaps expected. I had the impression that their celebrated general culture was not superior to that expected in a boy who had attended a good European high school. I was already acquainted with most of their cultural discoveries, and I found Oppenheimer's ostentation slightly ridiculous."[11]

While his students imitated him, Oppenheimer had a tendency himself to fix on particular people to idealize and emulate. At Caltech, he met the physical chemist Linus Pauling. They were the same age and became fast friends, planning to collaborate on scientific work. Oppenheimer was fascinated by Pauling and began to copy him. Pauling's biographer wrote that Oppenheimer "not only adopted some of Pauling's lecturing style; he began wearing an old fedora around campus, much like the one that Pauling wore." He bestowed lavish gifts on the chemist, from an expensive ring to items of great personal value, such as his childhood mineral collection. He wrote poems for Pauling that the recipient found "both obscure and troubling, mixing classical allusions with lines about mineralogy, Dante, and pederasty." Their friendship ended when Oppenheimer made a pass at Pauling's wife, which she found both flattering and resistable. Years later she told her husband, "You know, I don't think Oppenheimer was in love with me. I think he was in love with you."[12]

There were rumors circulating at Berkeley that Oppenheimer had had a homosexual affair with Harvey Hall, one of his first graduate students there. This gossip later came to the attention of FBI agents investigating Oppenheimer's political activities.[13] Whether or not these rumors were true (and there is no evidence to support them), it seems fairly likely that sexual ambivalence was one aspect of Oppenheimer's youthful problems of identity. He was, at any rate, frustrated with his lack of success in what he called his "negotiations" with the opposite sex. During his first year at Berkeley, he told his brother that girls were not worth worrying too much about. "The obligation," he asserted, "is always on the girl for making a go of conversation: if she does not accept the obligation nothing that you can do will make the negotiations pleasant." It was futile, he said, to be concerned about whether one was pleasing to women.[14]

During this early period at Berkeley, Oppenheimer developed a strong interest in ascetic philosophies and in ideas of discipline, transcendence, and renunciation, becoming particularly interested in Hinduism. He had begun to read Hindu texts in translation while he was an undergraduate at Harvard, and Rabi remembered him talking about Hindu classics while in Zurich. At Berkeley, Oppenheimer read more widely and deeply in this tradition. In 1931 he began to learn Sanskrit; he told his brother that he was "enjoying it very much and enjoying again the luxury of being taught."[15] In letters to Frank, he dwelled on themes of discipline and on the goal of transcendence, drawing not only on Hinduism but on a range of ancient, classical, and medieval models of the ascetic and contemplative life.[16] Oppenheimer archly advised his brother to cultivate "a little leisure, a certain detached solitariness and a quiet discipline which uses but transcends the discipline of our duties." In practical terms, he advised him to spend "not too hellishly many hours" on school and to eat "once in a while for your humility a green vegetable." With the help of such a regime, he said, one could hope to achieve "the assurance and the certainty, and to some extent that delectatio contemplationis which is the reward and reason of our way of life."[17]

Although he embraced his younger brother in speaking of "our way of life," Oppenheimer was worried about his sibling's decision to follow in his footsteps and become a physicist. He lectured Frank paternalistically on the responsibility of choosing a vocation: "I am sure you are right in preferring physics as a science to study and learn; but should you prefer it as a science at which to work, a vocation? By all means, and with my whole blessing, learn physics, all there is of it, so that you understand it, and can use it and contemplate it, and, if you should want, teach it; but do not plan yet to 'do' it: to adopt physical research as a vocation. For that decision you should know something more of the other sciences, and a good deal more physics." He told his brother that he was worried about "the possibility that you are more and more deeply committing yourself to a vocation which you will regret."[18]

At the heart of "vocation," Oppenheimer suggested, was "discipline." Of this he wrote, "I do value it—and I think you do too—more than for its earthly fruit, proficiency." Listing in his support "the bhagavad gita, Ecclesiastes, the Stoa, the beginning of the Laws, Hugo of St Victor, St Thomas, John of the Cross, Spinoza," Oppenheimer wrote that "through discipline, though not through discipline alone, we can achieve serenity, and a certain small but precious measure of freedom from the accidents of incarnation and charity, and that detachment which preserves the world which it renounces." Oppenheimer emphasized that it was the shaping of one's inner life that mattered above all. He announced that he wanted to be in the world, but not of the world.[19]

In 1933, Oppenheimer started attending Thursday-evening readings of the Bhagavad Gita in the original, under the guidance of the renowned Berkeley Sanskritist Arthur Ryder. Oppenheimer later described the experience and his view of Ryder as a moral paragon: "Ryder felt and thought and talked as a Stoic . . . a special subclass of people who have a tragic sense in life." Ryder gave him "a feeling for the place of ethics" and an understanding of vocation, teaching him that "any man who does a hard thing well is automatically respectable and worthy of respect." Oppenheimer found, expressed in the Bhagavad Gita and embodied in Ryder, an ascetic ethos on which to model his own self-fashioning. The Hindu notion of dharma meshed with Oppenheimer's interest in ascetic discipline and with his understanding of the pursuit of science as a calling, or vocation. Oppenheimer often used the language of the Gita to express ideas not substantively different from the Protestant work ethic.[20]

It is also possible, however, that Oppenheimer was intrigued by the Hindu classic's dynamic tension between detachment from and engagement in the world. He was ambivalent between the urge to seek refuge from a world in which he felt ill at ease and the competing desire for active involvement and intimacy with others. He was, so to speak, uncomfortable in his own skin—hence, perhaps, the appeal of what he called "freedom from the accidents of incarnation." Rabi later expressed disapproval of Oppenheimer's broad cultural interests and commented that if Oppenheimer had "studied the Talmud rather than Sanskrit," it "would have given him a greater sense of himself."[21] But for Oppenheimer, the point was that, unlike the Talmud, the Gita carried a message that appeared to transcend its particular religious and cultural tradition.

Felix Adler had sought to extract from Judaism and Christianity a universal morality that could be separated from religious doctrines and religious ritual. Oppenheimer's search for moral meaning beyond these Western traditions could be seen as continuous with Adler's universalistic ideal. However, Oppenheimer found Ethical Culture itself to be too narrow and constraining. Rabi observed that "too great a dose of ethical culture can often sour the budding intellectual who would prefer a more profound approach to human relations and man's place in the universe."[22] Oppenheimer's interest in the Gita and his prescription of renunciation, hardship, detachment, and inner peace in letters to his brother were a reaction against Adler's project. Whereas Ethical Culture emphasized moral action in the world aimed at improving human welfare, what Oppenheimer proposed verged on mystic renunciation of the world.

The ideal of hardship and the strenuous life that he prescribed to his brother stood in opposition also to his parents' ideal of cultivation, which expressed the German-Jewish attachment to *Bildung*. Instead of the flowering of the self through aesthetic cultivation, Oppenheimer now suggested the subjugation of the self through ascetic discipline. But there is no need to assume that Oppenheimer practiced what he preached. He did not live a life of ascetic self-denial. Instead, his study of Sanskrit and the Gita served as evidence of his cultivation. It was primarily an aesthetic and decorative accomplishment and a display of virtuosity. Hawkins recalled with a laugh, "I once was sitting in his living room before the war in Berkeley, and to the left on the bookshelf was a whole string of classics. I saw Plato and pulled down a volume, and I said, 'You know, I've just been studying this volume.' And he said, owlishly, 'I've read the Greeks, I find the Hindus deeper.' Wow! One-upmanship! He had that side to him, too."[23] Instead of a path to the disciplining of the self, Oppenheimer's interest in Hinduism functioned as ornamental cultural capital. Similar tensions between ascetic discipline and aesthetic cultivation, and between engaging in the world and renouncing it, were at the heart of Oppenheimer's scientific and intellectual identity.

OPPENHEIMER'S INTRODUCTION TO POLITICS

Oppenheimer's letters to his brother in the early to mid-1930s reveal a focus on inner self-discipline, as well as an attachment to aesthetic modes of self-fashioning. Though in some ways competing, these preoccupations came together in their inward focus on the self. In the later 1930s, however, this solipsistic and introspective orientation was to be shattered by new relationships and a growing political consciousness, leading Oppenheimer to new kinds of engagement in the world. Before the mid-1930s, he was highly insulated from political events. Harvey Hall later told the FBI that during those years, Oppenheimer "had very limited outside acquaintances and interests." Beyond physics, Oppenheimer was interested in "art subjects and classical concerts and movies," but not politics. Oppenheimer testified at the 1954 security hearings that in this period, he "was not interested in and did not read about economics or politics. I was almost wholly divorced from the contemporary scene in this country. I never read a newspaper or a current magazine like Time or Harper's; I had no radio, no telephone; I learned of the stock-market [crash] in the fall of 1929 only long after the event; the first time I ever voted was in the presidential election of 1936." Oppenheimer's interest in progressive politics was catalyzed by a number of influences. His father was helping relatives in Germany, including his own sister, to immigrate to America in order to escape from the Nazis. When Julius died in September 1937, Robert became their sponsor. He was deeply affected by what he learned of the Nazi persecution of the Jews, feeling what he later described as a "continuing, smoldering fury about the treatment of Jews in Germany." In addition, although his own wealth insulated him against the Depression, he was made aware of its effects by the plight of his graduate students, who were facing enormous problems finding jobs and for whom Oppenheimer had difficulty securing funds from the department.[74]

Oppenheimer's growing political involvement was closely interwoven with his relationships and social life. His first mature love affair began in the spring of 1936, when he fell passionately in love with a young woman named Jean Tatlock, the daughter of a Berkeley professor of medieval literature. Oppenheimer later said, "We were at least twice close enough to marriage to think of ourselves as engaged." Tatlock was, as Oppenheimer put it, an "on again, off again" member of the Communist Party. She introduced him to many of the causes and groups that he would support in the coming years. However, their relationship was stormy, and they broke up in early 1939. Tatlock was subject to periodic episodes of serious depression. The last time they saw each other was during one particularly bad such episode in the early summer of 1943. Oppenheimer responded to her call for help and, watched by security agents, made a trip back from Los Alamos to Berkeley to visit her. In January 1944, she committed suicide.[25]

Oppenheimer took a strong interest in the Spanish civil war. For a few years, he donated about a thousand dollars annually through Communist channels to the Spanish Loyalist cause and attended fund-raising parties for the Loyalists. He began a subscription to the *People's World,* the newspaper of the Communist Party on the West Coast. In April 1938, he signed a petition by academics calling on President Roosevelt to lift the embargo on the Spanish republic. He also became involved in local activism, working, for example, as recording secretary for the Teacher's Union at Berkeley. He joined the San Francisco Executive Committee of the American Civil Liberties Union in January 1939. He also sponsored the union of progressive scientists, the American Association of Scientific Workers. And he took an interest in local issues such as the plight of migrant workers in California and the longshoremen's strike.[26]

There is no conclusive or firm evidence that Oppenheimer ever joined the Communist Party. But he certainly cooperated closely with it during this Popular Front period, when the party was seeking alliances with liberals and progressives and when the class struggle was subordinated to the struggle against Fascism. Through his political involvement, Oppenheimer became close friends with Haakon Chevalier, a young leftist professor of Romance languages. Chevalier idolized Oppenheimer; for him, the years from 1937 to 1943, when the physicist left for Los Alamos, were "dominated by the personality of Oppenheimer." Chevalier later alleged that Oppenheimer had been a member of the same Communist Party cell that Chevalier belonged to. Oppenheimer denied this; he always maintained that he had never been a party member, although he once said that he had "probably belonged to every Communist-front organization on the West Coast." Oppenheimer engaged with a variety of progressive causes associated with the Popular Front and was actively involved in the Berkeley leftist milieu, in which party and non–party members worked together and in which political outlook was more important than a membership card. Oppenheimer told the AEC hearings that he had acquired a favorable view of the Soviet Union from reading, in 1936, Sidney and Beatrice Webb's *Soviet Communism: A New Civilization?* This book "and the talk that I heard at the time," he said, "had predisposed me to make much of the economic progress and general level of welfare in Russia and little of its political tyranny."[27]

Oppenheimer's involvement with progressive causes during the late 1930s meshed with and reinforced his role as a charismatic teacher. Indeed, his former student Philip Morrison credited Oppenheimer's development into a great teacher to the change in worldview that followed his political involvement: "My experience was that when I got there in '36, he . . . had become a superb teacher. I know that in '34 he was not. I'd talked to people who saw the change. He changed greatly in his attitude toward teaching and to students, between '34 and '36." This was not due merely to the accumulation of experience and confidence in the classroom. Rather, "it was an actual change in deep personality. He was much less isolated, more concerned with politics, more concerned with the outer world, more concerned with other people. He was sensitized. Before that, he was a very self-indulgent person . . . He spent his time studying Sanskrit and so on. He didn't care much about other matters . . . And I don't think I would have done very well if I'd been with his earlier self. I know he would not have increased the number of students the way he did."[28]

Oppenheimer's circle of students tended to share his political views. According to Chevalier, "when they learned that Opje was a member of the Teachers Union the young teaching assistants joined the union too. Many of them were or became left-wingers, by gravitation or contagion, and they were always eager to hear his views on the rapidly evolving political events at home and abroad." Physicist Edward Condon, rather dismissively, described Oppenheimer as "kind of an active parlor pink" at that time. "And," he said, "that's how all these other boys who were his students and were great imitators of him in every way, they too got involved with various left wing labor business."[29] It would be wrong, however, to think that the politics of Oppenheimer's students were merely imitative. For many, their left-wing political commitments were formed before arrival at Berkeley.

According to Morrison, about half of Oppenheimer's ten or eleven graduate students were "seriously left students and wanted to do everything that you can do." It was, he said, "especially galvanizing because Oppenheimer's landlady . . . was a very energetic and enterprising lady of high position socially, had a good deal of money, espoused the cause of the Spanish Medical Bureau, which was an organization to collect funds and to send people to go to serve as nurses and doctors in Spain and above all to support them with supplies, and things of that sort, which we were all engaged in. And Oppie's house was a good place to meet for that."[30] There was often very little distinction between a political event and a social event. So there were frequent parties to raise money for republican Spain and later on behalf of Russian War Relief. Chevalier wrote, "These were invariably lively and successful affairs, with some celebrity as a drawing card . . . and the reliable contingents of pretty girls . . . who would pass through the crowd selling flowers or encouraging people to buy drinks." Chemist Martin Kamen called the Berkeley leftist milieu the "cocktail front."[31]

Oppenheimer's relationship with Jean Tatlock may have been the catalyst to his political engagement, but it was far from being the only factor.[32] The causes and groups with which Oppenheimer became involved in the late 1930s satisfied a longing for involvement and for camaraderie. The progressive and internationalist character of the Popular Front appealed to him. Oppenheimer's left-wing political views during the thirties were to some degree continuous with the social values of the Ethical Culture School. It seems that Oppenheimer, like many intellectuals, was attracted to the cultural agenda of the Left, above all to the ideal of constructing and belonging to a new form of community—one that promised to overcome traditional divisions between people, both of class and of ethnicity. Sociologist Nathan Glazer has argued that American Jews who joined the Communist Party were often drawn to it by this cultural agenda: "To be a Communist meant to shed the limitations of one's social reality, and to join a fraternity that transcended the divisions of the world. This was the attraction of Communism to many Jews who no longer thought of themselves as in any way Jewish."[33]

When Oppenheimer later reflected on his political involvements of the 1930s, he also cast them in terms of issues of community, culture, and identity. In 1948, he described his awakening to politics as a coming to maturity: "I'm not ashamed of it; I'm more ashamed of the lateness . . . Most of what I believed then now seems complete nonsense, but it was an essential part of becoming a whole man. If it hadn't been for this late but indispensable education, I couldn't have done the job at Los Alamos at all."[34] He said at the security hearings that his youthful political involvement gave him a way to "participate more fully in the life of the community . . . I liked the new sense of companionship, and at the same time felt that I was coming to be part of the life of my time and country."[35] Oppenheimer, then, described his political interests in terms of self-shaping, of personality, of "becoming a whole man." It was, for him, an aspect of self-cultivation. But in contrast to his earlier, inward-looking focus, Oppenheimer now understood such self-shaping to require active involvement in a political community and engagement with the world. So Oppenheimer's activist, left-wing political involvement was both an extension of the universalism of *Bildung* and Adler's secular religion of Ethical Culture and a rejection and transcendence of the limitations of these outlooks in favor of a more worldly, more active, and even more universalistic philosophy and orientation.

In August 1939, Oppenheimer met Katherine (Kitty) Puening Harrison, and they married the next year. Kitty's second husband had been Joe Dallet, the Dartmouth-educated son of an investment banker. Dallet had severed ties with his family, joined the Communist Party, and gone to Spain to fight with the International Brigades. Kitty had also joined the party. She was permitted to do so only after she had performed "a number of tasks which were extremely painful to me, such as selling the *Daily Worker* on the street and passing out leaflets at the steel mill." Kitty, who came from a wealthy and established German family, found the poverty in which she lived with Dallet "depressing," and they separated in 1936. But the following year, she had a change of heart. In June 1937, just before Dallet went to Spain, they had a reunion in Paris. Kitty promised to join him in Spain. A few months later, he was killed in action. She returned to the United States, enrolling at the University of Pennsylvania to study biology, and married her third husband, a British physician named Richard Harrison.[36]

Oppenheimer met Kitty at a garden party in Pasadena while she was still married to Harrison, who was then doing an internship at a Los Angeles hospital and was friends with a number of Caltech physicists. Kitty said, "I fell in love with Robert that day, but hoped to conceal it." The following year, Oppenheimer asked the couple to join him for the summer at his New Mexico ranch. Kitty accepted the invitation, while Harrison stayed behind to study for his examinations. Kitty said that during that time at the ranch, "Robert and I realised that we were both in love." On November 1, 1940, Kitty divorced Harrison and married Oppenheimer. She was already pregnant with their first child, Peter, who was born in May 1941. Their affair had created a minor scandal at Berkeley.[37]

It was via Kitty that Oppenheimer came to meet Steve Nelson, a prominent Bay Area Communist Party organizer. Nelson had been a friend of Dallet's and had stayed with Kitty for two weeks in Paris to help her come to terms with the news of her husband's death. Oppenheimer and Nelson met in autumn 1940 at an event held to raise money for Spanish refugees. Nelson recalled that it was Oppenheimer who introduced himself after giving a talk about the Spanish war, surprising him by announcing that he was about to marry a friend of Nelson's. Oppenheimer invited Nelson to his house for a get-together with university people who wanted to meet a veteran of the International Brigades. The two men subsequently met occasionally as Nelson worked on organizing people in the university; he remembered that "a number of Oppenheimer's graduate students in the field of physics were quite active." Nelson said, "Our contacts were more on their terms than ours. They lived in a more rarefied intellectual and cultural atmosphere." He was impressed by Oppenheimer, particularly when the physicist "casually remarked" that he had read all three volumes of Marx's *Capital* on a three-day train ride. As Nelson put it, "Oppenheimer had such intellectual presence that almost everyone deferred to him. I was very impressed with our discussions and began to admire him." But he added that they "never became close friends" and that Oppenheimer's connection with the Communist Party had been "tenuous at best."[38]

Kitty's Communist connections, via Dallet, were to become a subject of interest to military intelligence and the FBI during the war, and they were also brought up in the 1954 hearings. But Kitty herself was far from a dyed-in-the-wool leftist. Jackie Oppenheimer, Frank's wife (both of them were Communist Party members), intensely disliked Kitty and said, "All her political convictions were phony." Following his marriage to Kitty, Oppenheimer withdrew more and more from the political causes and groups in which he had been involved. He later said that after his marriage, "a certain stuffiness overcame me."[39]

Like many affiliated with the Popular Front, Oppenheimer became disenchanted with the causes of the 1930s as World War II loomed. This change began for Oppenheimer when he learned more about the purges and political repression in the Soviet Union. He was particularly influenced by the negative opinion of Russia expressed to him by the politically liberal physicists George Placzek, Marcel Schein, and Victor Weisskopf, who had visited the Soviet Union a year and a half earlier.[40]

The Soviet-Nazi nonaggression pact was disillusioning for Oppenheimer, as it was for many intellectuals who had engaged with Communists in the cause of building a united front against Fascism. But Chevalier suggested that Oppenheimer's first reaction was to try to rationalize the Soviet position: "He communicated with extraordinary effectiveness his own conviction that political events ... could be made to yield their significance if examined objectively, in the light of the factors that had conditioned them."[41]

According to Segrè, "Oppenheimer and his acolytes followed the political line of the Communist Party of the United States," and Oppenheimer would often repeat "with the faith of the true believer, the nonsense originating from Stalin's Cominform." Chevalier recalled that in late 1939 or early 1940, Oppenheimer suggested writing pamphlets for circulation among academic colleagues setting out the Left's position on issues such as the nonaggression pact. Oppenheimer's hope, said Chevalier, was that "a well-written, well-printed and dignified-looking four-page brochure dealing seriously with some of the most fundamental current political problems, could have an incalculable impact on the academic mind." William R. Smythe, a physics professor at Caltech, told a security agent in July 1943 that "there was at one time a small pamphlet put out to be sent to the various professors of the California Institute of Technology, which definitely indicated a Communistic influence. [Smythe] ... personally suspected that Robert Oppenheimer was the one who instigated the sending of these pamphlets [but] had no tangible proof or evidence that such was the case."[42]

Nevertheless, after the nonaggression pact, Oppenheimer began to move away from his previous political causes. Morrison described the pact as a "seismic fault in the whole relationship between certainly the academic lefties and the Communist Party." After it, "the politics was less clear." Oppenheimer himself began to feel that he "had had enough of the Spanish cause and that there were other and more pressing crises in the world."[43] When World War II came, he embraced the war effort with the same fervor and the same ideals that had been behind his interest in the Spanish cause. Hans Bethe recalled how at the 1940 meeting of the American Physical Society, Oppenheimer "talked for quite a long time" about the significance of the fall of France. "He told us how much France meant to the western world, and how the fall of France meant an end of many things that he had considered precious and that now the western civilization was really in a critical situation, and that it was very necessary to do something to save the values of western civilization." Oppenheimer told the assembled scientists that they would have to "defend western values against the Nazis" and that "because of the Molotov–von Ribbentrop pact we can have no truck with the Communists."[44]

WAR, POLITICS, AND THE ATOMIC BOMB

Oppenheimer's left-wing political views strained his relationship with senior figures in the physics department at Berkeley. Oppenheimer was recognized as a brilliant theoretician, and his success in building a prestigious school of theoretical physics had significantly added to the renown of the department. Lawrence, the founder and director of the Rad Lab, greatly respected Oppenheimer as a physicist and also, it seems, had considerable affection for him as a person. However, there were also tensions, hinted at in the letter Oppenheimer wrote to Lawrence in August 1945 giving his reasons for deciding not to return to Berkeley: "I think it would not have seemed so odd [to you] . . . nor so hard to understand if you remembered how much more of an underdogger I have always been than you."[45] Caltech president Robert Millikan told a security agent in July 1943 that "while Robert Oppenheimer was at the University of California he was regarded as an extreme radical and having some subversive tendencies." He added that he thought Oppenheimer "unquestionably would have been fired had it not been for his extreme brilliance in his particular field."[46]

Both Birge (the department chair) and Lawrence were at odds with Oppenheimer over what Birge referred to as "the type of person who should be added to the department": "The only sort of person that Oppie seems to want is just the sort that Lawrence and I do not care to have on our staff." Birge explained, "New York Jews flocked out here to him and some were not as nice as he was . . . Lawrence and I were very concerned to have people here who were nice people as well as good students."[47] The chemist Martin Kamen, who worked at the Rad Lab during the late 1930s and early 1940s, thought that Lawrence "was ill at ease among Bohemian types." "Bohemian," in that context, meant Jewish and left-wing. The label was often attached to Oppenheimer's brother, Frank, who was then studying at Caltech—although, interestingly, not so much to Robert himself.[48]

When Oppenheimer enrolled Bernard Peters for graduate studies, Birge objected that the young man did not have a bachelor's degree. Peters and his wife had come to America as refugees after he had escaped from a German concentration camp. According to Condon, Peters "had never really had any undergraduate university work when he went to Dachau in the middle of his first year," but he was a "brilliant fellow," and Oppenheimer took him on for that reason. Oppenheimer, said Condon, was "a great iconoclast" and "was a great one to break university rules and get away with it."[49] Tensions between

Oppenheimer and Birge became marked toward the end of the war, with the department turning down Oppenheimer's recommendations for postwar appointments—refusing to commit to a place even for the brilliant Richard Feynman.[50]

Despite the mutual respect that Lawrence and Oppenheimer had for each other as physicists, their political differences became increasingly marked. In contrast to Oppenheimer's leftist milieu, Lawrence paid court to the rich and powerful as he sought to raise funds for the increasingly expensive research at the Radiation Laboratory. Lawrence, like most senior faculty of the Berkeley physics department, was rigidly conservative in his politics. In the 1930s, the personnel manager for the Rad Lab was George Everson, who was very anti-Communist and anti–New Deal. In reaction to the prevalence of left-wing ideas among young researchers at the Rad Lab, Lawrence "effectively banned" political discussion in the lab.[51]

Herbert York, then a graduate student in the lab, described Lawrence's rationale: "Scientists, he said, especially young ones, should not waste precious working time on extraneous issues for which they had no special training." It is significant that Oppenheimer did not see political activism as extraneous. The fact that he wrote an invitation to a party benefiting the Spanish Loyalists on a Rad Lab blackboard (which Lawrence later erased) suggests that he saw no need to separate his intellectual and academic life from his political life. Morrison, similarly, saw these roles as continuous: "I used to talk a lot—especially to the academic community in various groups and clubs and so on—about the fact that modern physics was answering the great problems of the world and that attention to it was an essential part of the modern world, and no political movement could afford not to take into account what this might mean." Morrison was at the time a reader of J. D. Bernal and other British socialist writers, who emphasized the importance of science as the key force of production and social change in modern society. Oppenheimer would have encountered similar ideas in the Webbs' book, which called science the "salvation of mankind" and praised the Soviets' "devotion to science," as well as their planning of scientific research. There is, however, no evidence that Oppenheimer was influenced directly or persuaded by Bernalist ideas.[52]

It turned out to be the war that would connect science most decisively with larger social, political, and economic forces. American science was harnessed to the war effort first under the National Defense Research Committee (NDRC), created in 1940 by Roosevelt at the urging of the president of the Carnegie Institution, engineer Vannevar Bush. From the early fall of 1941, the Rad Lab was increasingly given over to war research for the atomic bomb project, organized under the Office of Scientific Research and Development (OSRD)—the organization (again directed by Bush) that in June had incorporated and replaced the NDRC as the umbrella for wartime scientific research. Work at the Rad Lab on the separation of uranium isotope was carried out under the OSRD's S-1 Section, the branch that under the supervision of Bush's deputy, Harvard president James B. Conant, coordinated atomic bomb research.[53]

It was during this time that Lawrence's disapproval of left-wing political activism hardened into prohibition. Oppenheimer had previously been able to combine his political activism and membership in a radical circle with the maintenance of good relations with Lawrence. The entry of the United States into the war and the militarization of scientific research meant that Oppenheimer could no longer hold together his position as both outsider political activist and academic insider. This tension, which had already become evident before Pearl Harbor, was evinced by Oppenheimer's oscillations with regard to his support for the American Association of Scientific Workers (AASW).

Late in 1938, Martin Kamen attended a scientific meeting in honor of Nobel Prize–winning chemist Harold Urey. There he picked up information about the AASW, which had recently been formed on the national level. Kamen agreed to pass the material on to Oppenheimer and psychologist Edward C. Tolman, who were known as the most active among the left-wing and liberal scientists on the Berkeley campus. Kamen, however, found Oppenheimer extremely worried about an un-American activities committee recently set up by the California State Assembly. According to Kamen, "Oppie felt that his participation in any attempt to help the [A]ASW might be its 'kiss of death,'" and Oppenheimer took a position against setting up a campus branch of the organization. It struck Kamen at the time that Oppenheimer was exaggerating the risks. But by the fall of 1941, Kamen said, Oppenheimer had completely reversed his stand: he was now speaking enthusiastically about forming a chapter of the AASW, which would give support to the Federation of Architects, Engineers, Chemists and Technicians (FAECT), a union affiliated with the CIO (Congress of Industrial Organizations) that was campaigning to organize a branch at the nearby Shell Development Company plant. Oppenheimer apparently urged Kamen—as well as Al Marshak, also of the Rad Lab—to come to a discussion at his home to "hear the case for the FAECT." Oppenheimer introduced two representatives of FAECT, who made the case for setting up a campus branch of the AASW to bolster the FAECT's organizing at Shell. Oppenheimer had not, however, considered Lawrence's predictably negative response to the notion of political organizing at the Rad Lab. When Lawrence put his foot down and demanded the names of staff members who had attended the meeting, Oppenheimer's reaction was clumsy. He would not identify anyone himself but instead said that "he would inform those involved and leave it to them to see E.O.L. [Lawrence] if they wished." This put Kamen in a bind. If he confessed his involvement to Lawrence, he would irrevocably damage his relationship with the Rad Lab chief. But if he didn't, and Lawrence learned of his involvement, the consequences would be worse. He decided to go to Lawrence, who lectured him about Communist influence on labor organizations and complained about the trouble caused by Oppenheimer's "fuzzy-minded efforts to do good."[54]

On November 12, 1941, Oppenheimer wrote a letter to Lawrence, distancing himself from the AASW: "I . . . assure you that there will be no further difficulties at any time with the A.A.S.W. . . . and I doubt very much whether anyone will want to start at this time an organization which could in any way embarrass, divide or interfere with the work we have in hand . . . All those to whom I have spoken agree with us; so you can forget it."[55] Kamen reflected bitterly that "there were some of us who were not allowed to 'forget it.'" Throughout the war, Kamen was under surveillance by military intelligence, and in early July 1944 he was expelled from the Rad Lab as an alleged security risk. Blacklisted, he was forced to find work at the San Francisco shipyards.[56]

As he became involved with war work, Oppenheimer came under pressure to renounce his political commitments and solidarities. It was his entry into the atomic bomb project, more than events such as the nonaggression pact or the invasion of France, that precipitated his break with the Left. His extreme anxiety about Lawrence's criticism and his apologetic letter of November 12 were due precisely to the fact that Lawrence had just recently brought him into the bomb project. Oppenheimer's relationship with the project had begun at Berkeley in early 1941, when he and other theoretical physicists collaborated with Lawrence in the development of the electromagnetic process for the separation of uranium isotope. And it was Lawrence who thrust Oppenheimer forward, bringing him to the attention of the scientific leadership of the bomb project. Lawrence wrote to physicist Arthur H. Compton, a fellow member of the S-1 Executive Committee, in October of 1941 vouching for Oppenheimer, stating that the latter had "important new ideas" and assuring Compton, "I have a great deal of confidence in Oppenheimer." Lawrence requested firmly that Oppenheimer be included among the attendees at a conference on fast-neutron research to be held at the General Electric Research Laboratory on October 21 in Schenectady, New York.[57] Compton was sufficiently impressed by Oppenheimer's performance there, and by his subsequent correspondence with the young theoretician, that in January 1942 he appointed him as head of fast-neutron research at Berkeley. This was part of the program taking place at laboratories across the country under the overall direction of Gregory Breit. Partly because of his rigidity in restricting communication between laboratories, Breit's leadership had become widely unpopular. When Breit resigned in May 1942, Compton gave Oppenheimer his job. In the following months, with the help of experimental physicist John Manley, Oppenheimer took on the task of coordinating fast-neutron research.[58]

He had already begun to sever his ties with the Left. On December 6, 1941, the day before Pearl Harbor, Oppenheimer attended his very last party for Spanish war relief. It is unclear when he made his last contributions to that cause through the Communist Party. Oppenheimer said it was in early 1942. The report of the security hearings put it in April, coinciding with Oppenheimer's formal entry into the atomic bomb project. It was on April 22, 1942, that Oppenheimer completed his first security questionnaire; in May he began full-time work.[59] It is equally unclear when he last saw Steve Nelson. Oppenheimer told the security hearings that, after their initial encounter at the meeting for Spain, the only times he met with Nelson were a few occasions when Nelson and his wife visited the Oppenheimers socially at their home. This was not challenged during the hearings. However, in his autobiography, Nelson asserted that Oppenheimer contacted him shortly before leaving for Los Alamos, which would suggest that their last encounter took place sometime in early 1943. According to Nelson, they met at a Berkeley restaurant. Oppenheimer "appeared excited to the point of nervousness. He couldn't discuss where he was going, but would only say that it had to do with the war effort." Saying good-bye to Nelson, Oppenheimer added that "it was too bad that the Spanish Loyalists had not been able to hold out a little longer so that we could have buried Franco and Hitler in the same grave." "That," wrote Nelson, "was the last time I ever saw him, except on television."[60]

Throughout 1942, Oppenheimer was a rising star within the atomic bomb project. Compton was impressed at the progress Oppenheimer made in the fast-fission work: "Under Oppenheimer," he said later, "something really got done, and done at astonishing speed." Oppenheimer proposed a joint meeting of theorists and experimentalists to work out the problems of fast fission. The physicists met in Chicago on June 5 and 6; in the latter part of July, there was also an important meeting at Berkeley, which Oppenheimer chaired, focused on theoretical work. Meanwhile, the government was making arrangements for the atomic bomb project to be placed under the control of the Army. On June 17, President Roosevelt approved the proposals made by Conant and

Bush, the administrative heads of civilian wartime scientific research, that the Army take over direction of the program. The Army Corps of Engineers was the only organization with the capacity to organize a project of such magnitude. Moreover, the expenditures could now be hidden within the massive wartime defense budget. In September, General Leslie R. Groves was chosen as overall head of the bomb project, now known as the Manhattan Project after its Corps of Engineers administrative designation, the Manhattan Engineer District. Groves had much experience in leading large-scale construction programs— including the building of the Pentagon—and had a reputation as a tough and efficient organizer.[61]

Groves and Oppenheimer first met in early October 1942, when Groves visited the laboratories at Berkeley. Groves was impressed by a report Oppenheimer gave, and "the two men hit it off at once."[62] Particularly interesting to Groves was Oppenheimer's proposal for centralizing fast-fission research in one main laboratory. Oppenheimer, Groves, and Colonel Kenneth D. Nichols discussed the matter further on a train from Chicago to New York. And Groves invited Oppenheimer to Washington to go over plans for the new lab with Compton and Bush. On October 19, Groves gave the go-ahead for the new lab. Oppenheimer went immediately to Harvard to brief Conant.[63] He also played a key role in choosing the site for the new laboratory. He was familiar with northern New Mexico, which had many geographical advantages for a secret project, and he suggested looking at Los Alamos.[64]

As well as being instrumental in the choice of location, Oppenheimer played a central role in staffing Los Alamos, recruiting most of the senior scientists himself. In December 1942, he carried out a "personal raid" on the Radiation Laboratory at the Massachusetts Institute of Technology (MIT) in Cambridge, Massachusetts. Hans Bethe was one of the outstanding physicists there whom Oppenheimer enticed to Los Alamos. Oppenheimer "tended to wade into already established laboratories as if he were wielding the sword of scientific liberation." When talking to new recruits about the Los Alamos project, "he was burning with an inner fire." He later said that "the notion of disappearing into the New Mexico desert for an indeterminate period and under quasi-military auspices disturbed a good many scientists, and the families of many more." But, he said, "there was another side to it," and this was what Oppenheimer emphasized on his visits to the universities. Not only might the project "determine the outcome of the war," but "it was an unparalleled opportunity to bring to bear the basic knowledge and art of science for the benefit of the country . . . This job, if it were achieved, would be a part of history." Finally, he said, "this sense of excitement, of devotion and of patriotism . . . prevailed. Most of those with whom I talked came to Los Alamos."[65] Oppenheimer thus presented a romantic portrait of Los Alamos as a heroic scientific endeavor and the mesa as an ideal small university campus.[66] As he recruited these top-flight physicists to the project, he portrayed it as at once a wilderness scientific retreat, an opportunity to make manifest the truth and power of the new physics, and a way to connect their vocation as physicists with the global struggle against Fascism. But joining the project also meant joining a military organization. Oppenheimer's colleagues began to feel that he had not fully grasped what that meant.

In the fall of 1942, as Oppenheimer was at the forefront of the effort to establish and staff the new laboratory, he came to be recognized as the de facto head of the new enterprise. According to Bush, "Oppenheimer was chosen in November of 1942."[67] Physicists joining the project saw Oppenheimer as its scientific leader. But his directorship of Los Alamos was not formalized until February 25, 1943, in a letter from Groves and Conant, and it did not become final until his security clearance was granted in mid-July. By that time, the laboratory had been in full operation for about three months.[68] This delay enabled the general to test Oppenheimer's mettle and willingness to do his bidding before making a firm commitment. Oppenheimer's security problems helped to strengthen Groves's grip on him. Oppenheimer had been under investigation by the FBI since 1941, when the bureau put him on a list of potential subversives who might be placed in custodial detention.[69] G-2 (military intelligence) agents were very worried about Oppenheimer's close political and personal connections with Communists, and his security clearance was entirely due to Groves's support for him. Colonel John Lansdale Jr. was in the Army's Counter Intelligence Group, based in Washington, D.C., and was investigating what the Army believed to be Communist infiltration of the Berkeley Radiation Laboratory. According to Lansdale, the information they received from the FBI caused them "a great deal of concern." However, Groves's position "was (a) that Dr. Oppenheimer was essential; (b) that in his judgment—and he had gotten to know Dr. Oppenheimer very well by that time—he was loyal; and (c) we would clear him for this work whatever the reports said."[70] As a result, Oppenheimer was personally indebted to Groves for his position in the project.

Groves also overcame opposition to Oppenheimer's appointment from quarters other than his own security personnel and G-2 military intelligence. In the initial stages of the project, Oppenheimer was regarded by his scientific colleagues as an unlikely choice, a man unsuited by character, temperament, and ability to the task awaiting him. His selection as director of the new laboratory was not a result of his having climbed to the pinnacle of his profession, nor of his being a recognized leader of the scientific community. Oppenheimer's most significant role in professional physics before the war had been in training a new generation of theoretical physicists at Berkeley and Caltech. He had published approximately fifty articles and notes in the *Physical Review,* but he was not among those who published on the topic of fission before scientific interchange was halted because of the war. His influence was confined largely to the California schools at which he taught.[71] And the fervor with which Oppenheimer went about recruitment in fact alarmed the wartime scientific establishment. Henry Smyth, who ran the physics department at Princeton, protested to Conant when Oppenheimer poached Robert Wilson from his department, together with Wilson's group of young cyclotron physicists and a promising young theorist, Richard Feynman. While Conant supported efforts to find the best scientists for Los Alamos, he could not afford to gut the other wartime projects for which he was responsible and thus alienate the scientific leadership in the rest of war research. Conant wrote to Groves voicing doubts about Oppenheimer: "We are wondering if we have found the right man to be leader."[72]

According to Groves's deputy, Colonel Kenneth Nichols, the general "recognized that he would encounter opposition both in the scientific community and in Army security if he were to select Oppenheimer" as director of the new bomb lab. After all, as Nichols pointed out, Oppenheimer did not have a Nobel Prize—an honor that "contributed to the prestige of the other project scientific leaders," including Lawrence, Fermi, Urey, and Compton. Groves later wrote, "There was a strong feeling among most of the scientific people with whom I discussed the matter that the head of Project Y [Los Alamos] should also be [a Nobelist]," and "because of the prevailing sentiment that [Oppenheimer] would not succeed, there was considerable opposition to my naming him.[73]

Even among scientists at Berkeley, Groves's decision seemed surprising. Luis Alvarez recalled that despite Lawrence's support for Oppenheimer's appointment, even "some of Robert's closest friends were skeptical. 'He couldn't run a hamburger stand,' I heard one of them say." Alvarez himself was initially skeptical. Rabi thought that Oppenheimer "was absolutely the most unlikely choice for a laboratory director imaginable." For one thing, "he was a very impractical sort of fellow. He walked about in scuffed shoes and a funny hat, and, more important, he didn't know anything about equipment." Eugene Wigner agreed that Oppenheimer "did not look like a leader. He held himself slightly apart from others," and his "sensitive temperament and unusual habits" did not suggest the robustness of character required for such a leadership role. Such doubts were not dispelled by Oppenheimer's performance in the planning phase during late 1942. Samuel Allison, a physicist from the University of Chicago brought in by Oppenheimer to help with the planning, recalled:

> Just before Christmas of 1942, Oppenheimer asked me to come and help plan the preliminary layout . . . On the Mesa he and I sat down and planned the laboratory. He showed me what he called an organisation chart for a hundred personnel. I looked at it and felt sure that something was wrong, but I didn't know what. The best I could do was to poke at random. "Where are the shipping clerks?" I asked. He gave me a thoughtful sympathetic look. "We're not going to ship anything," he answered. I completely underestimated the size of the installation but not so much as he did.[74]

Manley was also concerned that a definite organizational structure had not been settled. He found Oppenheimer "about as unresponsive to such mundane matters as an experimentalist would expect a theoretical physicist to be, perhaps more so." At one point, Manley's urging seemed to have had an effect. In January 1943, he took a plane from Chicago to Oakland, flying through a storm for much of the journey. He finally arrived at the University of California, Berkeley, feeling tired out, and climbed up to Oppenheimer's office at the top of the physics building, Le Conte Hall. Manley recalled, "I had scarcely opened the door when he shoved a piece of paper at me, saying 'Here's your damned organizational chart.'"[75]

However, this did not put an end to administrative problems. When Robert Wilson made a visit to Los Alamos in March 1943 to inspect construction, he found the site in a very poor state and building work behind schedule. Following the trip, Wilson and Manley met with Oppenheimer in Berkeley to inform him of the project's "state of chaos": "Manley and I nagged at Oppy all day about his indecisiveness. We insisted that he had decisions to be made . . . We wanted to know who was to be in charge of what, not just vague talk about scientific problems nor even vaguer ideas about democracy. There were immediate problems to be faced, and from our point of view Oppy was not facing them."[76] When the two experimentalists pressed him to get on with organizational planning,

> Oppenheimer became extremely angry. He began to use vile language, asking us why we were telling him of these insignificant problems, that

it was none of our business, and so on. Both of us were scared to death. We were frightened because, if this was the leader and, if the leader was going to have a tantrum to resolve a problem, then how was anything going to get sorted [out]? So we withdrew, John and I, and discussed some more, and decided that we would take more initiative and not look for so much leadership from Oppy.[77]

Given such early worries, why did Groves regard Oppenheimer as the only man for the job? David Hawkins believed that Groves picked Oppenheimer "for reasons that were mysterious because many people senior to Oppenheimer were felt to be the more proper choice, but Groves somehow knew that Oppenheimer was the man for the job . . . It gives Groves credit for a level of insight that is not often attributed to him." Oppenheimer himself said blandly that Groves simply had "a weakness for good men." "My own feeling," Groves said, "was that he was well qualified to handle the theoretical aspects of the work, but how he would do on the practical experimentation, or how he would handle the administrative responsibilities, I had no idea." Groves felt that there were no other candidates who were not already running other important sites of the project. The other credible choice was Lawrence. Groves said, "I had no doubt that Ernest Lawrence could handle it . . . However he could not be spared from his work on the electromagnetic process." Nevertheless, Lawrence and Compton told Groves that if Oppenheimer failed, Lawrence would take over and pick up the pieces. Groves's biographer has suggested that the selection was based primarily on Oppenheimer's acceptance of the idea of a militarized laboratory: "Oppenheimer's unique compliance with what for [Groves] was one of the linchpins of bomb lab planning may have been the deciding, if not the overwhelming, consideration in determining that the physicist was 'the best man, the only man' for the director's job."[78]

The idea of Los Alamos as a military organization with scientists as commissioned officers was supported by Conant and was at the heart of Groves's conception of the project's organization. Oppenheimer was an initial supporter of the idea of a military laboratory. He even said, "I would have been glad to be an officer." Wilson remembered Oppenheimer justifying this stance by describing the war as a people's war. Oppenheimer thought that to join the Army was to place oneself side by side with the fighting masses: "Oppy would get a faraway look in his eyes and tell me that this war was different from any war ever fought before: it was a war about the principles of freedom, and it was being fought by a 'people's army.'" Wilson confessed, "Now I can be as idealistic as the next guy, but I thought he had a screw loose when he talked like that." Wilson doubted that "scientists could function at all" if they were "unquestioningly following orders." But for the time being, he was unable to convince Oppenheimer of this. Oppenheimer even went to the Presidio in San Francisco to begin the process of enlisting as a lieutenant colonel, and he began to look the part.[79] At least until 1941, Oppenheimer's hair had been long, and so tangled that he had to comb it with a steel dog comb; Bethe used to tease him that it made him look "like a Bolshevik." In preparation for his new role, Oppenheimer got a crew cut. Alvarez remembered it as being "almost as short as a military officer's." Robert and Jane Wilson thought that the length of Oppenheimer's hair reflected the various roles he assumed throughout his life: as a "young, radical professor at Berkeley . . . his hair was all little black curls. And then he was much more subdued at Los Alamos, the curls were not so curly."[80]

Oppenheimer's scientific colleagues saw him as initially very much under the power of Groves and sought to counteract this influence. Rabi, for example, said that Oppenheimer "was not a strong character. He was indecisive, and definitely not a fighter. If he couldn't persuade you, he'd cave in, especially to group opposition. Groves, on the other hand, could provide him with strong backbone in the form of consistent policy."[81] Rabi and other scientists opposed to the idea of a military laboratory sought to win Oppenheimer over to their own policy, to transform him into a representative of their aspirations for the role of the scientist in the new laboratory.

In January 1943, a pivotal meeting took place at MIT between Oppenheimer and physicists Robert Bacher, Edwin McMillan, Alvarez, and Rabi. It was here that Oppenheimer admitted to the plan for a militarized laboratory that he had discussed with Groves as far back as October 1942. When they were told of this, Bacher says, "we were horrified." Both Bacher and Rabi threatened to have nothing further to do with the project if this plan were carried out. They were worried that military rank would trump scientific expertise and interfere with the free flow of scientific communication. They told Oppenheimer that "lieutenant colonels didn't have anything to say, and that if he tried to establish a scientific laboratory [with] a hierarchy that was composed of military people, that it just plain wouldn't work." Rabi said, "We *knew* the military" through the radar work at MIT: "We'd been engaged in making military things, had the military around us. We knew it wouldn't work. In the first place, none of us would come." Alvarez said, "I don't think science can be done under authoritarian arrangements."[82]

This sustained opposition by his colleagues forced Oppenheimer to back down. He wrote to Conant informing him of the scientists' view "that the Laboratory must demilitarize: the arguments here were first that a divided personnel would inevitably lead to friction, and to a collapse of Laboratory morale, complicated in our case by social cleavage, and, more important, that in any issue in which we were instructed by our military superiors, the whole Laboratory would be forced to follow their instructions and thus in effect lose its scientific autonomy." A few weeks later, Oppenheimer was given a letter by Groves and Conant that not only formalized his own position as head of the new laboratory, but stipulated that, at least in its early stages, the laboratory would be civilian-run.[83] Groves remained bitter about this incident. He focused the blame on Rabi for swaying Oppenheimer on this critical issue. A 1946 War Department report, "Complications of the Los Alamos Project," had an entire section titled "The attitude of Dr. I. I. Rabi." Rabi's "influence" and "determined position" in demanding a civilian laboratory, the report stated, "was such that many of the troubles in the operations of the Los Alamos Laboratory stemmed from his original stand."[84]

Oppenheimer had initially asked Rabi to be associate director of the new laboratory. Rabi turned down the offer, feeling that the radar work he was doing at MIT was more vital to the war effort. Nevertheless, he made frequent visits to Los Alamos as a consultant, an exception to the quarantine of Los Alamos from the rest of the scientific world that Groves grudgingly allowed.[85] Oppenheimer became quite reliant on Rabi as an adviser, particularly in formulating a position with which to oppose Groves's desire for authoritarian military control of the lab. In early February 1943, during the beginning of construction of the Los Alamos Laboratory, Oppenheimer sent a note to Rabi asking for his recommendations before the general's visit to the site, which was to take place in less than a week. Rabi tutored Oppenheimer on how to be a leader of such a large-scale, bureaucratic, and demanding project. He told Oppenheimer, "The main idea is to get going immediately, to keep up the morale, to test the organization and to assemble a smoothly running team." In advance of Groves's visit, Rabi sent Oppenheimer a detailed memorandum providing, for example, predictions of when construction of the physical site would be finished and concrete proposals for assembling teams of scientists from various universities. He added, "Keep them flying." The experience of Rabi and Bacher from war projects at MIT was vital. Bethe said, "We found [the] MIT Radiation Lab [to be] the best model. So that was already adopted in the organization into six or eight divisions, and then each division into several groups . . . This general structure we just took over from the Radiation Lab." The importation of this structure to Los Alamos, he added, was Rabi's idea. Armed with such support from his colleagues, Oppenheimer was in a much stronger position to hold his own with the general. He continued to rely on Rabi's help throughout the project; according to Bethe, "Rabi made Oppenheimer more practical." Without Rabi's intervention, the project "would have been a mess.[86]

Groves was concerned about Oppenheimer's lack of either administrative experience or knowledge of engineering practice. He persuaded Oppenheimer to appoint as associate director a physicist with an industrial background. Edward Condon, from the Westinghouse Research Laboratories, fit the bill. Groves assumed that because of his background in industrial research, Condon would be the general's natural ally and would help to overcome what Groves saw as the congenital aversion of academic scientists to strict organization and a mission-directed program of research. Condon would take over procurement and management of personnel and relations with the military, leaving Oppenheimer to "think [through] the scientific problems and to establish the schedule of scientific and technical work." However, Condon proved to be just the opposite of what Groves had hoped for. Groves said, "Dr. Condon turned out to be not an industrial scientist, but an academic scientist with all of the faults and none of the virtues."[87] What angered the general most was Condon's opposition to the policy of compartmentalization of information, which Groves saw as essential to the project's security. In late April 1943, less than a month into the lab's full-time operation, Condon resigned. In his resignation letter, he cited the "extraordinarily close security policy." He said he found "the discussion about . . . the possible militarization and complete isolation of the personnel from the outside world" to be "morbidly depressing." Groves said, "I could never make up my own mind as to whether Dr. Oppenheimer was the one primarily at fault in breaking up the compartmentalization or whether it was Condon." He strongly suspected that it was Condon.[88]

Oppenheimer had begun to play a mediating role between the military establishment and the civilian scientific community. Both Groves and the scientists sought to fashion Oppenheimer into a representative of their interests and their vision for the project. For both camps, their hopes in Oppenheimer were based not so much on who he was and what he had accomplished, but on who they thought he might become with their guidance. Oppenheimer was embroiled in new relationships and subject to new pressures that would remake him as a person. When he moved to Los Alamos on a permanent basis on March 15, 1943, he left behind his life at Berkeley and stepped into a self-contained world.[89]

However, this break was not absolutely clean. Between 1941 and 1943, Oppenheimer found himself straddling two increasingly incompatible social worlds: leftist ties became an embarrassment to him as he realized that he was under the glare of the project's security apparatus. Sometime in January or February 1943, Chevalier and his wife came to dinner at the Oppenheimers' house. When Oppenheimer went to the kitchen to prepare the usual martinis, Chevalier followed him. Chevalier mentioned a recent encounter with a friend of his, with whom Oppenheimer was acquainted. This was a British chemical engineer named George Eltenton, who was involved with FAECT at the Shell Development Company. (Oppenheimer later testified that Eltenton had been present at the autumn 1941 meeting at Oppenheimer's home in Berkeley about setting up a branch of the AASW in support of FAECT.) Apparently, Eltenton had phoned Chevalier and asked to meet him. Chevalier told Oppenheimer that Eltenton had inquired into the possibility of Oppenheimer's passing technical information, via Eltenton, to the Russians. Both Oppenheimer and Chevalier would later deny that Chevalier was *asking* Oppenheimer to consider leaking information; either he was warning Oppenheimer about Eltenton, or he was announcing that he had been confused by Eltenton's suggestion. Oppenheimer told the security hearings in 1954 that he had exclaimed that what Eltenton proposed was "treason," that Chevalier had agreed with this judgment, and that the topic had been dropped.[90] Later in 1943, however, Oppenheimer would revisit that conversation, and it would gain a new significance in his life.

In the spring of 1943, Lieutenant Colonel Boris Pash, chief of counterintelligence for the Fourth Army Western Defense Command, began investigating reports that a Soviet espionage cell was operating in the Radiation Laboratory at Berkeley. Pash came to suspect that the cell included some former students of Oppenheimer's, namely, Giovanni Rossi Lomanitz, Joseph Weinberg, and David Bohm; that it was organized by Steve Nelson; and that Nelson had asked Oppenheimer for information on the bomb project.[91] Lomanitz, Weinberg, and Bohm were all members of FAECT, which Pash regarded as a Communist front organization attempting to infiltrate the Manhattan Project through the Rad Lab.[92] Pash's organization also believed it had evidence that all three were Communist Party members.[93] Lomanitz usefully reminds us, however, that things were neither so black and white nor so cloak-and-dagger: "I attended some meetings, because at that time meetings were much more open, free, and easy. There wasn't any great distinction . . . I was at some discussion meetings where members of the Communist Party spoke. Who was officially a member or what it took to be officially a member, I can't tell you to this day. It just wasn't all that conspiratorial."[94]

At the end of March 1943, Pash received a report from the FBI that Nelson had made an attempt to gain technical information on the bomb project from a scientist named "Joe." Pash set about trying to determine "Joe's" identity. The transcript of a conversation in Nelson's house (which had been bugged by the FBI) provided only the first name of the scientist and the fact that he came from New York and had two sisters living there. This led to a broad investigation of Rad Lab personnel. The first suspect was Lomanitz, whose first name, Giovanni, might (it was thought) be anglicized as "Joe." Eventually, however, security officers pinpointed as their suspect Joseph Weinberg, an Oppenheimer student who had come to Berkeley from City College, New York.[95]

The bugged conversation began with a discussion of a man that Nelson and Joe called "the professor." On the basis of a number of telling circumstantial details provided in the conversation, the FBI believed this professor to be Oppenheimer. Nelson mentioned, "I was quite intimate with the guy and there was a personal relationship also because his wife used to be the wife of my best friend who was killed in Spain. I know her well." He bemoaned the professor's increasing distance from the party and from leftist acquaintances: "He's very much worried now and we make him feel uncomfortable." Nelson had "spent a little time" at the professor's home but found him "jittery" and under "the impression that he was being watched." In Nelson's view, the man was confused: "You know what I mean. He's just not a Marxist . . . [and] now he's gone a little further away from whatever associations he had with us." Nelson saw this as connected to the professor's new role in the project and his desire "to make a name for himself." Nelson also blamed the professor's wife for "influencing him in the wrong direction." Joe agreed that "he's changed a bit . . . You won't hardly believe the change that has taken place."[96]

Nelson went on to ask if Joe would give him information on the project and "what kind of materials they are working on." Nelson indicated that he had approached the professor for information about the project but that he had been reluctant to provide anything specific, and Nelson had not thought it wise to press him.[97] Joe said that probably the professor didn't feel Nelson was senior enough to receive that information. Joe told him that the "basic idea" was published and available in the open literature, and he agreed to get Nelson reprints of articles. While telling Nelson "It's natural I'm a little bit scared," Joe went on to describe more specific features of the project, including the fact that "the material is uranium, a radioactive substance, as you know." He also provided estimated timescales for the project, which were regarded as highly sensitive information. Pash was therefore confronted with what appeared to be a very serious security breach, and he undertook an intensive investigation of the Rad Lab. Oppenheimer was not directly incriminated; the transcript suggested that the "professor" had resisted pressure to leak information. But it did seem that there were numerous threads connecting Oppenheimer to the persons involved, and Pash became increasingly convinced that Oppenheimer was at the center of a conspiracy.[98]

It was in June 1943 that Oppenheimer, trailed by security officers, visited Jean Tatlock in Berkeley. Pash saw this visit as suggesting Oppenheimer's continued links with the Communist Party. On the same trip, Oppenheimer hired David Hawkins to be his personal administrative assistant at Los Alamos. Army security suspected Hawkins of being a Communist.[99] On July 22, a security agent investigating Frank Oppenheimer reported his opinion that Frank was a member of the Communist Party and that Robert Oppenheimer was "very possibly a member of the Communist Party, and has definitely indicated sympathy towards that group."[100]

Around this time, Oppenheimer's relationship with young FAECT members at the Rad Lab was coming under scrutiny. Lomanitz was working at the laboratory and taking over work from scientists going to Los Alamos. But in late July, he found his draft deferment had been canceled. Lomanitz turned for help to both Lawrence and Oppenheimer.[101] Oppenheimer contacted the New York headquarters of the Manhattan Engineer District to ask them to intervene to keep Lomanitz on the project. However, all appeals on Lomanitz's behalf were unsuccessful, and he was drafted into the regular army. His work at the Rad Lab was taken over by Condon, who had recently resigned from his position as Oppenheimer's right-hand man at Los Alamos.[102]

In August, Colonel Lansdale visited Oppenheimer at Los Alamos and interviewed him about Lomanitz. Oppenheimer told Lansdale that he had encouraged Lomanitz to join the project at Berkeley and had also warned the young man, in the strongest terms, to "forgo all political activity if he came on to the project." According to Lansdale, Oppenheimer "had previously stated that he knew Lomanitz had been very much of a Red." Lansdale told Oppenheimer not to make any further requests on Lomanitz's behalf, because the young man "had been guilty of indiscretions." When told that Lomanitz had not given up "political activities," Oppenheimer's response was, "That makes me mad."[103]

Oppenheimer was clearly trying to ingratiate himself with Lansdale and was anxious to appear firm regarding security. Lansdale told him that "from a military intelligence standpoint we were quite unconcerned with a man's political or social beliefs, and we were only concerned with preventing the transmission of classified information to unauthorized persons." Oppenheimer replied that he thought a stronger approach was required with regard to the

Communist Party: "He stated that he did not want anybody working for him on the project that was a member of the Communist Party. He stated that the reason for that was that 'one always had a question of divided loyalty.' He stated that the discipline of the Communist Party was very severe and was not compatible with complete loyalty to the project. He made it clear he was not referring to people who had been members of the Communist Party, stating that he knew several now at Los Alamos who had been members." Perhaps Oppenheimer thought Lansdale's tolerance was feigned, designed to set a trap for him, and Oppenheimer was therefore making a show of being tough and "on-side." But Lansdale saw a deeper significance to the physicist's comments: "This officer also had the definite impression that Oppenheimer was trying to indicate that he had been a member of the party, and had definitely severed his connections upon engaging in this work."[104]

Lansdale had mentioned to Oppenheimer that security officers were particularly concerned about what they saw as Communist infiltration of the project through FAECT. Oppenheimer remembered that Eltenton was a prominent member of the union. On August 25, while in Berkeley, Oppenheimer went to see Lieutenant Lyall Johnson, the project's security officer for the university. Oppenheimer gave him Eltenton's name as a man the security agents should keep an eye on. Johnson passed the information on to Pash, who arranged for himself and Johnson to interview Oppenheimer the following day—this time with a tape recorder hidden in the room. Oppenheimer immediately began to talk about Lomanitz, offering to have a stern talk with him. Pash steered the conversation to what he considered a "more serious" matter, that of Eltenton. It was then that Oppenheimer came forth with what he would later admit was a "cock-and-bull story." It was an account in which neither he nor Chevalier was implicated. "I think it is true," he said at the time, "that a man, whose name I never heard, who was attached to the Soviet Consul, has indicated indirectly through intermediate people concerned in this project that he was in a position to transmit, without any danger of a leak, or scandal, or anything of that kind, information which they might supply." The approaches, he said, "were always through other people, who were troubled by them, and sometimes came and discussed them with me." Oppenheimer said that he did not want to give the names of the people approached, since that would "implicate people whose attitude was one of bewilderment rather than one of cooperation." But he did name Eltenton as a man "whose name was mentioned to me a couple of times" and who was "involved as an intermediary." He added, "Whether he is successful or not, I do not know, but he talked to a friend of his who is also an acquaintance of one of the men on the project, and that was one of the channels by which this thing went. Now I think that to go beyond that would be to put a lot of names down, of people who are not only innocent but whose attitude was 100 percent cooperative."[105]

Rather than revealing the kitchen conversation with Chevalier, Oppenheimer had painted a picture of a sustained espionage campaign against the project. As he was pressed for details, Oppenheimer said that "two or three" people were approached, "and I think two of them are with me at Los Alamos—they are men who are closely associated with me." These men, Oppenheimer said, were approached not by Eltenton himself but by "another party." Naturally, Pash asked him who that intermediary was. Oppenheimer replied, "I think it would be a mistake, that is, I think I have told you where the initiative came from and that the other things were almost purely accident and that it would involve people who ought not to be involved in this." A few questions later and Oppenheimer had blurted out that the intermediary between Eltenton and the men on the project was "a member of the [Berkeley] faculty, but not on the project." After a few minutes of sustained questioning, then, Oppenheimer had elaborated his story into a portrait of a conspiracy involving a number of people. He had also set project security on the trail of a faculty member apparently at the heart of this conspiracy. It was months later, when Groves ordered Oppenheimer to identify the faculty member, that he named Chevalier. It remains unclear whether Oppenheimer admitted to Groves at this time that the story of the three approaches had been a fabrication and that he himself was the only person approached.[106] It was due to his view of Oppenheimer's importance to the project and of the necessity of maintaining Oppenheimer's trust in him personally that Groves allowed months to go by before giving that order.[107] Meanwhile, the project's security officers were calling for Oppenheimer's head.

Pash wrote a memorandum to Groves on September 2 with his interpretation of the significance of Oppenheimer's naming of Eltenton. In Pash's view, Oppenheimer suspected that he himself was under investigation for radical activities, and his revelation about Eltenton was just an attempt to protect himself and "retain the [Army's] confidence" by preempting the findings of any investigation. Pash concluded, "It is not believed that he should be taken into the confidence of the Army in the matters pertaining to subversive activities."[108]

Not only had Oppenheimer triggered a large-scale search for a fictionalized (or at least massively exaggerated) espionage conspiracy, he had also increased the suspicions that were hanging over him. Los Alamos security officer Captain Peer de Silva wrote to Pash with his own opinion that "J. R. Oppenheimer in playing a key part in the attempts of the Soviet Union to secure, by espionage, highly secret information which is vital to the security of the United States."[109]

Historians Kai Bird and Martin Sherwin recently gave an interesting and novel interpretation of the Chevalier affair, one that is more charitable to Oppenheimer than most existing accounts. They suggested that Oppenheimer's story of the three approaches may not have been entirely fabricated. Eltenton later told the FBI that his Soviet contact, Peter Ivanov, had suggested approaching Ernest Lawrence and Luis Alvarez, as well as Oppenheimer. It is possible, Bird and Sherwin surmised, that the idea of these other approaches had been passed on via Chevalier to Oppenheimer. If that was the case, it would seem to provide a source for Oppenheimer's tale of the three approaches. What was actually said in the crucial kitchen conversation will never be known with any certainty. It nevertheless remains clear that the account Oppenheimer gave security officers was substantially fabricated—in particular his statement that the approaches were always to other people, rather than to him. These fabrications not only protected Chevalier, but also distanced Oppenheimer from the events. It should also be noted that Oppenheimer's account to the security officers suggested that the same intermediary contacted all three of the scientists on the project. But Oppenheimer would have had no reason to assume that Chevalier had in fact approached Lawrence and Alvarez. Apart from the fact that they were colleagues at the same university, and possibly had a casual acquaintanceship through Oppenheimer, these men were strangers to Chevalier, with opposing politics and outlook. Alvarez was on leave from Berkeley, having gone to work at the MIT Radiation Laboratory in 1940 and moving on to Chicago in 1943. And after Oppenheimer's political run-ins with Lawrence, if Chevalier had mentioned the idea of approaching Lawrence, Oppenheimer would no doubt have impressed on him just how unreceptive his deeply conservative colleague would be. Further, Oppenheimer had told Pash and Johnson that two of the people approached by the intermediary were now at Los Alamos. Lawrence's only relationship with Los Alamos was as a consultant, and Alvarez did not begin work at the facility until the following year. When interviewed by the FBI in 1946, and during the 1954 security hearings, Oppenheimer came up with no factual basis for the "three approaches" story, saying it was merely "concocted" or a "cock-and-bull story."[110]

BETWEEN GROVES AND THE SCIENTISTS

In the first few months of Oppenheimer's directorship of Los Alamos, his position was already highly compromised. De Silva recognized that the Army held the physicist in its grip: "It is the opinion of this officer that Oppenheimer is deeply concerned with gaining a worldwide reputation as a scientist and a place in history, as a result of the DSM project [Manhattan Project]. It is also believed that the Army is in the position of being able to allow him to do so or to destroy his name, reputation, and career, if it should choose to do so." De Silva saw that the Army could take advantage of his uncomfortable position in order to control him. De Silva thought that Oppenheimer should be told that the Army held his reputation in its hands and had the power to bring him down. "Such a possibility, if strongly presented to him, would possibly give him a different view of his position with respect to the Army, which has been, heretofore, one in which he has been dominant because of his supposed essentiality." And through its control over Oppenheimer, the Army could discipline the Los Alamos scientists: "If his attitude should be changed by such an action, a more wholesome and loyal attitude might, in turn, be injected into the lower echelons of employees."[111]

The summer of 1943 was a low point for Oppenheimer. His security troubles, together with the new administrative burdens and responsibilities that he faced at Los Alamos, were causing him to have doubts about his capacity to lead the project. According to Bacher, "as the work got started out there, and especially after I came in residence, I found that Robert Oppenheimer was deeply concerned about many things and seemed very worried about how he was doing as director . . . There was a period during this time when he felt he could not continue as a director." He even considered resignation, although "he was very much upset" about this possibility. During that summer, Bacher "spent about two hours a day with him, discussing things. Sometimes, after work at night, we'd talk for an hour or more." These talks involved keeping up Oppenheimer's spirits and morale, providing emotional and moral support. The supporting roles played by Bacher, Rabi, and others were crucial in transforming Oppenheimer into the powerful figure he ultimately became. Bacher said, "Within a relatively short time, he was as different from the professor I had known before the war as you could possibly think of anybody being." This transformation involved, crucially, the ability to stand up to Groves and to represent to him the scientists' collective point of view. According to Bacher, Oppenheimer "maintained a position of great influence with Groves," in spite of the fact that "it would be hard to think of two people who were more dissimilar. The directors of the laboratories who were more nearly people who you'd think could deal easily with Groves, were precisely the ones who didn't."[112] Oppenheimer's ability to be an effective representative of the scientists in the administration of the project depended on his status as an irreplaceable and essential person. The Los Alamos scientists needed a charismatic "natural leader"—and that is what Oppenheimer, with their help, became.

CHAPTER FOUR

King of the Hill

A CHARISMATIC LEADER

Arno Roensch, a scientific glass-blower, looked back at his time at Los Alamos during the war and remembered above all "seeing Oppie and his pork-pie hat and seeing him in the hall and having him nod to you and, I don't know, he was the project as far as I was concerned." At Los Alamos there was assembled an impressive array of the scientific stars of the day, yet Oppenheimer stood out as having a unique importance in the life of the laboratory. "Somehow he was the glue that held them all together and that feeling always stayed with me," said Roensch. "It's synonymous with Los Alamos."[1] For a great many participants in the wartime atomic bomb project at Los Alamos, the place and the time were indelibly associated with the person of J. Robert Oppenheimer. For them the project, in all its complexity, excitement, and solemn significance, was embodied by the enigmatic figure in the porkpie hat. Not only did Oppenheimer symbolize the meaning of the project, but he was, in their view, a powerful causal factor in its success. It is in seeking to describe this unique role that participants and historians have identified Oppenheimer as a "charismatic" leader.

Oppenheimer has been frequently described as a "born leader" or a "natural leader." Hans Bethe said simply, "He was a leader." Bethe went on to describe the inspirational quality of Oppenheimer's guidance: compared with Los Alamos, none of the other wartime laboratories had "quite the spirit of belonging together, quite the urge to reminisce about the days of the laboratory, quite the feeling that this was really the great time of their lives." He attributed this spirit to Oppenheimer personally. Los Alamos, Bethe said, "might have succeeded without him, but certainly only with much greater strain, less enthusiasm, and less speed." Robert Wilson pointed to Oppenheimer's "combination of skill, wisdom, and moral stature" when he said that Oppenheimer was "our leader in every respect." When Oppenheimer visited the laboratory late in his life, his successor, Norris Bradbury, introduced him as "Mr. Los Alamos." Oppenheimer, Bradbury said, had "built Los Alamos by the sheer force of personality and character." "He was the leader," Bradbury said. "He was the boss. It was Oppie's project." According to Laura Fermi, the wife of the famous physicist, Oppenheimer was "the real soul of the project." Enrico Fermi told Emilio Segrè, "When anyone mentions laboratory directors, I think of directors and directors and Oppenheimer, who is unique." Even Edward Teller, one of Los Alamos's most malcontent inhabitants, remarked, "Oppenheimer was probably the best lab director I have ever seen." According to Teller, Oppenheimer "was the constituted authority at Los Alamos. But he was more: His brilliant mind, his quick intellect, and his penetrating interest in everyone at the laboratory made him our natural leader as well."[2]

When Oppenheimer left Los Alamos after the war, scientific colleagues read out a tribute that vividly credited features of the site and the laboratory's social order to Oppenheimer as the charismatic leader:

> He selected this place. Let us thank him for the fishing, hiking, skiing, and for the New Mexico weather. He selected our collaborators. Let us thank him for the company we had, for the parties, and for the intellectual atmosphere . . . He was our director. Let us thank him for the way he directed our work, for the many occasions where he was the eloquent spokesman for our thoughts. It was his acquaintance with every single little and big difficulty that helped us so much to overcome them. It was his spirit of scientific dignity that made us feel we would be in the right place here. We drew much more satisfaction from our work than our consciences ought to have allowed.[3]

In its repetition of "Let us thank him," the tribute is reminiscent of a Christian prayer. But at the end, there is the implication of something more sinister— indeed, Mephistophelean: a sense of Oppenheimer as a seducer.

Participants' accounts suggest that in order to understand the complex social organization of Los Alamos, one must come to grips with embodied personal authority, with the special personal qualities of the project's leader.

Equally, as I will argue, Oppenheimer's charismatic role was a collective accomplishment, arising in response to social and organizational problems. The
organizational order of Los Alamos and the personal identity of Oppenheimer
were constructed together. If Los Alamos was shaped by Oppenheimer's personality, this was a recursive process: Oppenheimer was equally shaped and
transformed by Los Alamos. Identity, authority, and organizational order
were emergent properties of the ongoing social interaction of scientists, technicians, military personnel, and the other men and women engaged in the
project and the community life of Los Alamos.

Oppenheimer's charismatic authority was constituted as a partial solution
to the intense normative uncertainty that characterized everyday life at Los
Alamos. His charisma was a resource mobilized in attempts to define the situation, to identify what type of place this was, and consequently to specify norms
of appropriate conduct. Los Alamos was a hybrid organization involving
diverse groups with contradictory understandings of, interests in, and agendas
for the place.[4] The definition of the situation was chronically unclear and
contested; as one participant said, "Everyone had his own Los Alamos."[5]
Collective identity, the structure and legitimacy of forms of power and
authority, the specification of appropriate channels for communication, and
the delineation of legitimate forms of discourse were all at issue throughout
the war. Oppenheimer emerged as uniquely able to speak for, signal, embody,
and give legitimacy to a particular understanding of the place and its moral
order. His personal identity was centrally at stake in the collective construction of Los Alamos, a process that was both collaborative and conflictual.
Appreciating Oppenheimer's charismatic role requires a fine-grained portrait
of quotidian life at Los Alamos—a description of the experience of life both
inside and outside the laboratory.

"AN ISLAND IN THE SKY"

Los Alamos was a quintessentially modern site, severing people from traditional ties of community, assembling them under the auspices of a military-
industrial organ of the state, directing them toward a specific instrumental goal.
Participants in the project were faced with the tasks of making sense of this
radically new setting and constructing social bonds within it. To its new
residents, Los Alamos was largely a tabula rasa, a community whose identity
remained to be defined. For the project's military command, General Groves
and the Army Corps of Engineers, the problem was simply to construct a utilitarian and rationalistic planned compound on the mesa, on the model of an

Army camp. But the absence of a coherent tradition not only created a free field for the imposition of legal-rational authority, it also allowed the construction of new forms of identity and solidarity. Charismatic authority was a response and a solution to the problem of constructing a social order ex nihilo. Equally, Los Alamos provided a theater in which a charismatic self-presentation could be enacted. If Oppenheimer was, as many have suggested, a consummate actor, he found at Los Alamos the ideal stage. As a tabula rasa, Los Alamos had the potential to enable such a utopian, free shaping of self and community. However, the dislocation and isolation of its residents from broader attachments allowed their identities to be shaped, in a one-dimensional way, toward identification with the hegemonic institutional goals of the Manhattan Project.

Situated on a remote mesa named Pajarito Plateau, at seventy-three hundred feet above sea level, Los Alamos was nicknamed "the Hill" by the scientists and their families. The place seemed to many of these participants to be "a world unto itself, an island in the sky." It was entirely separate from the urban communities from which most of them had come, and they arrived there knowing little or nothing about what they would find.[6] The island metaphor expressed the new residents' profound sense of dislocation from their familiar social worlds. It is also reflective of the detachment of the military-scientific settlement from the locale and the communities amid which it was set.

The site had been selected in November 1942.[7] Most of the land was managed by agencies of the federal government—the Bureau of Indian Affairs, the Forest Service, and the National Park Service; the rest was divided between homesteaders and a boys' boarding school, the Los Alamos Ranch School, whose buildings provided the hub for the new settlement. The Manhattan Engineer District appropriated the land through compulsory purchase, bringing the school and the history of ranching and homesteading on the mesa to an abrupt end. As one historian of the region, Hal Rothman, put it, "Los Alamos had been dropped into a world to which it bore no relation."[8] To the extent that the new arrivals could find meaning in their situation, it was precisely in this unfamiliarity. Los Alamos was to them a world outside of and far removed from the regular flow of their lives; it was similar in that sense to a vacation. Françoise Ulam, the wife of Los Alamos mathematician Stanislaw Ulam, called it "a mountain resort as well as an Army camp"—"just like a camp out." The physicist Otto Frisch described it as "a first rate holiday place."[9]

Others were reminded of fabulous fictional locations. A recurrent image was "Shangri-La," the magical Tibetan monastery in James Hilton's highly popular novel of escapist fantasy, *Lost Horizon,* and Fritz Capra's 1937 film adaptation. It was also the name chosen by President Roosevelt for his re-treat (which Eisenhower renamed Camp David). So, for example, Phyllis Fisher wrote to her parents in October 1944 that her husband was taking them to an unknown location, which she jokingly called "Shangri-La." This image proved to be surprisingly apt, for, as she noted, it was "the name of a strange, hidden, magical mountain community."[10] At one party early in the project, Edward Condon, at that time associate director of Los Alamos, "picked up a copy of *The Tempest* and sat in a corner reading aloud passages appropriate to intellectuals in exotic isolation." To some Europeans, the relevant imagery was provided by Thomas Mann's *The Magic Mountain*.[11]

Such descriptions of the place as "magic" or particularly "spiritual" drew on American culture's romanticizing of the Southwest and New Mexico. In 1941, only two years before construction of the Los Alamos Laboratory began, the state of New Mexico adopted the slogan "Land of Enchantment" for its tourist campaign. This kind of imagery derived from the quasi-colonial idea of New Mexico as a romantic domestic Orient. With its ancient pueblos and picturesque Spanish-American villages, the state appeared to be caught in a time past, outside the boundaries of modern American life. Ruth Marshak (the wife of physicist Robert Marshak) remarked, "Too much cannot be said for the poetic gesture which placed that fantastic settlement, Los Alamos, in that fantastic state, New Mexico."[12]

The "poetic gesture" was widely seen to be Oppenheimer's. In the 1930s, he and his brother, Frank, had entertained friends and colleagues at the ranch not far from Los Alamos that they had leased and later bought. Oppenheimer had developed a taste for horse-riding and for chili, and a Santa Fe silver belt buckle became part of his attire at Berkeley. He had once said, "My two great loves are physics and desert country . . . It's a pity they can't be combined."[13] Through his prior connection to and enthusiasm for the place, Oppenheimer helped give legitimacy to the idea that this was an appropriate setting in which to do science. But underlying the invocation of the kitsch imagery of the Land of Enchantment was the fact that neither the residents of Los Alamos nor the project had any organic connection with the place itself. As Rothman aptly put it, they "could see the land they inhabited only as a stage, its scenic mountains as backdrop."[14]

To new arrivals, Los Alamos often appeared to be literally no place at all. Most recruits to the project arrived by train. The nearest station was Lamy, a tiny place a few miles from Santa Fe. Elsie McMillan, the wife of a Rad Lab physicist, described her reaction when she arrived at Lamy: "What desolation! Were we to live in this nowhere?"[15] Charles Bagley, an explosives engineer attached to the Special Engineer Detachment (SED), had a similar response: "Jesus, what's out there? Nothing."[16] Jacob Wechsler, another SED (as the detachment's members were called), told of being put on a train at the unit's headquarters in Oak Ridge, Tennessee, and being shuttled across the country without knowing where they were going nor where they were when they arrived. Sitting in the back of the truck that took them from Santa Fe thirty-five miles up the winding dirt road to Los Alamos, Wechsler put on his infantry gas mask to protect himself against the clouds of dust. The truck was waved through the gate into Los Alamos by military police and stopped near the base headquarters. Cameras and guns were checked in, and the men were then shown to the crowded barracks. "Nobody would say anything about why we were here, what our duties or chores were to be, what they called the base or anything else . . . We had a little indoctrination that said that there had been special clearance investigations for us to get here. We didn't even know what that meant."[17]

A strong initial feeling of disorientation and of arrival as being a sharp break from one's life hitherto was common among military personnel and civilians entering the project. This sense of separation was particularly powerful because of the inability to tell friends and relatives where one was going and because of the need to lie about it. Ruth Marshak recalls her realization "that when my husband joined the Manhattan Project it would be as if we shut a great door behind us. The world I had known of friends and family would no longer be real to me."[18]

The office of Dorothy McKibben at 109 East Palace Avenue in Santa Fe was the first stop for new arrivals. There they were issued a temporary pass, valid for twenty-four hours. Soon after arrival at Los Alamos, therefore, they were required to go to the Pass Office to receive a new one. Here they were fingerprinted and photographed. "The procedure," said Ruth Marshak, "struck me as similar to that which a criminal undergoes when he visits a police station."[19] Newcomers were briefed on security regulations and told not to reveal to anyone the location of the project, the scale of the site, the size of the population, or the names of any of the scientists. They also had to sign an acknowledgment of the Espionage Act. They were then given a new temporary pass, valid for two weeks, at which time they were to report back to receive their permanent passes.[20] Famous scientists were given new names and were assigned bodyguards. Enrico Fermi was to be called Mr. Farmer, Niels Bohr was Mr. Baker, and so on.[21] The New Mexico driver's license of Robert Serber stated "Not Required" under "Name." Under "Address," it read "Special List B."[22]

Administratively and politically, Los Alamos was an entity separate from the rest of northern New Mexico. Judicially, everyone and everything inside the post fence was "outside the clutches of New Mexico law [and] strictly the responsibility of the Army." Even the size of this new population was kept secret; accordingly, there was no official census at Los Alamos until April 1946.[23]

In order to maintain secrecy, the town was supposed to be as self-sustaining as possible. The post had a commissary selling general wares and a post exchange (PX). Clothes were often purchased from catalogs; all goods and mail for Los Alamos were to be addressed to Box 1663, Santa Fe, New Mexico. Many services were set up on an ad hoc basis within the post or laboratory. For medical care, the residents of Los Alamos were served by a small Army hospital with six beds.[24]

The post was surrounded by a barbed-wire fence. In order to enter or exit, all personnel had to show their passes. The only visitors allowed were those who had special business either with the laboratory or working on construction. Residents could not have guests or casual visitors from outside. Travel from the town was severely restricted; special permission was required to go more than one hundred miles, or further than Albuquerque. The most popular destination was Santa Fe. Few of the residents owned cars, however, and because gas was rationed, even a trip to Santa Fe was a special treat. Laboratory employees were given one day off per month for a shopping trip to the town, for which the Army provided bus transportation.[25] Such trips were defined not as a right but as a privilege, one that could be withdrawn at any time. The chemist Joseph Kennedy, who sat on the laboratory's Security Committee, proposed to the Governing Board that "a memorandum be sent to all personnel cautioning them that the privilege of visiting Santa Fe would have to be removed if the F.B.I. or G2 found any evidence of a leak from this source."[26] Personal trips beyond the local area were allowed only in special circumstances—for example, if there was a death in the family.

Communication in any form with persons outside the project was severely restricted. Mail censorship was introduced early in the project. Mail was allowed to be sent only via authorized drop boxes on the site, and both incoming and outgoing letters were censored. Residents did not have access to telephones for personal calls. A few senior project administrators had telephones in their houses, but these were for emergency purposes only. The laboratory had its own switchboard, operated by members of the Women's Army Corps (WAC). Long-distance calls from this switchboard were monitored. The monitor would announce at the beginning of the call that she was on the line; she had instructions to break the circuit "if there is any obvious breach of security" and to report "any conversation which seems doubtful."[27]

Secrecy and censorship were meant to sever or profoundly weaken any social ties extending beyond Los Alamos. According to Elsie McMillan, "security rules forced us not to make friends with outsiders. We might get too garrulous or let something slip." Jane Wilson, Robert Wilson's wife, described a chance encounter in the streets of Santa Fe with a college friend. Not having spoken for more than a year to anyone who was not part of the project, she was excited to see someone "from the outside world." However, even this encounter was against the rules, and she found herself unable to converse: "My conversation was a succession of fluid grunts. A moment's slip and I, by nature blabbermouthed, felt that I would find myself hurtling into the gaping entrance to hell. It was a relief to say goodbye. Then, like a child confessing that she has been naughty, I reported my social engagement to the Security Officer. Everything had to be reported to the Security Officer. Living at Los Alamos was sometimes like living in jail." She also felt confused and stifled by censorship regulations governing personal letters: "For fear of saying the wrong thing, one said as little as possible. Letters home were inclined to be terse and in my case, anyhow, painfully self-conscious. I couldn't write a letter without seeing a censor pouring over it." The inability to express oneself intimately in letters also blocked an important medium for personal reflection, an effect compounded by a prohibition against keeping personal diaries.[28] All these restrictions helped to produce the peculiar intensity and one-dimensionality of life at Los Alamos. Because the participants were entirely caught up in the immediate life of this small community, the project became the totality of their experience.

A COMPANY TOWN

Another feature of everyday life on the project was a feeling of impermanence and instability. The rapid expansion of the laboratory and community exceeded all predictions. Oppenheimer's initial understanding of the nature and scale of the project proved entirely unrealistic. He originally saw it as a type of retreat for a small group of senior physicists, a collaboration that would draw upon the academic physics community's intimacy, collegiality, and sense of vocation. At the end of November 1942, he wrote to James B. Conant (one of the senior civilian administrators overseeing the atomic bomb effort) that "the technical details of this work will in large part have to do with atomic physics so that any man whose experience has been in another field will necessarily be of more limited usefulness." He therefore thought that "in a tight isolated group such as we are now planning, some warmth and trust in personal relations is an indispensable prerequisite, and we are, of course, able to insure this only in the case of men whom we have known in the past."[29]

It was not long before this conception of a small, homogeneous staff began to seem inadequate. In the spring of 1943, Oppenheimer was trying to persuade the talented scientific organizer Robert F. Bacher to come to Los Alamos from MIT. Bacher insisted on the need for a strong engineering program: engineers, he said, "had to permeate the place." And Oppenheimer asked Bacher to recruit "a group of men whom you would call physicist engineers." Between late spring and early summer 1943, Los Alamos began trying to recruit chemists, metallurgists, and engineers.[30] The laboratory grew even more rapidly the following year, when large numbers of technicians, engineers, and junior scientists were brought in via the Army Corps of Engineers' Special Engineer Detachment. SEDs were technical workers from industry or college students; they were enlisted into the Army and channeled into the Manhattan Project, often after further technical and scientific education at universities under the Army Specialized Training Program.

Oppenheimer's earliest conception of weapon-design work was that it would require as few as six scientists—as he put it to Lawrence, "three experienced men and perhaps an equal number of younger ones." In late 1942 and early 1943, Oppenheimer was operating with an estimate of around a hundred scientists, or, including support personnel, several hundred persons.[31] The real figures turned out to be in the thousands. Providing housing and provisions for the expanding population was a constant source of difficulty: a post administrator stated in February 1945 that the increase in population "taxes practically all our facilities beyond capacity."[32] Construction on the project began in January 1943; scientific personnel began to move in on a permanent basis toward the end of March, and the population reached an estimated 3,500 by the end of the year. In December 1944 there were 5,675 people living at Los Alamos, and by June 1945 the total population had reached its wartime peak of approximately 8,750, including about 1,750 dependents.[33]

Los Alamos was a highly artificial, demographically anomalous community. It was overwhelmingly composed of young people. The average age of the scientific personnel was twenty-nine; only one scientist was over fifty-eight years old.[34] Oppenheimer himself was only thirty-eight when he became director. The youth of the population, combined with the uprooted character of the community, made for a sense of freedom and vitality. The life, wrote Jean Bacher (Robert Bacher's wife), was "peculiarly uninhibited and completely unrelaxed." On the weekends, the employees "let off steam—steam with a collegiate flavor. Large dances, which often turned into binges, were popular."[35]

The carefree atmosphere was accompanied by, and encouraged by, the enforced intimacy. This was true not only for the GIs, WACs, and single civilians, who lived in barracks and dormitories, but also for families. At Los Alamos, as in other small towns, "private matters were of public concern."[36] The informality and intimacy of life contributed to a sense of commonality and camaraderie. It reinforced the communitarian ethic of a group of people united by a common mission.

Los Alamos was a company town, in which the goals and values of the laboratory and the Manhattan Project were dominant. Community life at Los Alamos developed in a one-sided way, oriented completely toward the instrumental goals of the mission that had brought the residents to the site. At a Governing Board meeting in May 1943, David Hawkins, who organized liaison between the laboratory and the post, "emphasized the importance of putting across to the community the idea that this was a war project and the interests of the community must be subordinated to the progress of work in the laboratory."[37]

Particularly worrisome for the community's planners were nonworking wives, many of whom were raising families at Los Alamos. The problematic situation of the wives was exacerbated by the fact that Los Alamos was a planned community. The Army was charged with managing daily life in order to minimize any tendency for domestic problems to distract the scientists from their work. The more functions the Army took over in the management of everyday life, the more apparently superfluous was the domestic role of wives. A psychiatrist, called in by General Groves to assess the community, reported in the summer of 1944 that "the creation of jobs that provide emotional outlets for unoccupied wives has apparently done much to improve morale." For example, the establishment of a nursery school was "invaluable as a morale builder as well as an educational unit, permitting many mothers who would otherwise be discontented to become contributors to the productivity of the unit."[38] Ideally, all spheres of life, including the domestic sphere, were to be assimilated and subordinated to the overarching goal of producing the bomb.

As a planned community, Los Alamos had some characteristics of a utopian social experiment. More than one commentator has noted the "strong hint of utopian collectivism" that characterized the community.[39] An example of this was the linking of rent to salary, so that higher earners would pay higher rent.[40] Medical services were provided by the Army. Laura Fermi thought that at Los Alamos, the Army was running a "socialistic community."[41] Groves emphasized that the Army should, through its management of the town, ensure that domestic issues would not be allowed to distract scientists from the project's technical goals. He instructed Colonel Gerard Tyler, who in 1944 took over as post commander of Los Alamos, to "try to satisfy these temperamental people. Don't allow living conditions, family problems, or anything else to take their minds off their work."[42] At the same time, the scientists and their families, as members of the professional middle classes, were used to having a high degree of control over the conditions of their lives; they felt strongly that they had the right to personal autonomy, even while living on an Army post. Because of their value to the project, the scientists had the power to demand privileges. The Town Council, established in June 1943, was a way for civilians to air their grievances about living conditions. Reminiscent of student government, it was a gesture toward participatory democracy, anomalous in its wartime context. The young scientists and their families here debated with gusto the minutiae of post administration, from parking tickets to food in the PX to restrictions on dormitory visits by the opposite sex.[43]

It was hoped that the scientific and technical workers at Los Alamos would feel a moral bond to the project—a commitment above and beyond the call of duty. But the laboratory's administrators did not regard this attitude as something that could be left to develop organically in the community. Instead, they thought it had to be inculcated, or at least encouraged. Hence David Hawkins's concern with "the importance of increasing people's awareness of the war and of the fact that this is a war project."[44] In general, during the war, the Army paid considerable attention to the problem of generating "morale"— in other words, manufacturing appropriate motivations among soldiers, war workers, and the general population. Sociologist Morris Janowitz has argued that this emphasis by the military on morale, which was particularly evident during World War II, was a response to the increased reliance of the military on technology, and hence on personnel drawn from skilled and educated urban industrial populations. According to Janowitz, such personnel respond better to explicit motives than to simple discipline.[45] Certainly, this applied to the highly educated scientific and technical workforce employed at Los Alamos. The Army commanders at the post felt that they were forced to make considerable "concessions" to the "personal wishes" of the scientists in order to "keep the life of the working community running as smoothly and contentedly as possible."[46]

The concept of morale as employed at Los Alamos expressed a social ethic, emphasizing the virtue of integrating the individual into the life and norms of the community. For example, Oppenheimer's successor, Norris Bradbury, highlighted the "extraordinary harmony" and "sense of homogeneity" with which people worked together at Los Alamos during the war. By May 1945, the visiting psychiatrist Eric Kent Clarke thought Los Alamos a "closely-knit group with a feeling of accomplishment, and a sense of personal responsibility." "The compactness and isolation," he said, "actually has helped promote this." Despite the "feeling of impermanence," he felt that "there is now a sufficient volume of old residents to create an acceptance of the limitations, that is helpful in acclimatizing newcomers." He was pleased with Los Alamos's success in the "integration of the individual into the community." Clarke's philosophy of "mental hygiene" and "psychological adjustment" fit well with the project administrators' desire for a population single-mindedly oriented toward the goals of the project and bound together by a common ethos. Clarke thought that Los Alamos, with its close-knit community life and availability of healthy outdoor sports, was suited to "young, athletically-inclined extroverts." It was the "non-athletic, the introvert, the older people, the confirmed city-dweller," who he thought would be unhappy and dissatisfied. This "individualistic" group found "the isolation and necessity of living in such close physical contact with neighbors trying."[47] The isolated and close-knit nature of the community at Los Alamos, together with the collaborative nature of the scientific and technical work, meant that virtues of integration, teamwork, and sociability were primary.

So far, I have argued that Los Alamos was a community of uprooted and dislocated individuals, but a solidaristic one. Arrival there, I have suggested, was experienced as an autobiographical break, disconnecting project personnel from the social milieu from which they came and defining Los Alamos as a radically new situation. Once at Los Alamos, individuals were faced with powerful social-psychological pressures that recast their identity in conformity with the local communal identity of the laboratory and the town that supported it. The distinction between public and private at Los Alamos was only weakly maintained. All spheres of life, including the domestic sphere, were subordinated to the goals of the laboratory. Los Alamos was a peculiarly one-dimensional community, dominated by a single institution. Yet at the same time, the town was a place of radical normative uncertainty. The people who inhabited this place faced the problem of making sense of, and collectively defining, this new situation.

Were the scientists who worked in the laboratory supposed to think of the experience and the setting as simply continuous with the university environments from which they came? Was Los Alamos just an academic laboratory transplanted to a remote location? Or was it, on the other hand, more like an industrial research and development laboratory, along the lines of Bell Labs or the research branch of Westinghouse? Was it more like a military ordnance facility, such as the naval and army ordnance laboratories at Dahlgren, Virginia, and Aberdeen, Maryland?[48] Or was Los Alamos a radically new kind of place—similar in some aspects to each the above, but also highly different from any one of them? Participants faced the problem of finding a model on the basis of which to understand what was expected of them, what would count as proper behavior, what rights they could legitimately demand, and what privileges they would have to forgo. But the applicability of such models was always incomplete, and the relevance and/or legitimacy of any particular definition of the situation was open to contestation.

A MILITARY OR A CIVILIAN LABORATORY?

The key uncertainty facing participants was whether Los Alamos was primarily a military or a civilian facility. Two fences defined Los Alamos both geographically and symbolically. The outer fence, which divided the site geographically, legally, and politically from the state of New Mexico, marked out the Army post. A second fence separated the post—and hence the authority of its Army commanders—from the Technical Area, which was a formally civilian institution, operated under a contract with the University of California and with Oppenheimer as the lab's civilian director.

This apparently neat division of space and authority, however, was less simple in practice. Oppenheimer, like the military post commander, reported to General Groves of the Army Corps of Engineers. Indeed, Oppenheimer had been selected and appointed by Groves personally. The general maintained tight personal control over the project through telephone and teletype, and he very often visited Los Alamos, inspecting the site and attending key meetings. He exercised control through Oppenheimer and through the Albuquerque District Engineer, the Los Alamos post commander, and the Army and Navy liaison officers assigned to the laboratory. He was in close contact with, for example, the head of the laboratory's Ordnance Division, Captain William S. "Deak" Parsons. The role of the University of California was, in practice, limited to procurement for the laboratory and other business matters. Even when they signed the contract to manage the laboratory, the university's representatives were not told the project's purpose. So the fence around the Technical Area did not unambiguously demarcate civilian from military authority. As one Army report stated, "As the project developed, the lines between the responsibilities of the local scientific and military leaders became, of necessity, more flexible and overlapping."[49]

The boundary between military and civilian authority was further blurred by the growth in the proportion of laboratory staff who were military personnel—scientists and technicians in uniform. In August 1943, approximately 5 percent of personnel were military. In August 1945, civilians and military personnel were represented in equal proportions. The majority of these new technical personnel in uniform were SEDs. In July 1945, there were fourteen hundred SEDs working for the laboratory. In addition, about seventy WACs were employed there—the majority as clerks, librarians, and telephone operators, but several as scientific researchers and technicians.[50] The position of these personnel in relation to the military authority of the post and the civilian authority of the laboratory was unclear and contested. The HE (high-explosives) development for the implosion device (the design of the plutonium atomic bomb, "Fat Man") relied very heavily on SED manpower.[51] The GIs working in this program at testing ranges in the mountains and canyons around Los Alamos spent their working day beyond the reach of military commanders. Officers of the post often had trouble even knowing where their SED troops were at any particular point in the day, and most of these officers, unlike their troops, did not have passes to enter the Technical Area. General Groves was understandably disturbed by this obstacle to the authority of his commanders over their troops, and he moved to put officers of the Corps of Engineers directly in charge of the SEDs at the HE testing sites. George Kistiakowsky, the chemist running the HE program, strongly objected to this proposal. He told Oppenheimer that it was unacceptable, since the place of officers in the military chain of command would be unlikely to coincide with technical expertise.[52]

In general, the ability of SEDs to work effectively in the laboratory was given precedence over the enforcement of normal Army procedures. For example, Major T. O. Palmer, upon his appointment as commanding officer of the SED in August 1944, eliminated morning reveille and calisthenics for these troops. At a conference of senior post staff in early 1945, Palmer responded to the question of whether the SEDs were being "subjected to too much military work": he did "not see how the SEDs could have any fewer military duties." Clarke, the visiting psychiatrist, noted in passing that "the S.E.D. cannot be regarded as regular soldiers, having been segregated soon after enlistment for specific jobs because of specialized education. The work under civilian administrators places military regulations as secondary to the scientific program." This "anomalous situation" of the SED—between civilian and military authority—was, in his view, "a knotty one."[53]

As a former SED put it, "This was the least military of any military outfit I have ever been in. Now I understand that other units, the construction group and the MPs [military police], were really G.I., but not us. We were quite un-G.I." Another recalled, "Tech Area pressure kept the military from interfering and trying to make us G.I." Charles Bagley played down the distinction between military and civilian laboratory personnel, emphasizing the fact that he and many other SEDs were students when they were drafted or enlisted. They came to the project "right out of colleges, right out of the academic life." Another former SED said, "I was simply a civilian placed in military uniform." Still another reflected, "It was kind of frustrating to Major Palmer and all those who had been out where military was *the* thing to come out here where people were running around like a rag-tag militia." SEDs enjoyed the fact that work in the laboratory placed them temporarily outside the military environment. Roy Merryman recalled, "We kind of lived in the laboratory building, rather than sit around in the barracks on our bums. That was almost home to us." According to a former WAC who worked in the Tech Area, "it wasn't like anything else in the Army . . . Everyone had their job and no one was much concerned with rank."[54]

The uncertain coexistence of military and civilian forms of life inevitably produced conflicts and tensions on both sides. Civilians felt regimented and threatened by the restrictions of life on an Army post. Military personnel—particularly those in support functions, who did not have access to the Technical Area and from whom the project's goals were kept secret—felt that they were treated as secondary to the civilian scientists. Clarke reported that the "greatest single problem is the discrepancy between civilian and military life." The strongest complaint by military personnel working in the laboratory and on the post was that they received less pay for the same work. More generally, as Clarke stated, "the army group feel that they are penalized by being in uniform, are crowded out of their facilities, and in addition are patronized . . . The feeling prevails that the civilians are pampered lest they leave the project while the army group have no such similar opportunity."[55]

The relationship between civilian and military organization and authority was also complicated by the fact that formally, the laboratory's civilian status could be revoked. The original directive for Los Alamos was signed by both Groves for the military and Conant for the civilian scientific community. The directive divided the work of the laboratory into two periods, the first involving "certain experimental studies in science, engineering and ordnance," the second involving "large-scale experiments involving difficult ordnance procedures and the handling of highly dangerous material." In other words, the directive conceived of the program as being divisible into an initial phase of basic research and a later phase of weapon engineering. "During the first phase," the letter stated, "the laboratory will be on a strictly civilian basis." Upon commencement of the latter phase, however, "the scientific and engineering staff will be composed of commissioned officers." It was expected that at that time, the civilian scientists in the laboratory would be willing to take commissions as Army officers, and this was projected to occur sometime after the beginning of 1944.[56] Few senior scientists were eager to be enlisted, however. Bacher, for example, upon accepting his post in the laboratory, also submitted a resignation letter, to become effective on such date as the laboratory was militarized. This change never came to pass. But the civilian status of the laboratory was always a tentative one, with no guarantee of permanence. The fate of the laboratory was in this respect open. It was also entirely in the hands of one man—the overall executive head of the project, General Leslie R. Groves.

THE CHAIN OF COMMAND

At Los Alamos, to civilians and GIs alike, General Groves was the military personified. This perception was in many ways accurate, given Groves's strong personal control of the project. Rather than an impersonal bureaucracy, the Manhattan District is more correctly seen as having been a personal fiefdom. Formal divisions of power and authority in the project were beset with ambiguity and uncertainty, and above these formal relations, Groves's authority floated free, encompassing in its sweep the entire project. Historian Peter Bacon Hales pointed out that Groves was invisible in the project's organizational charts: "At the top was always the district engineer—first [Colonel James C.] Marshall, then [Colonel Kenneth D.] Nichols. Groves himself existed immaterially, at once everywhere and nowhere."[57]

A circular from Nichols, who functioned as Groves's deputy, stated, "Owing to the secret nature, urgency, scope, and importance of the projects assigned to this district, it is frequently necessary that instructions be issued and information requested without regard to normal organizational channels." Groves, in particular, was to be insulated from these usual channels:

General Groves does not wish to be designated to individuals outside the Manhattan District, particularly to investigating agencies, as having any direct connection with the Manhattan District . . . In general, if any

outside agency requests information concerning what is the next higher
echelon of command above the District Engineer, the answer should be
that the next higher echelon is the Office, Chief of Engineers, and no
particular individual designated.[58]

Groves's powers were sweeping; from his position outside regular organizational channels, he was able to overcome bureaucratic conservatism and
inertia. He prided himself on his ability to "cut through all sorts of intermediate layers of authority." Indeed, he himself proudly claimed that "there was
nothing more unorthodox than my operations either on construction or on
the Manhattan Project."[59]

Groves's extraordinary personal power derived, above all, from his ability
to claim a direct link to the authority of the president. Roosevelt, Groves
believed, "took a personal interest" in the project and had "gone over my
record when my name came in and had personally approved it." But Groves
met Roosevelt only once during the war, with Secretary of War Henry L.
Stimson, on December 30, 1944. They were allotted half an hour, but the
meeting took considerably longer. "That's the only time I saw him or talked
to him about the Manhattan Project," Groves said.[60]

Groves was obsessed with maintaining close personal control over the
operations of the Manhattan District. He was also the only individual with
a synthetic and overarching view of all dimensions and levels of the project.
Stimson informed Roosevelt that "Groves . . . is the only one who has a really
complete knowledge of the entire situation."[61] The broad range of Groves's
personal role was later described by Nichols: "Throughout the war . . . Groves
maintained direct access to both General Marshall and the Secretary of War
whenever he saw fit . . . In addition, he maintained direct control of construction and operations at Los Alamos . . . He also assumed control over many of
the international contacts, all intelligence efforts, and at the request of General Marshall, military planning for the use of the bomb." Nichols noted that
"an organization purist would say that this hybrid organization would generate friction, confusion, and might not be able to function smoothly." But in
Nichols's view, the arrangement "expedited decisions and results."[62]

The fact that Groves did not respect regular organizational channels meant
that he kept his subordinates in a state of uncertainty, thereby reinforcing
their dependence on him and strengthening his own personal power over the
network. However, despite his unique organizational power, Groves was
himself in a state of dependence: he relied utterly on the scientists' willingness
and ability to get the job done. When he tried to pressure Ernest Lawrence

by telling him, "Your reputation at stake," the Nobelist replied, "You know General, my reputation is made. It is yours that depends on the outcome of the Manhattan Project."[63]

COMPARTMENTALIZATION: SECURITY AND CONTROL

A key aspect of Groves's power was control over the circulation of information. A system of such control, termed *compartmentalization*, was already in place when the Army took over the atomic bomb project in the summer of 1942. But under military leadership, it took on new importance. The system required that, as far as possible, each task was to be performed in isolation from all the others. Those highest in the administrative hierarchy would have the greatest breadth of knowledge about the project, and only Groves, as overall director, would have a complete overview. The system was justified on security grounds: limiting each individual's knowledge of the project would make it impossible for an enemy spy to give away anything more than a fragmentary picture of what was taking place and, most importantly, of the technical details of how it was being accomplished. "Compartmentalization of knowledge," Groves said, "was the very heart of security."[64]

Compartmentalization also served other organizational functions. For Groves, it was a means of intervention in the *process* of scientific work on the project. It was a mechanism by which the practices of scientists could be transformed so as to be in keeping with the character of the Manhattan Project as a large-scale military-industrial system. When the Army Corps of Engineers took over control of the atomic bomb project, they found the state of scientific knowledge to be utterly inadequate as a basis on which to begin engineering development. It seemed to them unlikely that an academic approach would, on its own, lead to establishing the necessary levels of certainty on which to base planning. As Colonel Marshall described the situation, "When you get six or seven Ph.D.'s and three or four Nobel Prize winners around the table, you know, they are in the clouds." Marshall told the scientists that "if they didn't hurry up and make up their minds what they wanted to develop, we might not need a site; the war would be over." When Groves took over as head of the Manhattan District, it seemed to him "as if the whole endeavor was founded on possibilities rather than probabilities. Of theory there was a great deal, of proven knowledge not much. Even if the theories were correct, the engineering difficulties would be unprecedented."[65]

In Groves's view, scientific curiosity could lead to all sorts of blind alleys. If the necessary discipline could not already be found in the practices of the scientists, it would need to be imposed externally. This, then, was a key benefit of compartmentalization. This system of control over information, Groves said, "not only provided an adequate measure of security, but it greatly improved over-all efficiency by making our people stick to their knitting. And it made quite clear to all concerned that the project existed to produce a specific end product—not to enable individuals to satisfy their curiosity and to increase their scientific knowledge."[66] Compartmentalization defined scientists' work in instrumental or utilitarian terms, creating a structure of control and authority. Instrumental goals were established at the highest rungs of the Manhattan District, and the scientists were to labor to realize these externally determined ends.

By limiting scientists' tasks to particular technical problems and insulating them from knowledge of the overall framework of the project, compartmentalization strictly circumscribed the authority of scientists within the project. According to Arthur H. Compton, Groves "did not want any one man under him to have so much responsibility that he would become indispensable." By dividing authority and responsibility, Groves "avoided the troubles that he feared if some single scientist had been in a dominating position." Scientists were to be technicians, not strategists or policy makers. As Hales observed, compartmentalization was "a means to redesignate scientists and engineers as workers, equivalently obligated to management."[67] Groves was dismissive of scientists' resistance to the role defined for them in this system: "A lot of them were resentful because their opinions had not been asked for during the War; there had been no 'faculty' meetings. If I went to Los Alamos, for example, I would see Oppenheimer and his group leaders. We might go around and see things and talk to various individuals who were way down on the scale, but they weren't asked for their views on how to solve these problems or what we should do." For Groves, compartmentalization was about much more than secrecy; it defined a structure of authority: "We had strict discipline and we had to have it; not military discipline but real discipline, with everyone interested only in one thing and that was the achievement of the goal. As soon as the need for such discipline is removed there is a blow-up." Security regulations allowed Groves to regiment and standardize the work practices of the heterogeneous groups involved in the Manhattan Project, thereby creating a homogeneous organizational structure. Compartmentalization rendered the militarization of the whole project unnecessary. Without putting scientists in uniform, the general had discovered a way to firmly define their role and place within an organizational hierarchy and to limit scientific authority to the purely or merely technical.[68]

Security regulations functioned to break up associations, disrupting informal scientific culture and replacing it with the formal structure of the organization charts. In 1942, physicist Leo Szilard complained that compartmentalization caused "strain" and embarrassment between him and his "old friend" Teller and had led to their misunderstanding each other during discussion of an important scientific problem. According to Szilard, compartmentalization "poisons the discussion, even in those fields which are not explicitly excluded from discussion."[69] Szilard's colleague at the Manhattan Project's Chicago Metallurgical Laboratory (Met Lab), fellow Hungarian Eugene Wigner, similarly contrasted scientific fraternity with the formality of large-scale organizations, such as DuPont. The "outlook on life" and "attitudes toward the work" differed greatly between scientists and industrial engineers. Wigner thought that the engineer was motivated by "money and power," the scientist by the desire for "the esteem of his friends and collaborators" and "the satisfaction of having understood something." It seemed to Wigner that the DuPont men measured themselves and others by their place in the organization. By contrast, the moral order of science was violated by any privileging of the position over the person. So, he said, "if it is necessary for us to point to our authority embodied in the organization chart to someone, we may just as well terminate the relation with that collaborator—he would not do us much good."[70] Szilard and Wigner regarded compartmentalization as socially and scientifically corrosive, replacing scientific collegiality with an alien hierarchy.

Groves, however, was suspicious of such informal communication among scientists and of the social networks in which it was embedded. He was wary of foreign scientists—particularly Szilard, whom he regarded as disloyal.[71] The type of authority structure described by Wigner was inherently threatening to Groves. Without a scientific background, the general would have had a weak position in such an intellectual hierarchy. During one visit to Chicago, he defensively claimed that because he had attended Army schools for ten years after graduating from West Point, he had the equivalent of two Ph.Ds.[72] Colonel Nichols had a Ph.D. from the University of Iowa and had studied engineering at the Technische Hochschule in Berlin; he was, therefore, useful in being able to mediate these relationships for Groves. Groves said, "Because I suspected that Compton liked Colonel Nichols more than he did me, primarily because Colonel Nichols had a Ph.D. and looked very scholarly . . ., everything done with Compton was generally done through Colonel Nichols; that is, anything that was difficult."[73] The reputational authority structure of scientific community was an obstacle to Groves's personal control of the project.[74] If compartmentalization replaced such collegial authority with formal, organizational channels, so much the better, from Groves's point of view.

In the general's vision, Los Alamos was to be a place where scientific community and practice could be disciplined by military-industrial standards, in order to serve a military need. The decision to centralize research on bomb design in a new laboratory, situated in a secret location, was one of the first and most important moves that Groves made upon assuming control of the Manhattan District. Bringing the atomic scientists to Los Alamos would bring them directly under Groves's power and control, insulating them from other, opposing influences. Groves took a special and personal interest in Los Alamos. In contrast to other sites of the project—where his deputy, Colonel Nichols, had an important role to play—"in the case of Los Alamos, Groves made it clear that he personally would do all the direct supervision of the work." Groves later said that "from a practical standpoint, although not on paper, the chain of command was direct from me to Dr. Oppenheimer."[75]

However, Groves's desire to situate the scientific work securely under his control was potentially frustrated by an opposing agenda for the new laboratory. Oppenheimer and other senior scientists saw the lab as a chance to overcome the restrictions and difficulties of the compartmentalized and dispersed nature of the fast-neutron work that had been going on, without great success, in university laboratories across the country.[76] They argued that scientific freedom within the new laboratory was necessary if Los Alamos was to efficiently meet the instrumental goal of producing atomic bombs.

INFORMATION FLOW AND SOCIAL ORDER

An important motive for the creation of Los Alamos, from Oppenheimer's point of view, was to overcome problems of communication. Concentrating the fast-neutron research in one geographical location would be a vast improvement over the situation that he and John Manley faced in the latter part of 1942, in attempting to coordinate a gaggle of widely geographically dispersed fast-neutron laboratories.[77] However, less than a year into Los Alamos's operation, such problems of coordination and communication were being reproduced *within* this new laboratory. The senior scientists at Los Alamos were sensitized to problems of information flow, which arose not only from restrictions imposed by Groves, but also from the de facto compartmentalization that was an aspect of the increasing complexity and differentiation of the laboratory's organization.[78] As early as November 1943, Oppenheimer "felt the laboratory was now so complicated that he should call to the attention of the board the problem of relations between divisions." That same month, the laboratory established a Liaison Committee, chaired by Edward Teller, to coordinate communication with other sites of the Manhattan Project. By January 1944, however, this committee had been forced to turn its attention to the problem of so-called internal liaison.[79]

The scientists responded to these problems of communication by attempting to locate scientific work and intercourse in face-to-face interaction. In early May 1943, Bethe proposed to the Governing Board (the laboratory's top decision-making committee, composed of the most senior scientific and administrative staff) the idea of instituting "regular colloquia for the entire staff," to be held weekly or fortnightly. These might, he added, include reports from other sites of the Manhattan Project. Oppenheimer agreed with the thrust of Bethe's proposal and emphasized the "importance" that "should be given to regular meetings of the groups in which general laboratory affairs should be discussed, as well as specific problems of the groups." Bethe's proposal was accepted at the Governing Board meeting. The board decided that such "general colloquia" should be "held every two weeks on Tuesday night," and it made Teller responsible for organizing the meetings. However, somewhat ominously, the point was added that the meetings should be "carefully supervised."[80]

Almost inevitably, the Colloquium and its organizers came into conflict with the Army, to whose security rules the very existence of such meetings appeared to be an affront. Groves told Oppenheimer a few months later that both he and the post commander, Colonel Whitney Ashbridge, were "disturbed" about the "quite comprehensive review" of the program that Oppenheimer gave at the first colloquium. They felt that "from the beginning the colloquium has been operated very liberally according to war laboratory standards." Groves's personal scientific advisor, physicist Richard Tolman, supported such concerns. He was "troubled" by the presentation at a colloquium of a report on the methods that chemist Harold Urey was employing at Columbia University for separating boron isotopes. Oppenheimer replied in defense of the meetings, stating that he was "committed to this policy" and that he believed it was "the right policy."[81] Nevertheless, he agreed to a compromise under which restrictions were placed both on what could be discussed at the Colloquium and on who could attend these sessions.

Groves responded to the institution of the Colloquium by toughening his position that the laboratory should be a world unto itself, with the very minimum of contact, scientific or otherwise, with people or institutions beyond its fences. He told the Los Alamos Governing Board that "it was only with the greatest reluctance that he approved having anyone here who was allowed to leave, or having anyone visit here in a consultative capacity."[82]

Probably the most controversial aspect of the Colloquium arose from Bethe's initial suggestion that these meetings include reports from other sites of the project. At a Governing Board meeting on October 28, 1943, several leading physicists at Los Alamos took part in a lengthy discussion about this issue. The minutes stated optimistically that "the only specific restrictions imposed on the dissemination of information were on engineering details of the work in Chicago, and on production schedules that would determine the effectiveness of the end product as a military weapon." However, these restrictions were seen to be of great significance. Also, more far-reaching than these "specific restrictions" was the general climate of uncertainty about what could or could not be discussed. The result of such uncertainty was to make people even more cautious and tight-lipped than absolutely necessary. For example, in addition to restrictions on discussion of work at Chicago, Bacher, Bethe, and Oppenheimer responded to Tolman's comments by urging Teller not to discuss Urey's work on boron at future meetings.

During the October 28 meeting, Teller emerged as the most staunch defender of the original spirit of the Colloquium. He told the board that he felt "very strongly" that "imposing any limitations on the discussion in colloquia is contrary to the spirit with which the colloquia are supposed to be operated." He also felt that if restrictions were to be made, this should not be simply a matter for the division leaders on the Governing Board, but the policy should be at least announced to group leaders, one organizational rung down, on the Coordinating Council. This position pitted Teller directly against Oppenheimer, who advocated a more cautious and conciliatory path between scientific freedom and military regulation. Oppenheimer stated that "he could not formulate a policy except to say that the liberties permitted in this laboratory were much greater than in any other war laboratories and that he felt that discretion should be exercised so that those persons who were concerned about the scope of these liberties would not be alarmed. He felt that no information should be given which could not be justified by its connection with the work here." Oppenheimer was also "personally opposed" to the idea of discussing the policy with the Coordinating Council.

Teller did not hesitate to point out that Oppenheimer's proposed compromise violated the original intent of the Colloquium. The criterion that only information connected with the work at Los Alamos could be discussed was, in Teller's view, "too vague to be applied." He said that "a function of the colloquium was to bring a variety of matters to the attention of staff members in order to stimulate useful ideas. It would be practically impossible for one man to judge what the most productive questions for discussion would be." It was also unclear whether this criterion (that information was admissible only if relevant to the work at Los Alamos) was a change in policy. The general view was that "this statement does not represent a change but merely a formulation of existing policy" and that "the only basis for thinking that it is a change is Dr. [Vannevar] Bush's statement that all information would come to Los Alamos and none would go out." The other members of the Governing Board were leery of antagonizing Groves and Army security and felt themselves to be walking a tightrope between freedom and restriction: "It appeared to be the general feeling of the board that it would be wise not to act in such a way as to cause restrictions to be imposed by those who have the executive responsibility for the project. In particular it was felt that the laboratory would be harmed if documents were withheld from us on the ground that we discussed them too freely in colloquia." Teller left this meeting to attend another one; in his absence, the board recommended "that all work done in this laboratory can appropriately be discussed in colloquia. Of work done elsewhere which reaches us in the form of classified reports, all questions which have a presumptive bearing on the work of this laboratory may be discussed."[83] Teller, it appears, was overruled, but the board's decision left room for negotiation. What, for example, was to count legitimately as "presumptive" relevance?

These discussions exemplify a number of important features of life at Los Alamos. First, they show the way in which the senior scientists, in shaping the social order of the laboratory, oriented themselves in relation to their perception of the likely reaction of Groves and the military authorities. Groves was thereby able to exercise control in absentia. Second, the controversy provides an example of the normative uncertainty surrounding institutional practice at Los Alamos. The policy on information flow was unclear and contested, and different formulations were possible. As Teller recognized when he "asked that if restrictions were to be imposed they be made specific," this uncertainty could itself be restrictive. This was particularly the case if the board was inclined to err on the side of caution in relation to the military authorities. Third, Oppenheimer's role in the controversy demonstrates his intermediate position between the scientists and Groves's military authority. To Groves, Oppenheimer defended the Colloquium; to the scientists, he stressed the need to play within the Army's rules and to accommodate the interests of security. Oppenheimer embodied and personally mediated the conflicts and tensions at the heart of the social organization of the project.

In response to the Colloquium, Groves sought to impose a firm cordon of security around Los Alamos. However, any such cordon would necessarily be porous to a degree. Los Alamos required both information and materials from other sites of the Manhattan Project, so it could not be entirely isolated.[84] The Governing Board was constantly trying to get Groves to allow it more access to information about Oak Ridge production schedules. This information, essential for planning the Los Alamos program, came via Groves, and he did not want to have another set of estimates that would compete with his own and provide a basis upon which to challenge his authority. After all, one important way in which Groves was able to intervene at Los Alamos was in pressing the Governing Board to keep up with schedules. Issues of the flow of information were issues of authority and organizational control.[85]

The Colloquium was thus affected by general restrictions on liaison of Los Alamos with other sites and on the flow of information across the Manhattan Project. Moreover, the meetings were open only to laboratory staff members, defined as those with a bachelor's degree or higher. In general, technicians, including most SEDs, could not attend.[86] Many of the SEDs, for example, were enlisted during their undergraduate education, before receiving a degree. For these personnel, access to information was on a strictly need-to-know basis. McAllister Hull Jr., an SED directing the casting of explosive lenses, said, "That was the way I was taught. I was taught need-to-know . . . The badges said what kinds of things you could talk about. I suppose if you had a white badge . . . you could talk about anything . . . So there was that kind of compartmentalization. When somebody says there was no compartmentalization because people went to the seminars, well, who was invited to the seminars? . . . That's the point. I didn't get to the seminars. I didn't need to know this theoretical stuff that was going on in order to cast things out at S-Site." The extent of compartmentalization was asymmetrical, differing across the levels of personnel in the laboratory: "There was compartmentalization in terms of the technical people, but . . . for all of the real physicists there wasn't so much or wasn't any, with seminars, etc."[87]

SED Jay Wechsler, working on electronics under Otto Frisch, described in detail the experience of junior personnel: "We did not have any meetings. Otto attended meetings, apparently at a high level. At that time, I had nothing to do with meetings at a high level at all." Whatever overall picture of the project Wechsler acquired was through a gradual, informal, and intuitive process of piecing together what information he could pick up from scattered conversations. After receiving his clearance and badge, Wechsler went to work in the Technical Area:

> Otto showed me this lab that was essentially empty and said that it would be a physics lab. And we were going to be working on things. Being curious, I asked him, "Well, what?" What were we doing and so on . . . and I wasn't getting any real response. And he pulled out some piece of equipment that he wanted me to help [with,] getting ready and hooking up the electronics for and so on. It was a very, very strange beast. I couldn't quite understand it. He began to explain to me what it was, and I could not associate what we were trying to do with anything . . . I got to know some of the people next door. I was talking to one of them in the next lab and I asked him some questions. And he wasn't answering anything very much . . . I had set up a glass vacuum pumping system with some diffusion pumps and mechanical vacuum pumps and got the equipment out of some of the supplies. The kinds of things that I thought that I needed to work for Otto. But I still didn't know why. It just was not making an awful lot of sense. He just kept calling it a piece of equipment that he had shipped over . . . but nothing was hanging together . . . I didn't even understand how the piece of equipment worked, let alone what we were going to use it for. It was more like instructions—we need to be able to do certain electronic measurements and there are some defects in the electronics and there are some defects and leaks in the vacuum system and you need to be finding out where they were and . . . figuring out ways to fix it—and it was very cryptic. Very cryptic.

The pieces began to fit together when Wechsler went to the report library and found a copy of *Who's Who in Science*:

> I looked up Otto's name, and then I began to see that he had worked in fission and [was] Lise Meitner's nephew, and things started kind of coming together a little bit. And then I get thinking about the piece of equipment I was working on, and I realized that this was probably a special ionization chamber or something of that nature. So I came back to the lab and Otto was at the desk, and he was working on some things and I was sitting over where I had all my equipment. I was looking at him and I was thinking, "Well, this is a pretty famous guy, and I was trying to put two and two together, and finally he turned around and

he said to me, "What are you looking at?" And I said, "You." He said, "Why are you looking at me?" I said, "I just found out who you are." And he said, "So, now get back to work." I said, "Well, I think I even know what we're trying to do." And he said, "Oh? Then you'd better realize you'd better get back to work."

Wechsler concluded, "The more I got into it, the more I realized why nobody was saying anything. So I just didn't discuss it anymore at all."[88]

Access to information, then, stratified the laboratory. According to Wechsler, "it was clear that there were many areas that were being addressed and that there was some hierarchy up there who had figured how all this fit together." But at the level of the SED technician, "you sure weren't going to find anything much of that." Alongside formal channels of information flow, however, there existed a range of informal channels. According to SED Charles Bagley, "there was a lot of talk that went on in the barracks." McAllister Hull had friends working in the Theoretical Division, from whom he gained information about broader aspects of the project. "But," he said, "officially I didn't know a lot of these things that I in fact knew quite well."[89]

THE GOOD HOST

The Colloquium symbolized an informal, face-to-face, and collegial social order in the Los Alamos Laboratory, in continuity with the norms of the university. According to David Hawkins, attending was a "relief from the hard work, and it was for your general education, often dealing with subjects quite unrelated even to physics." For example, John von Neumann gave a lecture on the mathematical theory of games. To Hawkins, the Colloquium's value lay in helping to maintain an academic atmosphere.[90] Equally importantly, in the context of an expanding and internally divisionalized organization, the Colloquium served to make visible the laboratory and the program as a whole, to represent tangibly the social and epistemic coherence of the laboratory. The lab, represented abstractly on paper in organizational charts, was given corporeality in the gathering together of its scientific personnel in one hall.

The Colloquium was a means of disseminating information, but it was also recognized as being more than that. It was suggested at a meeting of the Governing Board in late May 1943 that the Colloquium be used to re-instill "habits of work" after the interruption involved in constructing buildings and setting up equipment. Oppenheimer said that the Colloquium contributed to the laboratory's "effectiveness, morale and security." Bacher, head of the

Experimental Physics Division, told the Governing Board that the "most important value of the colloquia" was "integration." In general, issues of epistemic integration and information flow were often discursively framed in terms of "morale." Philip Morrison argued that the Colloquium "unified" the scientific workforce and "made us all feel more responsible for the whole outcome." According to Bethe, the most important benefit was that "everybody in the laboratory felt a part of the whole and felt that he should contribute to the success of the program." Victor Weisskopf later praised the Coordinating Council in similar terms: it was a "very open council. Problems were really openly discussed and you had the feeling that you knew what was really going on." He qualified this thought by adding, "You hadn't, I actually think"; the sense of overview provided by the meetings was more illusion than reality. "But the point is you had the feeling and that was of such importance." This feeling perhaps arose because "you had an opportunity to protest and tell your opinion and that contributed very much to the morale of the place." All of this he attributed to Oppenheimer, who "as the chairman did that extremely well."[91]

The integrating effects of the Colloquium and other meetings were associated with Oppenheimer personally. It was he who usually presided at the colloquia and made the introductory remarks. Above all, he was credited by the scientists with establishing the Colloquium and with defending the values associated with it. This was despite the fact that the Governing Board minutes from the October 28, 1943, meeting reveal Teller to have been the more adamant defender of freedom of discussion. Weisskopf ascribed the Colloquium and its integrating effects to Oppenheimer personally: "Oppenheimer insisted on having these regular colloquia against the opposition of the security-minded people, who wanted each man only to know his part of the work. He knew that each one must know the whole thing if he was to be creative." Hawkins said:

> The battle that Oppenheimer had with Groves [was because] Groves's pattern with regard to military security was the well-known formula for "need to know." You're not told anything unless you need to know it. But of course that means that the need is available, is defined. And in research the need to know is not defined. So Oppenheimer said, Look, I cannot run a laboratory unless there is complete openness among all the parts of the laboratory. Everybody who will be considered a scientific staff member will have full access to all information, and Groves was horrified by this . . . but finally he had great respect for Oppenheimer. He knew that his whole future depended on Oppenheimer.[92]

Oppenheimer's individual role was emphasized also by Bethe, who wrote, "Oppenheimer had to fight hard for free discussion among all qualified members of the laboratory. But the free flow of information and discussion, together with Oppenheimer's personality, kept morale at its highest throughout the war." Luis Alvarez insisted that "the laboratory's fantastic morale could be traced directly to the personal quality of Oppenheimer's guidance." Rudolf Peierls also credited the comparative openness of discussion at Los Alamos, and the laboratory's consequent morale, to Oppenheimer personally: "Inside the laboratory he was able to maintain the completely free exchange of information between its scientific members."[93] Donald Hirsch, an SED who arrived the day of Trinity—the first atomic bomb test—was told at a briefing the following day that the laboratory was working on atomic weapons. "Oppenheimer's orientation meeting," he said, "startled us with his openness and we discovered later that was the hallmark of the way Oppenheimer ran the Hill. It was not a military establishment, in no way." Such openness was regarded as essential not only for morale but also—in direct contradiction to Groves's view—for efficiency. Morrison estimated that if there had been strict compartmentalization, the project "would have taken another six months . . . because everybody [would have had] to pass the blueprints back and forth," and because of the constraints that compartmentalization would necessarily have imposed on initiative in the lower levels of the organization.[94]

At Los Alamos, Oppenheimer became the personal embodiment of the virtues of academic-scientific forms of organization, as opposed to military regimentation. Hawkins remembered a Coordinating Council meeting at which the Army's security officer, Captain Peer de Silva, complained about the lack of respect shown to him by a young SED who had sat on the edge of de Silva's desk. Oppenheimer replied, "In this laboratory anybody can sit on anybody else's desk." De Silva, Hawkins said, "was slammed." Oppenheimer was as good as his word; one young scientist found that "his office was always open and each of us could walk in, sit on his desk, and tell him how we thought that something could be improved." This instantiated what Hawkins called the "spirit of the laboratory," the principle that "anybody in the lab who is involved in serious work may have an idea that's useful. It's a democratic principle." In this way, for Hawkins and many other participants, Los Alamos epitomized "the democracy of science." Even one of the scientists who was later most critical of Oppenheimer, Harold Agnew, said, "Here everybody was equal. There was no question about it. There were no special privileges for anybody. That was a tribute to Oppie who understood that's the way it worked . . . To me it was just a wonderful place, wonderful experience."[95] In a situation of competing and conflicting norms, Oppenheimer gave voice to, and signaled in more subtle ways, a definition of the situation that made it legitimate to act in ways contrary to militaristic or bureaucratic codes of behavior.

In addition to being credited with institutional reforms, Oppenheimer was viewed as himself *embodying*—instantiating in his character and displaying in his person—these same values and functions. Over and over, Los Alamos scientists drew attention to how Oppenheimer in effect "knew it all." He was the one person at Los Alamos, it was repeatedly said, with the intellectual range and ability to possess an overview of the scientific work. Bethe recalled that Oppenheimer "knew and understood everything that went on in the laboratory, whether it was chemistry or theoretical physics or machine shop. He could keep it all in his head and coordinate it . . . There was just nobody else in that laboratory who came even close to him" in knowledge. Oppenheimer's breadth of knowledge and understanding of the project, Bethe said, were "clear to all of us, whenever he spoke." And as the laboratory's formal organization and technical program became increasingly complex, it was a relief to know that Oppenheimer "had it well organized in his head." According to Peierls, "his quick perception enabled him to remain in touch with all phases of the work." He could walk into a technical discussion in an area about which he might be presumed ignorant and make a decisive intervention—if not because of his factual or theoretical knowledge, then because of his ability to cut to the heart of any kind of problem. It was said that Oppenheimer "once joined a metallurgy session during an inconclusive argument over the type of refractory container to be used for melting plutonium. Although this was hardly familiar ground to a theoretical physicist, after Oppenheimer had listened for a time, he summed up the discussion so clearly that the right answer, though he did not provide it, was immediately apparent."[96]

But other commentary on Oppenheimer's intellectual scope and its integrating power is not so easy to understand in these terms, instead gesturing at the mental role of his physical presence. Weisskopf noted "some almost super ESP kind of connection" by virtue of which Oppenheimer managed to be on the spot when and where exciting developments were taking place. Stanislaw Ulam described one such occasion: "I saw Robert Oppenheimer running excitedly down a corridor holding a small vial in his hand, with Victor Weisskopf trailing after him. He was showing some mysterious drops of something at the bottom of the vial. Doors opened, people were summoned, whispered conversations ensued, there was great excitement. The first quantity of plutonium had just arrived at the lab." Oppenheimer, said Weisskopf, "was intellectually and even physically present at each significant step; he was present in the laboratory or in the seminar room when a new effect was measured, when a new idea was conceived. It was not that he contributed so many ideas or suggestions; he did so sometimes, but his main influence came from his continuous and intense presence, which produced a sense of direct participation in all of us."[97] Weisskopf emphasized "how tremendously important it was for the morale at Los Alamos . . . if you come to the final experiment and the director is there." Oppenheimer "always went to the important discussions at seminars, in spite of his administrative load."[98] This was a display of human concern and personal involvement. His presence made a difference, both intellectually and morally.

Robert Wilson remarked on the intellectually transformative power of Oppenheimer's physical presence:

> In his presence, I became more intelligent, more vocal, more intense, more prescient, more poetic myself. Although normally a slow reader, when he handed me a letter I would glance at it and hand it back prepared to discuss the nuances of it minutely. Now it is true, in retrospect, that there was a certain element of self-delusion in all that, and that once out of his presence the bright things that had been said were difficult to reconstruct or remember. Nor, as I left, could I quite decide what it was we had agreed to do. No matter, the tone had been established. I would know how to invent what it was that had to be done.[99]

Oppenheimer's body, gestures, and physical presence were equally essential to the apparently effortless way in which he was able to assert authority over his colleagues and to direct the laboratory with a minimum of friction. According to Wigner, the scientists at Los Alamos "disliked being visibly directed. Oppenheimer understood that. He knew their strengths and weaknesses without asking and treated them with some sensitivity." Wigner provided an example of this graceful exercise of authority:

> Oppenheimer . . . smoked a pipe and he gave some deft direction with his pipe alone. When a subordinate reached him with a grievance or request, Oppenheimer received him with the pipe in his mouth. He listened carefully to the man, all the while making clear that his pipe also required some attention. In this way, Oppenheimer quietly sent his subordinates an important message: that a successful project understands the personal needs of its members but does not cater to them. He did all this very

easily and naturally, with just his eyes, his two hands, and a half-lighted pipe.[100]

Oppenheimer's capacity to weave together the manifold intellectual threads of the project was inseparable from his ability to bring moral cohesion to the scientific workforce. The English physicist James Tuck, in describing Oppenheimer as a "great gentleman," pointed to his ability to "sit above the warring groups and unify them." According to Peierls, "he guided the discussions . . . in the same spirit of a cooperative search for the answer in which he had guided discussions with his students." Now, however, the "students" included Nobel Prize winners. It was therefore important that this guidance be experienced not as domination, but as facilitation. In Bethe's words, "he never dictated what should be done. He brought out the best in all of us, like a good host with his guests."[101]

Peierls recalled that Oppenheimer "managed to deal with people in a manner which made them feel that they were respected, and gave them the confidence that their views and their needs were taken into account." Above all, he "was able to delegate responsibility and to make people feel that they were being trusted." A. L. Hughes, the assistant director of Los Alamos, said, "We trusted him. He was completely honest." Oppenheimer's "remarkable capacity for seeing the other point of view" allowed him to identify and overcome people's doubts and worries. The head of metallurgical work at Los Alamos, Cyril Smith, recalled that when he was faced with a difficult scientific dispute with a colleague, an informal, five-minute discussion with Oppenheimer was all that was required to give "the necessary perspective" so that he "knew exactly what to do." Teller was impressed by Oppenheimer's chairmanship of the 1942 summer conference in Berkeley on the theoretical physics of the atomic bomb, in which role Oppenheimer "showed a refined, sure, informal touch." Oppenheimer was seen personally, and even physically, to catalyze the emergence of a unity and coherence that already existed in potential. Hawkins recalled that if there was "an incipient disagreement" during a Governing Board meeting, "one would listen patiently to an argument beginning, and finally Oppenheimer would summarize, and he would do it in such a way that there was no disagreement." "It was," Hawkins said, "a kind of magical trick that brought respect from all those people, some of them his superiors in terms of their scientific record, brought them to acknowledge him as the boss . . . So that's why . . . there was never any disagreement that he was the leader of that enterprise."[102]

Oppenheimer was celebrated for knowing all the science of Los Alamos, but also all the scientists—and not just the scientists. "The indefatigable Oppie," journalist Robert Jungk wrote, "knew not only all the scientists, but also most of the laborers by their first names." One young SED wrote home to his parents after the bombing of Hiroshima, describing the "informality" of Los Alamos, which, he said, was "unparalleled in any other organization that I have seen." For example, he told them, "several times Dr. Oppenheimer has called me for something or other . . . and every time, when I would answer the phone with 'Doty,' the voice at the other end would say, 'This is Oppy'."[103]

Oppenheimer set the tone, and the laboratory followed his example. This moral example was communicated, crucially, through his physical presence. "His porkpie hat," wrote historian James Kunetka, "became emblematic" of his presence throughout the laboratory. He showed himself, and by doing so, he showed his concern and his integrative knowledge. His appearances all around Los Alamos were like a squire's passage through his domain: a display of mastery over, as well as of belonging to, the place. "Each Sunday," a group leader wrote, "he would ride his beautiful chestnut horse from the cavalry stable at the east side of the town to the mountain trails on the west side of town greeting each of the people he passed with a wave of his pork-pie hat and a friendly remark. He knew everyone who lived in Los Alamos, from the top scientists to the children of the Spanish-American janitors—they were all Oppenheimer's family." When the Oppenheimers' daughter was born, the "whole town" came to give its blessing: "The sign 'Oppenheimer' was placed over baby Tony's crib and people filed by in the corridor for days to view the boss's baby girl."[104] Oppenheimer was also celebrated for his concern over the lives of scientists' spouses and families. Like a secular saint, he was celebrated for tending the sick and consoling the bereaved; he was said to be "a little aloof, but still a warm and comforting presence." At the marriage of secretary Marge Hall to the young physicist Hugh Bradner, Oppenheimer took the role of the "'father' who gave the bride away."[105] Knowing everything about Los Alamos, then, meant knowing human and moral things as well as natural and technical things. The man who was supposed to be personally responsible for the integrative and morale-enhancing Colloquium was the same man who was supposed to know essentially everything about the lives of Los Alamos people—to know them as emotional and social beings as well as the bearers of scientific thought.

However, underlying this appearance of integration were sharp demarcations of status and privilege within the laboratory, and these distinctions werc also constitutive of Oppenheimer's authority and role. As previously discussed, the SEDs' experience of the laboratory's social and informational order was markedly different from that of the civilian scientific staff members. What Hawkins called the "democratic principle" of the Colloquium was only for the initiated. As McAllister Hull put it, "One of the things Oppie did was to insist on the seminars, to say we've got to do it this way, we've got to share, we can't have this compartmentalization. But that's just [at] the level that Oppie operated on. Now you get below that, where people are manufacturing things, building things, then it was 'need to know.'"[106] The distinction between openness and compartmentalization, then, mapped onto status-laden distinctions between thinkers and manufacturers, mental work and manual labor, knowledge and skill, science and engineering. Hawkins thought that a key value of the Colloquium lay in maintaining academic norms and reminding the staff, "We mustn't forget that we are scientists, not just engineers."[107] These restrictions on who could attend the Colloquium also defined what the Colloquium represented.

By being excluded from the Colloquium, the SEDs were rendered institutionally invisible (as they have been made historically invisible in most published accounts of Los Alamos).[108] To the extent that the laboratory social order came to be defined, for the senior scientists, by the Colloquium, it was defined by a setting from which the SEDs were absent. Their absence and invisibility were essential to the self-reflexive mythos of Los Alamos at the time: that it was a social order continuous with the university. Arguably, the SEDs were the hybrid scientist-engineers that Bacher had predicted would be essential to the success of the project, and they were particularly important in the implosion work. But they also represented the model of the "scientist in uniform" that Rabi, Bacher, and others had insisted would be unworkable. Although the SEDs were essential, their presence threatened to muddy the civilian-academic definition of the situation mapped out by Los Alamos's scientific elite and associated with Oppenheimer's personal charisma. As the Colloquium came to symbolize the moral order of the laboratory, the exclusion of the SEDs allowed the lab to be conceived as scientific rather than engineering-focused, academic rather than industrial, civilian rather than military, and a small, face-to-face community rather than a large, impersonal organization. In other words, this exclusion fed into the symbolic accomplishment of all those features of Los Alamos's moral order with which Oppenheimer's charismatic authority was associated. The invisible labor of the SEDs and other hidden workers (such as the female "computers" who carried out highly routinized implosion calculations)[109] was the social substructure supporting Los Alamos's "democracy of science" and Oppenheimer's charismatic role.

RITES OF PASSAGE

Oppenheimer's Los Alamos persona was constructed in counterpoint to that of Groves. This was particularly clear in the contrast between the meanings attached to the physical presence of each man in the laboratory. Whereas Oppenheimer's tours of the lab were regarded as integrative, Groves's inspections, on his frequent visits to the site, were experienced as disciplinary control.[110] They were also taken as opportunities to playfully test Groves's authority, as scientific personnel attempted to catch the general out and, if possible, subject him to ridicule. Physicist Charles Critchfield recalled an incident when Groves toured the chemistry labs at Los Alamos. The head of chemistry, Joe Kennedy, asked the general whether he would like to see the bomb's initiator. Groves replied, "Of course." Kennedy knew that the little black ball in the cardboard box was loaded with fifty curies of polonium and was hot enough to burn one's fingers if touched. Mischievously, he asked Groves if he would like to pick it up. Groves replied, "'It's probably hot isn't it?' and so he certainly wasn't that stupid, but . . . Joe didn't like him." SEDs and WACs also found Groves's visits an opportunity to comically test and challenge his authority. Bagley recounted one such occasion, when Groves inspected the cluttered SED barracks: "The aisles were narrow because we were double-decked on both sides and there was just enough room to walk through, and we had barracks bags hanging over the ends of the bunks . . . Because General Groves was going to pay a visit, . . . we had to have everything neat. So we loaded down the barracks bags and we hung them up with everything we could think of, and they stuck out in the aisle . . . Groves was corpulent, to say the least. He couldn't get through. He had to turn sideways and wiggle through the barracks bags, down the aisle." The young men delighted in having created this "barricade."[111]

Groves was uncomfortable as a public speaker. Even while expressing his admiration for the general, Agnew admitted, "I've never heard Groves speak [publicly] except at the end of the war, we got an E Award . . . and he gave a little introductory speech and that was all, he probably read it. But he had no pizzazz." WAC Eleanor Roensch said that "sometimes General Groves—I didn't like General Groves—he required that all of us be marched out. We would march out and go to Theater Number 2 this one time, to listen to some important speech that he wanted to give us and when we got there we were all crammed into this Theater. There were so many of us and we had to stand and we waited and waited and waited and finally General Groves appeared and he made some speech, something like be sure to write home to your folks." Bagley remembered the same incident: "One time he gave us a pep talk, and I guess he looked at the military manual. Now he was talking to a bunch of people that [for the] most part were graduates or a couple of years, anyway, into college. He said, 'I want you fellows to write home to your parents. This is the time of the year, now don't forget to write home to your parents.' Now that got to be a big joke around there." On another occasion, Groves told an assembly of SEDs that he had "tried to hire civilians first. He said, we looked around and we found out that there weren't that many civilians, so we went to the Army Specialized Training [Program] and we scraped the bottom of the barrel and we got you guys. I'd say, what a stupid man to say that to a bunch of guys. We always referred to ourselves as the scrapings of the bottom of the barrel." Bagley added, "In my opinion of him, he did not consider the enlisted man a worthy person, just from such foolish things like that."[112]

The image of Groves as clumsy, uncomfortable, and occasionally crass was accentuated at Los Alamos by the contrast with Oppenheimer's displays of sophistication and self-assurance. Agnew said, "Groves didn't have this panache, this pizzazz, the porkpie hat . . . and a lot of money. And all of that went together and Oppie was the darling." Agnew also pointed out that "Groves physically was a little rumpled looking. He spent the whole three or four years I would say riding on trains, so he looked a little disheveled at times." The disparity was particularly apparent when the director and the general toured the laboratory together. Hull recalled a revealing incident:

Just about the time I'd gotten the techniques down pretty well, Oppen-heimer comes up with Groves. Now Oppie always thought he knew everything about anything in the lab that was going on, and mostly he did. But what that meant was that if he came into your lab, as he did in mine with Groves, he didn't ask me to explain what was going on. He proceeded to explain what was going on. So I stood back and folded my arms, and Oppie and Groves stood over one of these big bowls . . . and of course there were three rubber tubes with hot water going into the casing. So, Oppenheimer is talking here and Groves is standing here, and Groves stands on the hottest [tube]. It pops off from the wall, and a stream of water just below boiling point shoots across the room. And if you've ever seen a picture of Groves, you know what it hit. And I contained myself from laughing, because I valued my stripes.

And Oppenheimer looked over and said, "Well, just goes to show the incompressibility of water." And at that point I broke up. Groves was so embarrassed he didn't look around, so I retained my stripes.[113]

This anecdote suggests not only the laboratory workforce's perception of Groves, but also the ways in which Oppenheimer, as he escorted Groves around the lab, performed his intellectual and social mastery of the site. While Groves may have been the object of humor here, the solidarity between Oppenheimer and his workforce—which the anecdote also suggests—was ultimately to Groves's advantage.

Groves later said, "One of the big complaints made about me after the War was that scientists didn't like me. I think the answer to that is: who cares whether they liked you or not? That wasn't the objective; it was to have things running well." Groves's view was expressed in the November 1946 report "Complications of the Los Alamos Project," which stated that "frictions which developed between the scientific and military personnel" were in fact useful to the development of the project, since they generated cohesion among the scientific workers at Los Alamos:

> The success of the project as a whole was in a measure because of these difficulties rather than despite them. The scientists and technicians who came to the project, inexperienced as they were in large and complex undertakings, inexperienced in cooperating, under pressure, with large and varied groups, had of necessity to collaborate not only with the military but also with their own numerous and diversified members: physicists, chemists, metallurgists, engineers, etc. Their task, and the time-schedule which was imposed on it, made this collaboration imperative. There seems to be no doubt, in retrospect, that the common attitude of opposition and objection to the military, on the part of scientists and technicians, drew all groups of the latter closer together and tended to fuse them into a far more cooperative whole than might otherwise have been possible.[114]

Rather than being in opposition to each other, Oppenheimer's and Groves's leadership roles were functionally complementary. In contrast with Oppenheimer's harmonizing style, Groves's leadership consisted of more coercive managerial power. Los Alamos scientists observed that Groves was used to getting his way by intimidating his subordinates into a state of blind obedience.

The apparent absence of coercion in Oppenheimer's leadership was made possible by Groves's assumption of this more severe role. As physicist Raemer Schreiber put it, Groves was the "fall-guy" for anything the Los Alamos residents and scientists were not allowed to do. After all, he said, "you've got to have somebody to be mad at."[115]

The notion that Oppenheimer's authority was opposed to the general's was, therefore, largely an illusion. Groves had just as much of an interest in supporting Oppenheimer's personal role as did the scientists, because their allegiance to Oppenheimer functioned to accommodate them to bomb work and ultimately to the general's own authority. Oppenheimer was, after all, Groves's handpicked lieutenant, and it was Groves who was always firmly in charge. As they supported and celebrated Oppenheimer's charismatic role, the scientists were both making and seducing themselves into the new technoscientific culture of nuclear weapons. This seduction depended on the accommodation between values of scientific freedom and collegiality and the more coercive structures of large-scale military-industrial organization.

AUTHORITY AND COLLEGIALITY

A fundamental tension at Los Alamos was the one between the collegial equality of its senior scientists and the mission-directed and hierarchical structure of laboratory organization. For example, the decision regarding who was to lead the Theoretical Division could not be made solely on the basis of scientific reputation. With a workforce in this division consisting of men such as Hans Bethe, Edward Teller, Victor Weisskopf, and Rudolf Peierls—among many others at the top of the field—the imposition of vertical lines of authority, with section leaders reporting to group leaders who reported to a division leader, was likely to strain relations among scientific equals.

The most important example of strain arising from the hierarchical and mission-directed organization of the laboratory concerned the place of Edward Teller in the Theoretical Division, which was headed by Bethe. Teller was a personality who did not fit easily into the structure of the laboratory. Marjorie Ulam recalled that she learned early on that "Teller seemed to be the enfant terrible and had violent disagreements with Oppenheimer." In Serber's opinion, Teller was "a disaster to any organization." Throughout his career, Teller reveled in that sort of judgment and portrayed himself as deeply uncomfortable in large-scale technocratic organizations (though not uncomfortable about nuclear weapons work in itself). "Before I participated in the Manhattan Project," he said, "I was anything but an organization man." Rather, "I worked on subjects that I liked because I liked them. I did not work on anything for the purpose, let us say, of my career . . . I was not and did not desire to become part of an organization." However, "one cannot work on something like atomic energy without becoming part of an organization."[116]

Until he left for Chicago on February 1, 1946, Teller would sometimes vent his frustration at Los Alamos life to his friend and colleague Maria Mayer, a physicist who was working on the Manhattan Project at Columbia University. To Teller, Los Alamos was a "physicist-reservation," and he joked, "Don't shoot them except in season."[117] He chafed at the enforced conviviality of life at Los Alamos and the requirements of collaborative work in the formal organizational structure of the laboratory. In one letter, he quipped to Mayer that he was being "virtuous": "With getting up early, being not later for appointments than expected and being, I hope, on the whole a good boy."[118]

Teller regarded himself as temperamentally unsuited to the emotional requirements of Los Alamos's style of large-scale collaborative work. In his letters to Mayer, he often made reference to his efforts to maintain the emotional self-control necessary to fit into the social structure of the laboratory: "So far I kept alive successfully. Yesterday I passed up one of the best opportunities to get mad. Now I am duly proud of it."[119] Teller apparently found the close-knit, small-town atmosphere of Los Alamos to be oppressive: "In this place it is a real blessing to be, for a short time, devoid of company. You know how dense the population is in this part of the country."[120] For many of Los Alamos's scientists, the communal life and the availability of outdoor activities were a compensation for the restrictions of life there. Skiing and hiking, as well as an interest in the "Land of Enchantment" mystique of New Mexico, were important aspects of a developing community identity at Los Alamos. Teller, however, was scornful of such interests and activities. He was thankful for the coming of spring, when "the snow is relegated to the place to which it belongs, namely to the scenery." He was also less than impressed with the Native American culture, which many of the other scientists romanticized. When his wife, Mici, dragged him along to watch an "indian dance," he found it "picturesque and just as dull as I expected it to be."[121] After a particularly heavy snow in 1945, Teller wrote that Mici had joined a group on a skiing expedition: "Only those are left behind who don't ski and therefore don't count." But "it is almost allways [*sic*] true that I do not mind to be left alone."[122] Teller perceived himself to be a marginal individual at Los Alamos, and he cultivated the self-image and role of the individualist opposed to the pressures of organization and to the compromise, self-control, and conviviality that it required.

Teller's discontent was focused on his place within the organization. He resented working under Bethe in the Theoretical Division, and he also resented the burden of what he felt were mundane and routine calculations, which he was required to carry out for the theory of the implosion. He objected in general to being on an intermediate rung in the hierarchy of a mission-directed laboratory. In his opinion, Bethe "overorganized" the division. "It was much too much of a military organization," he said, referring to it also as "a line organization." According to Bethe, as soon as the laboratory structure was set up, "Edward essentially went on strike . . . He continued to work, but from then on he seemed rather disinterested in working on the direct business of the laboratory." He was dissatisfied with his place in the hierarchy; Bethe said, "I believe maybe he resented my being placed on top of him." Teller also took umbrage when Bethe appointed Weisskopf to be deputy division leader. According to Weisskopf, "Edward maintained that he was the better physicist and should have been given the job. I didn't try to deny the allegation but pointed out to him that Hans probably had chosen me because I was better in dealing with people."[123] In general, Teller's situation in the Theoretical Division instantiated the problem of vertical organization of scientists whose professional reputations were essentially equal. According to his biographers, "at Los Alamos Teller was officially only a scientist with a third-rank authority, but the chances are that in his own mind—even if only subconsciously—he saw himself as the director's equal."[124]

In particular, Teller was upset that his brainchild—the hydrogen bomb, or superbomb—was not included in the laboratory's mission. In his work in the Theoretical Division, he was told to focus on implosion and that the "Super" would have to wait. The Super had been the focus of discussion during the 1942 summer theory conference at Berkeley. The idea of using a fission bomb to set off a fusion reaction in deuterium had captivated the participants at the conference, including Bethe and Oppenheimer. But when it came to organizing the program at Los Alamos, the Super was relegated to the back burner. Teller felt strongly, and in a personal way, the contrast between the character of discussion at Berkeley and the more controlled atmosphere of Los Alamos. According to Bethe, "in Berkeley it was very much the style that Teller and Oppenheimer liked, namely, that we talked constantly to each other. And this had to come to an end. We had to sit down in our offices and actually work something out, and this was against his style." The pressure of time, which lay powerfully behind the mission-directed and hierarchical nature of the laboratory, was particularly symbolic to Teller of the general constraint that he felt in his work. Bethe said, "He resented particularly that I was no longer available very much for discussions. I remember one occasion when I was terribly busy, and he came in to discuss some problem which sounded to me rather far away from our main problems, and so after an hour or so I looked rather conspicuously at my watch, which was one of those dollar watches you haul out of your pocket, and he didn't like it at all."[125]

To some extent, Teller projected these generalized feelings of discontent at Los Alamos onto the figure of its scientific director. But this discontent coexisted, it seems, with a continued and genuine appreciation for Oppenheimer. Teller focused his sense of anger and frustration much more strongly on men such as Bethe, who were Oppenheimer's deputies but above Teller in the laboratory hierarchy. Teller had long "liked and respected Oppenheimer enormously. He kept wanting to bring up his name in conversation." Bethe thought the two men were in many respects "fundamentally . . . very similar." Teller's frustration at Los Alamos did, however, strongly color his attitude toward its director. What others regarded as Oppenheimer's sprezzatura—his effortless superiority, his ability to "lead without seeming to"—to Teller appeared to be just a more insidious form of control. According to Teller, none of the other sites of the Manhattan Project that he visited, including Columbia, Chicago, and Oak Ridge, "were run as systematically, with as much direction or with as much psychological finesse as Los Alamos, and this, to me, was deeply repulsive." Oppenheimer, in Teller's view, was a "politician" who was charming to anyone, so long as they could be of use to him.[126]

When physicist Felix Bloch, who also suffered under the rigidity of the organization, resigned from the project, Teller wanted to drive him to the train station at Lamy. That same night, Oppenheimer invited Teller to one of his parties, something he had never done before. In Teller's view, the reason behind the invitation was that "Oppenheimer did not want me exposed to the evil influence of a deserter . . . I may be unjust, but the whole thing just looked like too much of a coincidence. He used friendships, he exploited friendships. Granted, he did not want me to leave Los Alamos, but obviously he manipulated people."[127] Rather than miss the party entirely, Teller dropped off Bloch and went to the party late.

Personal access to Oppenheimer and to the close circle around him was necessary for power and status in the project, and Teller did not wish to forgo that. His sense of being marginal and disempowered in his position in the organization was closely bound up with the feeling of being distanced from

Oppenheimer. Teller, according to Bethe, "resented even more that he was removed from Oppenheimer. He had come to like discussions with Oppie very much, but Oppie was terribly busy. In the end an arrangement was made that Teller would see Oppenheimer once a week, for one predetermined hour—I think from ten to eleven on Monday morning or some such thing, and so that was that."[128]

Perhaps most revealing was Teller's reaction when, in June 1945, his old friend Leo Szilard sent him a copy of the petition drawn up by Chicago scientists against use of the bomb. Against Szilard's wishes, Teller went to Oppenheimer to ask his permission to circulate the petition. Oppenheimer dissuaded him from doing so, and Teller, without mentioning his conversation with Oppenheimer, wrote to Szilard telling his friend that he disagreed with him on this issue.[129] Teller later regretted allowing himself to be so influenced by Oppenheimer: "Oppenheimer persuaded me at least not to take action at that time and I'm sorry, I was wrong. He should not have. I should not have. But I did."[130] Nevertheless, if Teller was indeed so influenced, this does suggest a degree of admiration and respect for Oppenheimer that at least coexisted in an ambivalent way with any more negative feelings.

In general, the Los Alamos scientists did not directly associate Oppenheimer's leadership with the more coercive aspects of the laboratory's organization. His perceived ability to direct on the basis of consensus rather than coercion was supported by his institutional separation from instrumental functions. The enforcement of schedules and policy decisions was delegated to division leaders such as Bethe, Kistiakowsky, Parsons, and Bacher, who acted as his deputies. When the laboratory was put on an extremely tight schedule in the final months of the project, the supervision and enforcement of the schedule was carried out by a "Cowpuncher Committee," chaired by physicist Sam Allison. Bacher also served on this committee; he took charge of a range of instrumental functions for Oppenheimer, and sometimes it was Bacher, rather than Oppenheimer, who conveniently took the blame for unpopular decisions. Teller, a year and a half after the end of the war, wrote to Mayer of his thoughts regarding Bacher, who had recently been appointed to the Atomic Energy Commission. Teller admitted that he "never particularly liked" the man, who was "at best a third-rate physicist" and essentially a careerist and opportunist, "a great master of how-not-to-stick-your-neck-out." Teller described Bacher as "a normal human being of the genus 'manager'" and as "a great administrator. He loves organization charts and he loves reports in proper shape and he is completely devoted to priorities." Teller added, "To Oppy he was the ideal yes-man."[131]

Ultimately, the differentiated structure of the organization allowed a resolution of the superbomb issue—if only for the duration of the war—and a way out of the stalemate in relations between Teller and Bethe. In June 1944, Teller and his group were removed from the Theoretical Division and given independent status, reporting directly to Oppenheimer. The restructuring of the laboratory later that summer, in response to the new emphasis on implosion, provided an opportunity for a more permanent resolution of Teller's position. He and his group were allowed to work on the superbomb in the new F-Division, led by Enrico Fermi, with whom Teller had a good rapport. F-Division was a collection of groups that were carrying out research deemed unlikely to have a payoff before the end of the war, and they were therefore left outside the main thrust of the laboratory. This allowed Teller and his small band of superbomb enthusiasts to avoid the strict mission control and discipline to which most of the laboratory was subjected in the final "weaponeering" stages of the project. This separation freed Teller to work on the Super; however, it also decisively defined this work as outside the central mission of the lab and as something for postwar rather than wartime development.[132]

FIRST AMONG EQUALS

Oppenheimer's leadership at Los Alamos was defined by the social structures in which it was embedded. The association of his leadership with consensus rather than coercion was substantially a product of the assumption of coercive roles by Groves and by Oppenheimer's scientific deputies, such as Bacher. It relied also on the combination of the laboratory's differentiated structure with practically unlimited material resources. This combination enabled avoidance of conflicts among senior scientists, as in the case of Teller's removal to F-Division. It allowed the reconciliation, to a degree, of the laboratory's vertical hierarchy with collegial equality among the lab's scientific elite. When Oppenheimer took up his position at Los Alamos, his reputational authority within the scientific community was not sufficiently great to give his decisions and opinions automatic authority above those of his scientific colleagues. This was precisely the initial worry of Bush and Conant: that because Oppenheimer did not have a Nobel Prize to provide authority above the men he was to direct, his leadership would not be adequate to move the project along. At the outset of the project, Oppenheimer was very far from being a natural leader of the scientists. Yet it was precisely in terms of natural qualities of leadership that we find his role at Los Alamos described. One can make sense of this paradox,

I suggest, by understanding how Oppenheimer's leadership was shaped, enabled, and constrained.

Oppenheimer was celebrated at Los Alamos for his ability to see the big picture: to synthesize the entire body of science involved in the project and, from this overall perspective, to bring order and cohesion to decision making and discourse. Famously, he could sum up opposing views in such a way that the argument would appear resolved—his "magical trick that brought respect" even from those who were "his superiors in terms of their scientific record."[133] Although not set apart by a Nobel Prize, he was seen to be able to "rise above" the scientific flock, due to this combination of moral and intellectual qualities. His authority derived from an ability to speak for and bring to bear a consensus that was seen to already exist in potential. His synthetic knowledge, together with his perceived moral qualities, allowed him to reconcile conflicting parties and made him the "natural" spokesman for an underlying, though not yet realized, consensus. It was this underlying collegial consensus, for which he was believed to speak, that was the root and source of his authority; hence the close association between Oppenheimer's leadership and organizational forms (such as the Colloquium) that expressed that collegial order.

Oppenheimer was himself aware of this collective source of his authority and the limits that it entailed. In early September 1944, Parsons, the head of the Ordnance Division at Los Alamos, sent Oppenheimer a memorandum requesting far-reaching "executive" powers within the laboratory to direct work on mechanical, electrical, and explosives aspects of the design of the implosion bomb.[134] This would have given Parsons an extraordinary position of power in relation to the leaders of the divisions working on implosion. But Oppenheimer denied that such power was his to bestow.

In his reply to Parsons, Oppenheimer outlined his understanding of the structure of power and authority in the laboratory and the real limitations placed on his own power as director. He told Parsons, "The kind of authority which you appear to request from me is something that I cannot delegate to you because I do not possess it. I do not in fact, whatever protocol may suggest, have the authority to make decisions which are not understood and approved by the qualified scientists of the laboratory who must execute them." If Oppenheimer was a master at reconciling disputes, it was because these were precisely the situations in which his authority to intervene was most legitimate and therefore effective. He qualified his comments on the limitations of his authority, saying, "I do not mean by this that in the case of a divided opinion it is not appropriate for you or for me to let our own views render the decision." But this rendering of the decision, he suggested, was always with reference to some potential consensus: "I should not consider making a decision which was not supported by responsible and competent men in the laboratory." Ultimately, he told Parsons, decisions had to be arrived at and justified through collegial discussion:

> You have pointed out that you are afraid that your position in the laboratory might make it necessary for you to engage in prolonged argument and discussion in order to obtain the agreement upon which the progress of work would depend. Nothing that I can put in writing can eliminate this necessity. All that I can say is that I will support decisions reached by you in consultation with the other members of the laboratory and that such support is in fact axiomatic and trivial as long as these decisions are reached after competent technical discussion and after the opinions of all vitally concerned have been given appropriate weight. I am not arguing that the laboratory should be so constituted. It is in fact so constituted.[135]

Oppenheimer's individual leadership was itself a collective accomplishment, a matter of interactional work by the members of the laboratory. Oppenheimer's ability to direct without seeming to do so, to rely on consensus rather than coercion, was a product of the interactional support of his role and his authority by other powerful members of the lab. John Manley described the give-and-take on which Oppenheimer's leadership relied: Oppenheimer "had no great reluctance about using people," but "it was an enjoyable experience because of the character of Robert to do it so adroitly. And I think that he . . . realized that the other person knew that this was going on . . . It was like a ballet . . . each one knowing the part and the role he's playing, and there wasn't any subterfuge in it."[136]

The collaborative nature of Oppenheimer's leadership, however, implied constraints on his power. His charismatic authority relied upon a network of support from his colleagues. His authority was collectively produced and collectively constrained. As Oppenheimer mediated between the various interests engaged in the project—in particular between the military and the civilian scientists—his leadership was shaped by these interests. Personal access to Oppenheimer, and influence or control over him, became a key resource by which other actors sought to influence the project. Oppenheimer was thus a nodal point at the intersection of the networks of power in the Manhattan Project.

The first atomic explosion took place in the desert at Alamogordo, New Mexico, on July 16, 1945. Months of preparation had gone into readying the equipment and the test site. General Groves urged that the test be completed in time for the news to be delivered to Truman at the Potsdam conference.[137] With such vast resources already committed to the project, and so much riding on the outcome of the test, Groves could not afford to allow for contingencies. It is an indicator of Oppenheimer's significance in the project that in the hours leading up to the test, from one o'clock in the morning until about five o'clock, Groves was with Oppenheimer "constantly." "Naturally he was nervous," Groves wrote to the secretary of war two days later. "I devoted my entire attention to shielding him from the excited and generally faulty advice of his assistants who were more than disturbed by the unusual weather conditions." Groves sequestered Oppenheimer from his advisers and, by restricting and controlling access to the director, cemented his own power in the local situation. Groves was in control of the test by being in control of Oppenheimer. "Every time the Director would be about to explode because of some untoward happening, General Groves would take him off and walk with him in the rain, counselling with him and reassuring him that everything would be all right."[138]

Here we see the volatility of charismatic authority, but we also see how Oppenheimer's person and his personal authority were shaped and stabilized by the other actors engaged in the project. If Oppenheimer was the right man for the job, it was because people worked to make him that. Shaping him into the natural leader was a collective task and a collective accomplishment, one in which competing actors and groups in the project each had their own stake. Oppenheimer himself was the focus of struggle between these competing interests. He embodied not only the success of the project, but also its tensions and contradictions.

CHAPTER FIVE

Against Time

MOMENTUM

Oppenheimer, we have seen, was credited with defending academic norms and values against military hierarchy and compartmentalization. The Colloquium symbolized the triumph of scientific collegiality and free exchange of information over military restrictions, and the threat that the laboratory would be militarized once the project became large-scale never materialized. However, there were tensions at Los Alamos other than this overt conflict between the military and the academic ways of life. The laboratory's civilian scientific leadership was charged with producing a weapon, in time for use in the war. The character of life and work at Los Alamos changed as the project grew and as the laboratory's program became increasingly synchronized and integrated with the industrial production of fissionable material and with military planning for combat delivery of the bomb. Habits that were appropriate for open-ended scientific inquiry gave way to a more disciplined, instrumental, and hierarchically directed focus on engineering problems.

David Hawkins noted that the bomb project built up "an institutional momentum which . . . grew great enough to almost limit or even predetermine many later decisions."[1] This momentum was expressed in a rigorous system of scheduling. Schedules formed a powerful underlying framework, structuring and disciplining the social life of the project. They allowed the coordination and control of an increasingly vast and complex enterprise, spanning geographically dispersed production and research sites.[2] In the organizationally and socially hybrid structure of the Manhattan Project, encompassing military, industrial, and academic forms of life, scheduling was a means of creating symmetry among different types of settings. Within sites, schedules bound the activities and regimes of individuals into a coordinated, collective task. They served as a powerful mechanism of social control that tied both the daily lives and the consciousness of the scientists to the overall goals of the institution. The regulation of time was a means by which the role of the scientist, within the new institutional framework of the bomb project, was constituted and defined. The domination of everyday life by the schedule served to narrow the scientists' perspective to the merely instrumental attitude of accomplishing the "job at hand." The pressure of time, expressed in the schedule, became a force against any moral examination or questioning of the project's goal.

Oppenheimer played a crucial role in easing the introduction of industrial modes of temporal organization into the laboratory. He helped to establish the legitimacy of the schedule and to habituate the Los Alamos scientists to this new form of discipline. It was an aspect of his leadership that he was able to seamlessly weave together academic norms of free inquiry with the military's imperative to produce a bomb on time. He attached his personal authority to the schedule, thereby lending *moral* force to the discipline of the laboratory's temporal regime. He also worked to suppress what few attempts there were to raise moral concerns about use of the bomb, and in so doing he kept the scientists focused on the instrumental problems of the weapon's design and manufacture. Oppenheimer's directorship of the laboratory was therefore bound up with key tensions between collegiality and bureaucracy, the university and the factory, science as a calling and science as an instrumental occupation, and soldierly duty and broader moral responsibility.

THE ATOMIC FACTORY

The Manhattan Project was an organizational network incorporating, in a web of contracts, numerous geographically dispersed sites: the plutonium-production factories at Hanford, Washington; uranium-isotope separation plants at Oak Ridge, Tennessee; and the atomic weapons laboratory at Los Alamos, as well as university laboratories at the University of Chicago, Columbia University, the University of California at Berkeley, and elsewhere. It was a hybrid network requiring coordination of the Army Corps of Engineers, naval ordnance facilities, the Army Air Forces, commercial-industrial conglomerates, and also academic sites—laboratories and universities.[3] Engineering the bomb meant engineering a complex social structure, combining these diverse organizations and directing them toward a well-defined instrumental goal. It meant bringing together laboratory science with the organizational techniques of American Taylorist and Fordist manufacture. The reach of the resulting network and infrastructure was suggested by Niels Bohr's comment on the successful separation of uranium isotope: "You see, I told you it couldn't be done without turning the entire country into a factory. They have done just that."[4]

Managing such diverse organizational and human elements required the alignment of divergent practices and settings, in particular the reconciliation of large-scale engineering and industrial production with a scientific state of the art characterized by a very high degree of uncertainty in basic processes and techniques. This required deviation from the usually conservative and risk-averse engineering practices of firms such as DuPont. But it also meant the transformation of scientific practice. The previous chapter discussed how the compartmentalization of information was a means by which General Groves could discipline the scientists' practices, rendering them compatible with military-industrial organization. The management of time was another means for accomplishing such a transformation and for imposing control over daily life in the laboratory. Deciding how working time was to be organized was a key element in establishing what kind of social order Los Alamos was to be.

The organizational identity of Los Alamos, as we have seen, was confused and contested. Since this was a laboratory within a military post, life at Los Alamos was obviously quite different from university patterns. But Los Alamos scientists saw, or constructed, a degree of continuity with academic norms and practices. Stanislaw Ulam wrote of his early days at Los Alamos: "The atmosphere of work was extremely intense at that time and more characteristic of university seminars than technological or engineering laboratories by its informality and the exploratory and, one might say, abstract character of scientific discussions." This informal, academic character was at odds with the sort of compartmentalized focus that Groves hoped to instill in his workers. Ulam found "a milieu reminiscent of a group of mathematicians discussing their abstract speculations rather than of engineers working on a well-defined practical project." And this character of the work was expressed crucially through the organization of time. Ulam was impressed that "discussions were going on informally often until late at night."[5] For Ulam, informality was tied to a voluntaristic form of organization. Los Alamos, in his view, was unlike a technological and engineering laboratory because one did not clock in and out but stayed on to discuss a problem into the night, perhaps sleeping later the next day. An academic setting was characterized by the lack of a rigid distinction between working time and free time. The autonomy of the professional scientist was powerfully expressed by the flexibility of his working day—one regulated internally, by his dedication to the task, rather than externally by a clock and a schedule.

Flexibility in the organization of daily time was particularly significant for the GIs who, as members of the Special Engineer Detachment (SED), worked as scientists and technicians in the Los Alamos Technical Area. The workday of the Technical Area was not easily coordinated with that of an Army post. As noted in the previous chapter, a certain esprit de corps also developed among the SEDs, due to the unusual license that they enjoyed within the Army. In particular, the voluntaristic spirit represented by this relatively unstructured organization of daily time was important in their view of themselves as colleagues of the civilian scientists, albeit colleagues in uniform.

A culture emphasizing self-motivation and seeing a problem through rather than working fixed hours permeated the laboratory. SED glassblower Arno Roensch recalled that if there was an urgent job, they would work in the shop into the night, and the SED commanders made allowances for this: "The lab was open day and night in those days. It was something, everyone felt so dedicated. There were no time limits . . . If you wanted to go back to work and work till one o'clock or two o'clock in the morning there was no problem. You just signed the out sheet in the orderly room and you could sleep in the morning."[6]

In the early stages of the project, considerable uncertainty surrounded scheduling. The Los Alamos scientists were frustrated at the difficulty of getting reliable scheduling information from the production sites. They blamed this on compartmentalization; and, as I suggested in the previous chapter, General Groves's insistence that he be the main conveyor of scheduling information to Los Alamos was motivated by his desire to maintain personal control of the organization. But there was, in addition to this, a more straightforward reason for Los Alamos's lack of clear scheduling information in the first year of the project. Groves explained that "this lack did not result so much from poor liaison as from the fact that during this period all schedules were vague, incomplete and contradictory. It was not only difficult but impossible to arrive at sensible schedules for bomb research and development, when we simply could not predict when the necessary U-235 or plutonium would be ready."[7]

This situation of uncertainty changed roughly midway through the Los Alamos program. As soon as it became possible to make reasonably reliable predictions about schedules for the production of fissionable materials coming from Oak Ridge and Hanford, the practices and routines of Los Alamos began to be coupled with those of the distant uranium and plutonium factories. This involved the "domination of research schedules by production schedules . . . Every month's delay would have to be counted as a loss to the war."[8]

The linking of these schedules provided the technological momentum of the project. The first large shipment of uranium-235 from Oak Ridge arrived at Los Alamos in August 1944.[9] The Hanford plutonium pile became ready for the first time on September 26, 1944.[10] This period was marked by a massive reorganization of the laboratory around the implosion method of bomb assembly and also by a large influx of SED manpower. The new sense of urgency led to increased regulation and greater routinization of life and work at Los Alamos. The laboratory's Administrative Board (which succeeded the Governing Board in mid-1944) expressed concern at numerous times in this period that workers were *not* sufficiently dedicated to the task of building the bomb.[11] But there was considerable equivocation by the scientific leadership of Los Alamos between coercing and more subtly co-opting the appropriate orientation from the lab's workforce.

On August 17, 1944, Oppenheimer commented to the Administrative Board that "too much time was being spent in the PX [general store/café] by employees." Oppenheimer "felt that exhortation should be used before more forceful methods, in order to instill a feeling of responsibility in employees." For example, he "wished to have time clocks installed only as a last resort." In the meantime, he "requested that a letter be distributed to all employees drawing attention to the present laxness."[12] Oppenheimer regarded appeals to moral responsibility as preferable to the imposition of external regulation. But he was also keenly aware of the laboratory's position within the broader nexus of the Manhattan Project, and of his own accountability to General Groves.

The allegation of time-wasting was revived a month later when Colonel Elmer E. Kirkpatrick Jr., of the Corps of Engineers, reported to the Administrative Board on his visit to Los Alamos. Kirkpatrick had had a long association with Groves on Army construction projects, and Groves had brought him to the project that September as the Manhattan District's deputy engineer. One of Kirkpatrick's key duties, as Groves's special representative, was making preparations to ensure the timely delivery of bomb components to the planned overseas operational base. Visiting Los Alamos soon after his appointment, Kirkpatrick commented prominently on his "strong feeling that a large part of the personnel were lacking in a sense of urgency, as evidenced by a laxness in working hours." He emphasized that this applied both to the Technical Area and the post. Kirkpatrick's comments were, it seems, taken extremely seriously, particularly because the "situation was recognized by the board as it had been discussed several times in the past." Promptly after Kirk-patrick's report, a siren was introduced at the lab. Starting Monday morning, October 2, 1944, the siren sounded at 7.25, 7:30, 8:25, and 8:30 a.m.; 12:00 noon; and 12:55, 1:00, and 5:30 p.m.[13]

The siren was understood as belonging to an industrial mode of organization. Bernice Brode, a physicist's wife, wrote of the scientists' being "cooped up in a factory atmosphere where the whistle blew at 7 and 7:30 summoning them to the grind."[14] According to Laura Fermi, despite the "impression of confusion," in fact "our life there was more than orderly, it was overregulated." She wrote, "Our daily schedule [was] adjusted to the sirens that announced beginning and termination of work." But this compulsion by the siren was re-inforced by Oppenheimer's embodied moral authority. His personal example played the key role of vouching for, and lending legitimacy to, the normative orientation toward time that was signaled in a mechanical way by the siren. Brode wrote, for example, "Even Oppie abandoned his former pattern of living. I remember the days when he would not accept a class before 11 in the morning so he could feel free to stay up late for parties, music, or ideas. But at Los Alamos, when the whistle blew at 7:30, Oppie would be on his way to T [Technical Area], and hardly any would beat him to it. When Sam Allison came to the site from Chicago, he shared Oppie's office for some time. Sam said his one ambition was to be sitting at his desk when Oppie opened the door." Groves was worried about Oppenheimer's health collapsing from overwork. "But," he said, "I never could slow him down in any way."[15]

Although the siren signaled the end of the workday at 5:30 p.m., working hours varied across disciplines. An experimentalist might work on an experi-ment all night, but, as Robert Serber recalled, "a theorist had no excuses and it was eight to five, five days a week." Many workers, however, both scientists and technicians, would stay late at the Technical Area. In general, the siren did not rigidly demarcate work time from leisure time or private time, a fact that was upsetting to many of the scientists' wives. Ruth Marshak saw the Technical Area as "a great pit which swallowed our husbands out of sight, almost out of our lives . . . They worked as they had never worked before. They worked at night and often came home at three or four in the morning. Sometimes, they set up Army cots in the laboratories and did not come home at all. Other times, they did not sleep at all . . . The loneliness and heartache of some scientists' wives during the years before the atomic bomb was born were very real."[16]

Marshak ascribed these long hours to "curiosity and zeal . . . [and] an inspir-ing patriotism." Such motivations were cultivated by the wartime atmosphere and were reinforced by Oppenheimer's charismatic example. Oppenheimer himself, said Los Alamos administrator Dana Mitchell, "gave you a sense of urgency and made you feel that what you did was important," helping to foster the scientists' very high degree of commitment to the institutional goals of the project. Historian Lillian Hoddeson noted that the scientists, though working on "this strongly mission-directed problem . . . experienced the joy of research and the sense that they were working on their own problem." She posed, in passing, the question of "how it is possible for a large laboratory to create an environment in which many or most of its scientists can experience such a sense of free inquiry while in fact they are working directly in line with the mission."[17] One example of how this subjective sense of freedom could coexist institutionally with strict overall direction was the fact that the working day itself was only weakly regulated. Despite the overt regulation of the siren (which in any case, it appears, was generally taken to define only the minimum required working day), scientists would organize their workday as required to accomplish the task at hand.

However, alongside this weak regulation of the day was a much stronger regulation of the overall temporal framework for the completion of tasks. Increasingly, work at Los Alamos came to be governed by a system of schedules. These overarching schedules set the rapid tempo into which the Los Alamos scientists choreographed their individual and group work routines.

IMPLOSION AND THE IMPERATIVES OF TIME AND SCALE

The increasing routinization of scientific work at Los Alamos can be seen most clearly in the program to develop the implosion method of assembly. The implosion program began in spring 1943 as an organizationally isolated, low-key effort involving just a handful of scientists. By late 1944, it was the centerpiece of the laboratory's mission. The laboratory's leadership saw inculcating a new orientation to time as essential in habituating the scientific workforce to such a large-scale endeavor. This new orientation was understood to require a break with academic norms and practices.

Until the summer of 1944, it was thought that the gun method of assembly would be used for both the uranium and the plutonium bombs. This method involved shooting together two subcritical pieces of fissionable material, inside a specially designed gun barrel, to form a critical mass. Work on this technique was organized within the laboratory's Ordnance Division, under the naval ordnance expert Captain William S. "Deak" Parsons.

However, from the beginning, alternative methods were considered. Among these was implosion, referred to in Serber's introductory lectures given at the laboratory in April 1943 (known as the "indoctrination course"). The fundamental idea was to squeeze a subcritical mass of fissionable material in upon itself with the simultaneous application of great pressure from all sides equally. Serber's presentation sparked the interest of the experimental physicist Seth Neddermeyer, formerly of Caltech.[18]

Neddermeyer enthusiastically developed the concept and presented his ideas to a meeting on ordnance problems on April 28, 1943. His presentation, however, met with little encouragement. Parsons called it "a touch of relief" from the usually "dead earnest" atmosphere of the laboratory.[19] But it was policy across the Manhattan Project to "buy time with money" by concurrently pursuing a variety of different potential technical solutions to any problem, whatever the extra cost. That way, the failure of any one research program would not slow down the project overall. In line with this policy, Oppenheimer was willing to allow Neddermeyer some free rein to follow his intuition.[20] Neddermeyer put together a small group of volunteers (Hugh Bradner, John Streib, and Charles Critchfield). This team worked in an arroyo away from the main part of the Technical Area, carrying out exploratory research in which they wrapped steel pipes with TNT and blew them inward, the goal being to compact the hollow pipes into a solid bar. The hope was that it would eventually be possible to compress a spherical shell of fissionable material so as to form a critical mass. The key problem faced by the implosion program was how to generate a symmetrical imploding shock wave. As ordnance expert L. T. E. Thompson put it in a memo, "a spherical shell under high external pressure, with impact load over one section, should begin to collapse, I think, in about the same manner of a dead tennis ball hit with a hammer."[21]

In June 1943, the team was incorporated as a group (E-5, Implosion Experimentation) into the Ordnance Division, with Neddermeyer as the group leader, but the implosion program remained institutionally isolated within the division. Working "in a little corner," Neddermeyer did not succeed in constructing broad networks of support for his idea. A poor politician, he was not personally suited to the entrepreneurial role of generating such networks. Los Alamos colleagues described him as "shy," "a very mild sort of guy and a very poor salesman." As Neddermeyer himself told Oppenheimer, "I'm no operator."[22]

The fortunes of the implosion program changed when the mathematician John von Neumann visited Los Alamos as a consultant in September 1943.[23]

Von Neumann had been working for Conant's National Defense Research Committee (NDRC), on the hydrodynamics of shock waves produced by shaped charges. He had been introduced to this problem by a Russian-born chemist from Harvard, George Kistiakowsky, head of the Explosives Research Laboratory in Bruceton, Pennsylvania. Kistiakowsky's approach was novel and unorthodox, treating high explosives as precision instruments. Von Neumann immediately saw a connection between implosion and his work for the NDRC.[24] He gave his personal endorsement to the implosion idea and helped to catalyze support among the laboratory's leadership. Von Neumann was important in his ability to move between cognitive and technical domains. He translated knowledge of high explosives into the language of mathematics and physical theory and hence into the language and sphere of interest of powerful actors such as Teller and Oppenheimer.

According to Neddermeyer's collaborator Charles Critchfield, von Neumann "woke everybody up." Among those stirred was General Groves, who reprimanded Captain Parsons for not keeping him informed regarding implosion. Bethe, Oppenheimer, and Teller also became excited, as they began to see the potential in implosion for a more efficient weapon. Teller called Critchfield and asked, "Why didn't you tell me about this stuff?" to which the latter responded that "Seth and Hugh [Bradner] and Streibo [John Streib] and I have been working on this and nobody paid any attention to it." As Neddermeyer put it, "I made a simple theory that worked up to a certain level of violence in the shockwave ... Von Neumann is generally credited with originating the science of large compressions. This is true with respect to the organized research of the project itself. But I knew it before and had done it in a naïve way. Von Neumann's was more sophisticated."[25]

Groves himself attended a number of key transitional meetings held by the Los Alamos Governing Board to plan a new, concerted effort on implosion. Groves was excited about implosion, but he demanded a clear and determinate time frame for its development. As the implosion work was foregrounded in the laboratory's concerns and was established as a key aspect of the development of the Manhattan Project as a whole, there was a powerful impetus to discipline and routinize the work. This would transform the character of the "HE Program," the experimental and engineering effort to develop implosion using high-explosive assemblies.[26]

Whereas Neddermeyer's group was small-scale, informal, and voluntaristic, there was now increased pressure for the formalization of the program's organizational structure. Almost immediately after von Neumann's visit, Oppenheimer expressed concern that the implosion work was "short staffed."

"Use more men and move faster," he told Neddermeyer. At its meeting on October 28, 1943, the Governing Board noted critically that the implosion experimentation work was "being carried on by a group of eight men whose relations with the rest of the engineering division are rather loose" and who "have not yet become accustomed to the idea of a large scale operation."[27]

Oppenheimer believed that expansion of the implosion work required placing the HE program under robust leadership. A serious hindrance to the success of the program, he told Conant, was the "reciprocal lack of confidence" between Neddermeyer and his Ordnance Division boss, Parsons. Unless new leadership and an influx of new staff were introduced to the work, he said, "I should very seriously doubt whether the implosion method could be developed in time." For this leadership, the Governing Board looked to Kistiakowsky, who began to visit Los Alamos as a consultant in the fall of 1943. Kistiakowsky was, as he himself put it, a somewhat "reluctant bride," at first skeptical of the work's chances for success—"I didn't think the bomb would be ready in time and I was interested in helping to win the war." Additionally, he "feared that difficult relations may result between myself and Seth Neddermeyer." Neddermeyer was likely to resent any interference in the implosion program, which had been his brainchild and personal project. Kistiakowsky was forced to swallow his reservations, however, finding himself under "pressure . . . [from] Oppenheimer and General Groves and particularly Conant, which really mattered, to go there on full time." Bowing to this pressure, he took a permanent position at Los Alamos early in 1944, bringing with him a team from the explosives division of the NDRC.[28]

Kistiakowsky soon found his initial doubts confirmed by his situation within the Los Alamos organization. He was caught "in the middle" of a "continuing angry conflict" between Parsons and Neddermeyer, whom he described as being "at each other's throats." Their conflict revolved around competing ideas about what sort of place Los Alamos was and how work there should be run. Parsons was "accustomed to developing mass products" and based his management of the division on a conservative military model. Neddermeyer, who was in temperament the exact opposite of Parsons, "believed that the implosion research should be done by a small group, in a consecutive set of experiments until the right way of doing it was achieved . . . Neddermeyer believed that this had to be discovered in a scientific, orderly fashion."[29]

Kistiakowsky was committed to a model of wartime research and development as necessarily large-scale and rapid. In his role as consultant in November 1943, Kistiakowsky had already recommended the reorganization of the HE program, establishing a clear division of labor. He also advocated centralized means of coordinating the expanding program of research—for example, a data-analysis project "to provide a centralized mechanism for correlation and interpretation of data obtained in the field by the operating crew." By February 16, 1944, when Kistiakowsky joined the program full-time, the E-5 group had been divided into the sections that he had suggested. Between March and September, the sections followed an "ambitious work schedule" that Kistiakowsky had prepared for the diagnostic work.[30]

During 1944, implosion research took place on a wide range of fronts. After von Neumann's visit, members of the Theoretical Division became engaged in developing mathematical models of the hydrodynamics of implosion, in April beginning to use IBM machines for the complex calculations involved.[31] On the experimental front, diagnostic studies to determine the degree of symmetry of the implosion were becoming a multidisciplinary effort cutting across a number of divisions and involving a variety of groups of physicists, metallurgists, machinists, chemists, electronics engineers, and explosives experts. By the spring and early summer, implosion was no longer an isolated project in one marginal group in the Ordnance Division, but the subject of coordinated work throughout the laboratory. Coordination and control of the work threw up difficult management questions among groups. The implosion program required systematizing a diverse array of groups of scientists and bodies of expert knowledge.

"QUEER DUCKS" AND ORGANIZATION MEN

The place of Neddermeyer's group within this growing network was increasingly problematic. By the summer of 1944, this group, known as E-5, was at the focal point of a large proportion of the laboratory's effort. However, the group was regarded as unprepared for this new role, and the blame was placed squarely on its leader, Neddermeyer. From the time that von Neumann focused interest on implosion, "it was evident," according to Bethe, that Neddermeyer "was not the right man." He was, said Bethe, "rather lackadaisical and was really not trying to put a big effort into it . . . Neddermeyer did not have the drive and had no intention at all to make it a large effort."[32] He had not come to terms with the expansion of the implosion program, nor with the character of the Manhattan Project as a large-scale technological system. Such a view of Neddermeyer was also expressed by Kistiakowsky when, in the early summer of 1944, the latter forced action on the issue of leadership of the HE research.

At the beginning of June, Kistiakowsky gave an ultimatum to Oppenheimer, demanding that Neddermeyer's role be clearly defined and circumscribed. Kistiakowsky noted that when he had voiced such concerns early on, Oppenheimer had reassured him that "the natural reluctance of Seth to surrender some of his authority and to submit to a much closer supervision than that exercised by Capt. Parsons previous to my [Kistiakowsky's] coming, will be overcome in time and friendly cooperation based on mutual confidence established." Kistiakowsky now felt that he had been let down in these expectations. The impasse between Neddermeyer and Parsons, which the introduction of a third party was supposed to have broken, was now reproduced in relations between Neddermeyer and Kistiakowsky. Neddermeyer, alleged Kistiakowsky, was "j[e]alous of his authority" and had become "essentially resentful of all attempts to interfere with the manner in which he runs E-5." According to Kistiakowsky, the result of the ambiguity of the chain of command between himself and Neddermeyer was that "our discussions of E-5 affairs, except those of purely technical nature, have usually ended either in a stalemate, with no action being taken by Seth or, when I was more insistent, even led to acrimonious altercations. In most instances I have chosen not to force through my opinions against his open resentment and have gradual[l]y converted my relation to E-5 to that of a technical consultant." Such a de facto division of authority was unacceptable to Kistiakowsky: "I do not think that as the *administrative* leader of E-5, Seth does a good job, however earnestly he tries."[33]

By the late spring to early summer of 1944, a clear schedule for the delivery of plutonium had been established. Quantities sufficient for a weapon were projected to become available by the summer of 1945.[34] There was therefore a considerable demand, both from within the Manhattan Project and from its political sponsors, for an increase in the pace of work at Los Alamos. It was in this context that tensions in Kistiakowsky's relationship with Neddermeyer came to a head.

The autonomy that Neddermeyer had enjoyed in 1943, and that Parsons had allowed him, was, in Kistiakowsky's view, increasingly problematic as the implosion program took on the character of a large-scale and centrally planned enterprise. In this emerging organizational structure, the parts had to be engineered to fit the whole. So Kistiakowsky's criticisms of Neddermeyer were focused on the local culture of the implosion experimentation group, which, he suggested, was incompatible with that of the larger laboratory. In particular, he implied that the intense mutual loyalty between Neddermeyer and his men was pathological, since it generated resistance to organizational and personnel changes. Neddermeyer, Kistiakowsky said, "is unwilling to bring new blood into E-5 and the few persons he has suggested as possible candidates acceptable to him appear to be 'queer ducks', if one is to believe other physicists opinions; on the other hand he develops a feeling of strong loyalty to men who are already working with him and is very slow to admit their limitations and weaknesses, with the net result that the quality of the technical E-5 staff is not nearly as high as it could have been." And Neddermeyer himself could not be altogether removed from the HE project, partly because the E-5 staff was "very loyal to Seth."[35]

Kistiakowsky also established a contrast between the growth and differentiation of the project as a whole and the integration of the HE program around the person of Neddermeyer. According to Kistiakowsky, Neddermeyer's "natural tendency, which he controls but rarely, is to do every job himself, the result being that he neither deputises his authority down the line, nor passes the jobs to specialists outside his group." Consequently, his time was taken up on "matters which could have been taken care of by others," and he did not have "enough time to give adequate technical supervision to his staff" or for "a more careful planning of the future needs and technical plans of E-5." Kistiakowsky framed his main criticism of Neddermeyer's leadership as an issue of time: "It is not that E-5 has broken down completely; actually lots of valuable data are being, and will be, obtained." But under Neddermeyer, "more time will be taken to obtain them than absolutely necessary and this may jeopardize the timely completion of the entire H.E. project because the work of E5 is the foundation on which everything else rests." As a member of Neddermeyer's group, Hugh Bradner, later recalled, "Kisti said, We're fighting a war. We can't afford to have somebody who takes time . . . And [so] Kisti would have *no* patience with [Neddermeyer] at all."[36]

Oppenheimer was unwilling to lose Kistiakowsky from the project and was more willing to sacrifice Neddermeyer. He essentially agreed with Kistiakowsky's criticism that Neddermeyer was behaving too much like an academic scientist. As Neddermeyer himself stated, Oppenheimer "became terribly impatient with me in the spring of 1944 . . . I think he felt very badly because I seemed not to push things as for war research but acted as though it were just a normal research situation." Kistiakowsky presented a number of options by which Neddermeyer could continue to guide the scientific research, but with control of administrative and personnel aspects handed over to either Kistiakowsky or someone even "more hardboiled than I am." Oppenheimer went further, giving Kistiakowsky complete overall control of the HE work, as an associate division leader in the Ordnance Division. Neddermeyer would become a senior technical adviser. Neddermeyer walked out of the meeting at which this was announced. Oppenheimer wrote to him that evening: "In behalf of the success of the whole project, as well as the peace of mind and effectiveness of the workers in the H.E. program, I am making this request of you. I hope you will be able to accept it." Neddermeyer, however, remained bitter that implosion was taken out of his hands.[37]

Kistiakowsky set about the task of transforming the HE program into a large-scale and disciplined operation. He redrew the "organization table," establishing a new Technical Steering Committee, which he would chair. This committee met weekly and was to have strict executive control over the HE program. "The decisions of this Committee," Kistiakowsky announced, "are binding on the leaders of the operating branches."[38]

Kistiakowsky faced the problem of coordinating the increasingly large-scale and differentiated structure of the HE program. The size of the program was illustrated by Kistiakowsky's (unsuccessful) request in May 1944 that several members of Group E-9 be allowed to attend Coordinating Council meetings (hitherto restricted to group leaders): "I should like to point out that in some respects the organization and activities of the Groups comprising H.E. Project are quite different from Groups in the rest of the laboratory. With few exceptions these other Groups have a small number of men in them, usually under ten; their activities are well centralized, being limited to a few adjoining offices or one building. On the other hand, Groups E-5 and E-9 have, or will have, more than fifty men a-piece." Kistiakowsky argued that in fact, "it could be said that the activities of these two Groups are equivalent to a full division." The size of these groups was indeed unusual; also atypical was the fact that their work locations ranged from laboratories in the Technical Area to "widely scattered field locations." As a result, the structure of the high-explosives work was highly decentralized, with sections having considerable autonomy from each other as well as from the overall leadership. Given this autonomy and the scale of their operations, Kistiakowsky argued, "the section leaders within these Groups are more nearly comparable to the group leaders in other divisions of the laboratory."[39]

It was during this period that research on the plutonium-gun idea faced its most serious setback and the laboratory's hopes came to be placed squarely on implosion. This shift was due to the results of a set of tests conducted by experimental physicist Emilio Segrè, between early April and mid-July 1944, on samples of plutonium from the Clinton reactor at Oak Ridge. Segrè found an extremely high rate of neutron emission in the samples, much higher than that exhibited by the small quantities of laboratory-produced samples with which Los Alamos had previously worked. This was determined to be a result of contamination by the isotope plutonium-240 of the Clinton plutonium. The clear implication was that the probability of spontaneous fission in the plutonium that was due to arrive from Hanford was even higher than had been predicted on the basis of the relatively pure laboratory plutonium. The relatively slow gun method, it now appeared, would not assemble a critical mass in time to prevent predetonation (premature detonation). The plutonium-gun program faced a crisis.[40]

The response to this situation was a rapid reorientation of the development of the plutonium bomb in favor of the implosion method of assembly. Initially, a number of alternative solutions to the crisis were considered. Gun researchers did not automatically abandon the gun principle as infeasible but considered the possibility of producing faster guns. Chemists and physicists looked into the possibility of separating out the problematic isotope. What enabled such alternatives to be so quickly ruled out was the particular way in which the laboratory had been organized. The power of the Segrè experiments to effect far-reaching technical and organizational change depended to a large degree on the fact that the implosion program had already been pursued with some priority since the previous fall. The shift was also facilitated by the Governing Board's strong, centralized control of the laboratory's program, reinforced by the overarching importance attached to speed of development.[41]

Oppenheimer told the Administrative Board on July 20 that "essentially all work on the 49 [plutonium] gun program and the extreme purification of 49 should be stopped immediately" and that "all possible priority should be given to the implosion program."[42] This decision led to the reorganization of the laboratory, bifurcating its work between the gun method for the assembly of the uranium bomb and the implosion method for the plutonium version. Oppenheimer created two new divisions focused entirely on implosion: the Weapons Physics Division, or G (for "gadget") Division, under Robert Bacher; and the Explosives (X) Division, under Kistiakowsky. The reorganization led to massive growth in the scale of the implosion program. By July 1945, the work under G and X divisions accounted for almost 35 percent of the laboratory's personnel.[43] The demand for personnel created by the intensive development of implosion, and the later engineering phase of the project, led the laboratory to recruit large numbers of SEDs from the Corps of Engineers.

The reorganization coincided with the anticipation, in late summer and fall of 1944, of the arrival of the Hanford plutonium. From now on, work rhythms at Los Alamos were to be tied to the industrial output of the uranium and plutonium factories. A strict system of reporting and scheduling was applied to the HE program by Kistiakowsky in August 1944, under orders from Oppenheimer and Parsons. Kistiakowsky announced to his men that "our reporting system must be considerably more systematized and strengthened in the future, and must include scheduling of anticipated progress." He required biweekly reports from each group. These reports were to be highly detailed, covering "every separate research program, which is either actively worked upon now, or for which definite future plans have been laid." The concern for planning and for meeting production output was also reflected in the requirement that the report present "a coherent picture of short and long range scheduling." Kistiakowsky stated that he wanted to have "thorough discussions of the outstanding problems with each Group Leader at bi-weekly intervals, and Group Leaders will have to be prepared to give me information on their plans and on factors which determine the rate of progress."[44]

Kistiakowsky's demand for regular reports proved more difficult in practice than in theory. According to the explosives chemist Hyman Rudoff, the idea of a fortnightly report was a compromise Kistiakowsky came up with to placate experimenters who were "loath to devote all that time to paper work" and "jibbed at the idea of a weekly report." However, "not all of the American workers were absolutely certain what a fortnight was, so he came up with what was, in essence, an equivalent. He asked for a semi-monthly report." This did not resolve matters. At a low point in the work, Kistiakowsky reprimanded his staff for not reporting frequently enough. When they protested that they did indeed submit bimonthly reports, Kistiakowsky replied, "Gentlemen, I am only a poor Russian immigrant, but even I know the difference between a semi-monthly and a bi-monthly report. Why don't you?"[45]

FREEZING THE DESIGN

The move toward routinization of the work in the new divisions was rendered problematic in part by the degree of technical uncertainty still inherent in the science and techniques of implosion. If a weapon was to be prepared in the next year, it would require taking urgent steps to stabilize, or "freeze," an implosion design. Captain Parsons, associate director of the laboratory since the August reorganization, played the key role in urging a quick decision on, and enforcement of, an implosion design. In November 1944, he took on the role of chair of the Intermediate Scheduling Conference, an interdivisional committee that began meeting in August with the task of coordinating the various groups involved in the design and testing of the implosion device. Immediately before coming to Los Alamos, Parsons had helped take the proximity fuse from the laboratory and testing ground into naval combat use.[46] At Los Alamos, he emphasized that the laboratory was just one part of a wider system, which would extend to the combat use of the atomic bomb.

Parsons regarded uncertainty as not merely a technical problem, but also a human and organizational one. Uncertainty was to be reduced, as far as possible, by clarifying and enforcing the instrumental mission of the laboratory. This would be achieved by a strong executive authority who could autocratically force the project from the experimental stage through to the delivery stage, a role for which Parsons saw himself as ideally suited. "In my opinion," he warned, "if the executive function I have outlined above is not made the responsibility of a qualified individual (not a committee or board) the necessary decisions and vigorous action will not be taken, and success in the experimental program will not be followed by successful fabrication and delivery of the weapon." Parsons requested "full authority . . . to carry out the policies and schedules by resolving conflicts in detail, coordinating the plans and schedules of the divisions and groups whose primary efforts are directed toward development, design and production of all mechanical, electrical and explosive parts of the gadget which will be carried in the airplane."[47] Oppenheimer balked at granting Parsons individually such far-reaching powers. But it was Parsons's vision of the disciplined and hierarchical organizational structure, which would be required to accomplish the laboratory's transition from research to engineering, that was to guide the final six months of the project.

Parsons's demand for the "freezing" of the design was connected in part with his concern about what he saw as chronic residual uncertainties in the implosion process. The problem of generating symmetry had dogged the implosion design from the beginning. The solution adopted by the laboratory was to employ explosive lenses, a configuration of explosives that would focus the blast into a symmetrically converging shock wave. The idea of using lenses was introduced to Los Alamos by the British physicist James Tuck, who had worked on similar devices before coming to Los Alamos in May 1944. The lenses adopted at the laboratory consisted of a fast explosive, Composition B, and a slow explosive, baratol. Manufacturing these lenses was a very delicate and difficult procedure, because any imperfections would distort the imploding wave.[48]

Parsons felt that the research and development phase was being unjustifiably prolonged by the scientists' fascination with the complex and, in his view, esoteric lens method.[49] The esotericism of the lens assembly was, Parsons believed, a product of physicists' domination of the laboratory and the consequent lack of importance attached to engineering protocols. He wrote to Groves that "in this endeavor the importance of an exciting, elegant research solution must not be placed above the prosaic materialization of this solution in a finished weapon."[50]

Instead, Parsons argued, the laboratory should focus on what he regarded as a simpler, non-lens model. He wrote to Oppenheimer in October 1944 expressing concern about intentions to rely on the lens model. A meeting on November 2 was devoted to whether to continue research on a non-lens version of the "gadget." Because no agreement was reached, Oppenheimer allowed work to continue on "both lens and non-lens problems," though he stated that he found this situation "undesirable."[51]

Parsons wrote again to Oppenheimer on February 19, 1945, outlining his objections to the lens program: "We embraced lenses as our first love in July 1944, renewing our pledge in September and December. With this highest priority assignable here (in thought, in field work, in explosives casting, design and shop work) . . . we have failed to make the schedules confidently predicted in September and December." He continued, "'To know all is to forgive all'— perhaps—but we are now asked to discard other alternatives and 'concentrate' on lenses, essentially because our previous attempted consideration on lenses has produced so few experimental facts that we must now decide without those facts." Parsons suggested that "this failure to meet all lens schedules is not a cause for recrimination but is data in itself." Faced with the lack of experimental data, he argued, "physicists, deprived of basic understanding[,] have lost their seven league boots." He pointed out the problems that existed not only in the conception but also in the craft of lens development: "I believe that experience to date shows that regardless of how simple an explosive lens looks in schematic design, its actual design, manufacture and final casting are steps which usually require several times as long as the most pessimistic 'rational' predictions. These are arts in which we can hope for no outside assistance except in machining before June, 1945." Parsons predicted that it would be "difficult in cold blood to look for an adequately tested lens implosion gadget in 1945" and argued that "the possible Summer 1945 gadget is a non-lens model." He added, "The non-lens implosion gadget as a limited objective (June–September, 1945) I believe could be engineered if there is good luck at every turn and if the philosophy is kept straight." On February 28, Parsons put forward the case against lenses at a meeting with Oppenheimer and other senior figures among the laboratory's scientific staff. But the committee decided against him and in favor of an all-out pursuit of the lens model.[52]

Kistiakowsky, Bethe, and Rudolf Peierls had been particularly strong advocates of the lens program.[53] It was the lensed design that was taken into the final two stages of development: the freezing of the implosion design, which took place between the fall of 1944 and the late winter to early spring of 1945; and the actual production of the bomb for the Trinity test, between March and July 1945. Kistiakowsky later emphasized that Oppenheimer's personal intervention had been crucial in resolving the dispute in favor of lenses: "In early 1945 we had a top-level meeting with General Groves present in which a kind of battle royal was fought, in a friendly way, between Parsons and me . . . Oppenheimer in the end decided for the lenses and that was that." Indeed, Oppenheimer is said to have told Parsons, "You might say some of your best friends are lenses," implying that nothing but prejudice lay behind Parsons's skepticism about the lens program.[54] However, although Parsons lost his specific battle for a non-lens device, his general vision of the organizational and attitudinal changes required for the transition to engineering was instantiated in the steps taken to freeze the (lensed) implosion design.

Taking the project into its "weaponeering" phase meant establishing both a definite bomb design and a clear organizational hierarchy and chain of command. The execution of this program was very much in line with Parsons's organizational philosophy. Parsons called for an executive authority to steer the technical work at Los Alamos toward its instrumental goal. Scheduling was a key means by which this control would be imposed upon the organization and the technical work, as Parsons made quite clear in his February 1945 memo on "homestretch measures." Decisions made by the executive would be expressed, he said, "in a binding directive, schedule, design or production order as the case may be." He emphasized the need for strong, top-down direction of the weaponeering program: "Ruthless, brutal people must band together to force the FM [Fat Man] components to dovetail in time and space. This is totally true of the first battle model, whether it is lens or non-lens." Backing away from his earlier expression of distaste for committees, Parsons advocated the establishment of a committee of senior members of the laboratory to strictly determine and enforce designs and schedules. "They must," he wrote, "feel that they have a mandate to circumvent or crush opposition from above and below, animate or inanimate—even nuclear!"[55]

In Parsons's model, then, engineering the atomic bomb meant engineering a sociotechnical system, requiring the disciplining and control of nature, technologies, and human beings. Scheduling was a key means by which these heterogeneous components of the system could be forced to "dovetail in time and space."

On March 1, 1945, Oppenheimer appointed the so-called Cowpuncher Committee, to "ride herd on" the implosion program. The committee's membership (Bethe, Kistiakowsky, Parsons, Bacher, Samuel Allison, Cyril Smith, and Kenneth Bainbridge) was substantially on the lines suggested by Parsons. This committee, meeting weekly, "relentlessly defined and redefined the assignments to individual groups, while constantly adjusting scheduled milestones." It oversaw and integrated the implosion work of the Explosives and Gadget divisions, including the fabrication and inspection of explosive lenses, the design and construction of detonators, diagnostic tests, chemistry and metallurgy, studies of nuclear physics, design of the inner metal parts of the implosion assembly, and coordination of the Trinity test, as well as administrative matters such as establishing shop priorities. Most importantly, the Cowpuncher Committee determined and enforced the project's schedule in the run-up to Trinity. Without this committee, Parsons later stated, "Los Alamos would still have been fumbling over minor engineering and procurement problems in the Fall of 1945."[56]

Under this regime, the implosion design was stabilized and the laboratory shifted gear from research to weapon engineering and testing. In April 1945, Kistiakowsky reported that "one can now state with a reasonable degree of assurance that all major research and design gambles involved in the freeze of the program of the X-Division have been won." With the routinization of the technical work at Los Alamos, its schedule could be seamlessly dovetailed with other production schedules: "Progress is more and more determined by the rate of supply of manufactured items." And in his monthly report for May, Kistiakowsky added that X-Division's activities "have lost all semblance to research and have become so largely production and inspection and testing that their brief summary here seems impractical."[57]

The Cowpuncher Committee relentlessly drove the laboratory in its work leading up to the Trinity test on July 16, 1945. Everyday life took on a peculiar intensity in the effort to ready the weapon for the test. During the early spring, recalled one Los Alamos resident, "the momentum of the Post and Laboratory quickened. There was an eerie quality in our immediate surroundings. The key men and women worked at a feverish pace and the auxiliary functions tried to keep up with them." McAllister Hull Jr., an SED who was in charge of the casting of explosive lenses, recalled, "I lived at S-Site for about three or four weeks before the shot [Trinity]. My colleagues brought me clean uniforms. We had three shifts. So I ate five meals a day, showered in the showers they used for chemical clean-up, and slept on my desk." He added, "We had one goal. That's to make this thing as fast as possible, to make it work." Hull oversaw a team of powdermen from DuPont, brought in to cast the explosives on a production-line basis for the test. In the period leading up to the test, "we worked like dogs, and we took chances and we didn't worry about anything."[58]

CREATING A CHAIN OF EVENTS

From 1944, the work at Los Alamos was integrated into a military operation aimed at the combat delivery of the weapon. In August General Groves, as he began to acquire firm production schedules for fissionable materials, gave the Air Force estimates for when the bombs would be ready: the implosion bomb in January 1945 and the gun-assembly bomb in June. Groves later said that he intentionally gave dates ahead of his actual expectations in order to "avoid any possible unnecessary delay in the use of the bomb."[59]

In September 1944, bomber pilot Colonel Paul W. Tibbets began to organize the 509th Composite Group, the combat unit that would deliver the bomb. Tibbets trained his team and carried out practice drops at Wendover Field, Utah. He was in close contact with Groves and Parsons at Los Alamos. During this period, Parsons insisted that the work at Los Alamos had to be closely integrated with military preparations for use of the bomb. Parsons saw his task in organizing delivery and liaison with the Air Force as the coordination and linking of different sociotechnical systems. This involved, as he put it, "the planning, training and logistic functions which seemed necessary to connect Y [the laboratory] with the 20th Air Force as a weapon producing, servicing and operating team." On September 7, he urged Oppenheimer to ensure that "success in the experimental program" would be "followed by successful fabrication and delivery of the weapon."[60]

Parsons argued that making Los Alamos into a "weaponeering" organization required making sure that its technical staff was committed to a military conception of the project. On September 25, he wrote a report for Groves in which he expressed his concern about "tender souls" who are "appalled at the idea of the horrible destruction which this bomb might wreak in battle delivery." Such "loose reasoning" might lead to "the expressed or unexpressed hope that 'We may never have to use this weapon in battle'" and thereby to the jeopardizing of the mission. It appears that Parsons was referring to early rumblings of discontent among scientists at the Chicago Metallurgical Laboratory (Met Lab), who were starting to discuss alternatives to military use of the bomb. Parsons was clearly worried that these ideas might spread to the New Mexico site, but he noted with satisfaction that as yet, "in its expressed form, this hope [of avoiding the bomb's use] is not encountered at Los Alamos."[61]

Transmitting Parsons's report, Oppenheimer was at pains to assure the general that the laboratory was committed to the production and use of a weapon:

> 1. I believe that Captain Parsons somewhat misjudges the temper of the laboratory. It is true that there are a few people here whose interests are exclusively 'scientific' in the sense that they will abandon any problem that appears to be soluble. I believe that these men are now in appropriate positions in the organization. For the most part the men actually responsible for the prosecution of the work have proven records of carrying developments through the scientific and into the engineering stage. For the most part these men regard their work here not as a scientific adventure, but as a responsible mission which will have failed if it is let drop at the laboratory phase . . .
>
> 2. I agree completely with all the comments of Captain Parsons' memorandum on the fallacy of regarding a controlled test as the culmination of the work of this laboratory. The laboratory is operating under a directive to produce weapons; this directive has been and will be rigorously adhered to.[62]

By September–October 1944, there was only one legitimate definition of Los Alamos's mission: it was to produce a weapon to be used in combat.

Parsons worked meticulously to ensure that no contingencies would arise to stand in the way of a clear-cut chain of events from production of the bombs at Los Alamos to their military use. He described his task as "the overall planning and technical initiative functions required to crystallize and integrate the combination of B-29 and bomb into a battle weapon." As early as September 1944, Parsons had been concerned about delivery problems. He pointed to, for example, the "notoriously unreliable engine" of the B-29 bomber. As historian Al Christman put it, Parsons's worry was that "worn out [B-29s] could become the fatal weak link in the chain of events."[63]

Parsons was equally concerned about the "mass-bombing psychology" of the Army Air Forces. Unlike mass raids, the success of which was judged statistically, the atomic bomb mission would need to guarantee the success of a single plane. In March 1945, Parsons was designated as head of the Los Alamos team, code-named Project Alberta, which was responsible for the combat delivery of the bombs. He wrote to Groves of his eagerness "to represent you in the initial battle delivery."[64] Project Alberta was assigned to the First Technical Detachment, attached to the 509th Composite Group. This unit would assemble the bomb components and oversee technical preparations on the Pacific island of Tinian, the base from which the 509th would attack Japan. Throughout, Parsons regarded both human and technological aspects of delivery as proper objects of engineering concern. Flight-testing and training, which he supervised over this period, "served not only to perfect the design but also to sift the young scientists and technicians and eliminate any who got jittery or tended to make mistakes under field conditions." A final precaution was delaying the arming of the bomb until the B-29 was on its way to Japan. Parsons himself would carry out this task on board the *Enola Gay* as it flew toward Hiroshima.[65]

NOT TO REASON WHY

An important effect of the laboratory's industrial discipline and social control was the suppression of moral and political dissent. The project's military leaders saw any moral qualms as dangerous interference in the chain of events from weapon production to weapon use. Groves attended a meeting in late March 1945 with Under Secretary of War Robert P. Patterson, at which the latter "asked whether there was any indication of anyone flinching from the use of the product." Groves was adamant that he had "heard no rumors to that effect" and explained his personal view "as to what a complete mess any such action would make of everything including the reputations of everyone who had authorized or urged or even permitted the work in the first place." The defeat of Germany, Groves argued, while removing the character of a "race" against a possible enemy bomb, "would not remove the necessity for going ahead."[66]

As early as March 1944, at a Los Alamos dinner party held by James Chadwick, leader of the British scientific mission, Groves had commented to the effect that "the real purpose in making the bomb was to subdue the Soviets."[67] Groves's remark made a particularly deep impression on one of the guests, physicist Joseph Rotblat: "I felt deeply the sense of betrayal of an ally ... this was said at a time when thousands of Russians were dying every day on the Eastern Front ... Until then I had thought that our work was to prevent a Nazi victory, and now I was told that the weapon we were preparing was intended for use against the people who were making extreme sacrifices for that very aim." When it became evident to him, toward the end of 1944, that the European war would soon be over and that there was now no possibility of a German atomic bomb, Rotblat decided to leave the project. As a pretext for detaining him, security officers accused him of being a spy. Nevertheless, he was able to refute these charges, and he sailed for Britain on Christmas Eve. But he was not allowed to tell anyone at Los Alamos his real reasons for quitting the project, and the official explanation was that he left due to worries about his wife in Poland. He was the only member of the project ever to leave on moral grounds.[68]

At Los Alamos, there was little overt opposition to the lethal use of the weapon. Los Alamos scientists played key roles in the planning of the atomic attacks. In late April 1945, Groves established a committee to help choose targets for the bombings. The committee, which included three Los Alamos scientists (John von Neumann, William G. Penney, and Robert R. Wilson), met at the laboratory in May to gather scientific opinion on the probable effects of the bomb and on suitable targets. In addition to the committee members, among the attendees at these meetings were Oppenheimer, Richard Tolman, and Parsons. Oppenheimer opened the meeting with a summary of the agenda: the height of detonation, report on weather and operations, "gadget" jettisoning and landing, status of targets, psychological factors in target selection, use against military objectives, radiological effects, coordinated air operations, bombing rehearsals, operating requirements for safety of airplanes, coordination with Twenty-first Bomber Command.[69]

These meetings were entirely focused on the military goal of the most efficient and most devastating use of the weapons. Humanitarian and moral concerns were entirely absent from the discussions. Hans Bethe and Robert Brode were brought in to provide calculations of the optimum height at which the bomb should be detonated for maximum destruction. Dr. Joyce C. Stearns of the Army Air Forces listed potential targets: Kyoto, Hiroshima, Yokohama, Kokura Arsenal, and Niigata. In addition, the group discussed the possibility of bombing the emperor's palace. It was agreed that the targets should be "in a large urban area of more than three miles diameter." Attacking a civilian population was favored because of the "psychological" effect. The group thought that Kyoto had "the advantage of the people being more highly intelligent and hence better able to appreciate the significance of the weapon. Hiroshima has the advantage of being such a size and with possible focusing from nearby mountains that a large fraction of the city may be destroyed." The assembled experts warned against a purely military target: "Any small and strictly military objective should be located in a much larger area subject to blast damage in order to avoid undue risks of the weapon being lost due to bad placing of the bomb." They also thought about following the atomic attack with an incendiary raid: "This has the great advantage that the enemies' fire fighting ability will probably be paralyzed by the gadget so that a very serious conflagration should be capable of being started." The chief worry about an incendiary raid was that radioactive clouds would make any such immediate follow-up hazardous, although it would be possible the very next day.[70]

This focus on immediate military questions contrasted with the situation at the Met Lab, where discussion of the moral and political implications of the atomic bomb led to the drafting of the Franck Report. Completed on June 11, 1945, it called for a nonlethal technical demonstration rather than military use of the bomb and argued for international control of atomic energy. The report was the outcome of discussions that had been taking place informally at the laboratory since the previous summer. In addition to the report, during June and July, Leo Szilard circulated his petition against military use of the bomb. Chicago scientist Ralph Lapp wrote that while this "dissent was futile with respect to altering the decision on the bomb," its real importance was in kindling the postwar scientists' movement for international control of atomic energy. A number of factors led to this ferment at Chicago. It owed a great deal to leaders such as Szilard and James Franck. In addition, the scientists at Chicago were not subject to the extreme restrictions faced at Los Alamos. The Met Lab was affiliated with the University of Chicago and was in an urban setting. As the official history of the AEC put it, "the Metallurgical Laboratory retained the essential features of academic research." Scientific work on reactor design had been completed, and plutonium production had been taken over by DuPont, so the Met Lab scientists were not under the same pressures of time as their Los Alamos counterparts. Moreover, their intense discontent with DuPont's management of the plutonium program made the Chicago scientists less inclined to trust the overall leadership of the project.[71]

At Los Alamos, any potential concern with the moral implications of the bomb, or with its long-term consequences, was eclipsed by the scientists' disciplined focus on technical problems. Victor Weisskopf wrote that he entered the project for fear of the consequences if the Nazis acquired an atomic bomb first. He "secretly wished that the difficulties would be insurmountable." However, he said, "imperceptibly, a change of attitude came over us. As we became more deeply involved in the day-to-day work of our collective task, any misgivings that we had at the start began to fade, and slowly the great aim became the overriding driving force: We had to achieve what we had set out to do." Weisskopf noted how the focus on the instrumental problems of the bomb's operation and effects left little mental space for moral reflection and tended to deaden any ethical sensibilities.

> We tried to determine the degree of destruction, the number of victims if the bomb exploded over a city, and the potential for radioactive damage to humans, animals and soil. All this required painstaking research in our laboratories and at our desks. Under the circumstances we were unable to confront the moral issues of our work even though we recognized them. There is no denying that constant discussions about the nature of the damage caused by fire and radiation sickness, and about the millions of deaths led to a growing numbness toward those terrible consequences.

Richard Feynman described a similar transformation of consciousness. The original reasons for the development of the bomb faded from his conscience as, he suggests, its creation became an end in itself. "What I did immorally was not to remember the reason and why I was doing it. So when the reason changed, which was that Germany was defeated, not a single thought came to my mind that it meant that I should reconsider why I was continuing to do this. I simply didn't think." Rotblat said, "Scientists with a social conscience were a minority in the scientific community. The majority were not bothered by moral scruples; they were quite content to leave it to others to decide how their work would be used."[72] The very fact that Rotblat was the only scientist to quit after the disappearance of the original rationale indicates that, in attempting to understand what drove the relentless and dedicated work on the bomb, we must look not to some motivating set of ideas or reasons, but rather to the social organization of the project.

The apparent inexorability of the bomb's development was produced by the system-building efforts of Groves, Parsons, and the Los Alamos scientists themselves. From early 1943, Isidor Rabi, on the basis of his experience managing the radar program at MIT, advised Oppenheimer of the importance of getting the project moving forward. Rabi warned, for example, that morale was "sinking" among research groups "standing idle" while Los Alamos was being constructed. It was imperative that they "be put to work immediately."[73]

The generation of impetus, then, was aimed at binding people to social order—conceived of in terms of "morale"—at the same time that it bound them to a schedule. This binding and ordering function of the schedule helps to explain how it was that, as sociologist William Ray Arney said, "the individual scientist at Los Alamos was a consummate team member at once vitalized by the work and submerged in it." The high level of social integration at Los Alamos—the strong identification of individuals with the aims of the organization—was to a large degree orchestrated through the management of time. There was very little private time. Most waking hours were devoted to working on the bomb. Workers were kept in a frenzy of activity, and this itself was crucial in maintaining the cohesion of the group. According to Emilio Segrè, "the pressure of work was immense and enhanced by the unavoidable deadlines and heavy responsibilities." Teller noted in a letter the frenetic and all-consuming pace of activity: "I should like to have a chance to think, to do some useless work, even to get bored ... Since the war started I have not been bored."[74] In particular, the structuring of time and the laboratory's "ownership" of time subdued thoughts of the moral implications of the atomic bomb.

Oppenheimer, as director, ensured that the instrumental mission of the laboratory was enforced. A key aspect of this enforcement was preventing scientific deviations from the established technical design. His successor, Norris Bradbury, observed that "Oppenheimer ... very resiliently kept people working on those two tasks [implosion and the gun method of assembly] and did not let the effort of the project be diverted, diluted into other goals not directly related to getting this particular job done[,] namely getting the atomic bomb done in the time that the war might last."[75] It was not a great extension of this role for Oppenheimer to move to deter any moral or political diversions from the laboratory's work or any dilution of the workforce's collective dedication to producing the bomb. He accomplished this through a combination of coercion and co-optation, both enforcing the hierarchically established mission of the laboratory and formulating the moral justification for that mission.

Robert Wilson—then a young group leader and experimental physicist, of Quaker background—wrote of how he called a meeting sometime in late 1944 or early 1945 to discuss the topic "The Impact of the Gadget on Civilization." "It was evident," said Wilson, "that the Germans would be beaten, and I wanted to raise questions about what our next steps should be." Oppenheimer had warned Wilson against holding the meeting, telling him that it would lead to trouble with the laboratory's security officers. When Wilson ignored this advice and went ahead, Oppenheimer came to the meeting, drawing on his authority among the scientists and his powers of oratory to direct the discussion: "Eloquent and persuasive as ever, Oppenheimer dominated the meeting. He argued that we should redouble our effort in order to demonstrate the reality of the nuclear bomb, so that the United Nations would be set up in an intelligent manner to deal with the problems presented by this new weapon."[76] There was no question at the meeting but that the bomb should be built. "It is significant," Wilson wrote, "that no one at that meeting . . . even raised the possibility that what we were doing might be morally wrong. No one suggested that we pack our bags and leave." Oppenheimer skillfully channeled the internationalist idealism of this group of young scientists into the task of building the bomb and used the meeting to reinforce their dedication to the work. "It is hard to express now the loyalty we felt for Oppy, our leader, and our confidence that he would do the right thing . . . Maybe we had no other choice than to put our trust in him, but in any case, we did," said Wilson. "With missionary zeal, we resumed our work."[77]

Philip Morrison remembered the way in which Oppenheimer conveyed a sense of the rightness of their task and of its urgency. Oppenheimer was unflinching in his support for use of the weapon: "Oppie had said spookily, I remember it so clearly: 'We must *use* Fat Man. We must bomb Berlin and Tokyo simultaneously.'" Morrison spoke of his experience as a junior scientist in thrall to Oppenheimer: "For nearly three years Oppenheimer labored ceaselessly: with him, for him, we worked as hard. It was our labor of love . . . I admired Robert Oppenheimer. He was of course my senior and my superior. It would be presumptuous to say I loved him or even feared him—Oppenheimer filled me with angst. Whatever the case, I listened to him. He had many arguments." The primary argument that Oppenheimer had in his arsenal, one to trump all opposition, was that only combat use of the weapon would demonstrate to the world its destructiveness, and that this destructiveness itself might mean the end of all war.[78]

Oppenheimer played a central role in supporting and lending legitimacy to the system of social discipline expressed in the project's institutional momentum. That this discipline was not questioned and was not experienced as coercive was due in part to Oppenheimer's ability to inspire his workforce with a picture of the laboratory's mission as being universalistic and moral. At a memorial service for Roosevelt, held at Los Alamos on April 15, 1945, Oppenheimer delivered a eulogy that was at the same time a rallying cry to continue the laboratory's work:

> We have been living through years of great evil, and of great terror. Roosevelt has been our President, our Commander-in-Chief and, in an old and unperverted sense, our leader. All over the world men have looked to him for guidance, and have seen symbolized in him their hope . . .

that the terrible sacrifices which have been made, and those that are still to be made, would lead to a world more fit for human habitation . . .

In the Hindu scripture, in the Bhagavad-Gita, it says, "Man is a creature whose substance is faith. What his faith is, he is." The faith of Roosevelt is one that is shared by millions . . . in every country of the world. For this reason . . . it is right that we should dedicate ourselves to the hope, that his good works will not have ended with his death.[79]

This secular sermon was a performance that asserted and instantiated Oppenheimer's charismatic leadership of the laboratory and community. Oppenheimer was at once a priest, calming and reassuring his flock, and a military leader, rallying his troops in time of war. And as he celebrated Roosevelt's leadership, his message was also that the scientists at Los Alamos should trust and have faith in their more immediate leadership and in the rightness of their mission.

Oppenheimer personally blocked the distribution at Los Alamos of Szilard's petition against use of the bomb. The petition stated, "We feel . . . that [atomic bomb] attacks on Japan could not be justified, at least not unless the terms which will be imposed after the war on Japan were made public in detail and Japan were given an opportunity to surrender." Communication between Los Alamos and Chicago was hindered by compartmentalization, so the logistics of sending the petition to Los Alamos were complicated. In the first attempt, Szilard gave the petition to Ralph Lapp, who was traveling to Los Alamos. Lapp was supposed to give it to another scientist to hold in a sealed envelope until Szilard could explain to Oppenheimer its purpose. But the petition was handed to Oppenheimer immediately, and he declared that it could not be circulated. Then Szilard asked Teller to try to have the petition distributed in the laboratory. But after speaking with Oppenheimer, Teller wrote to his Hungarian friend telling him that he could not support his protest. "The things we are working on," Teller said, "are so terrible that no amount of protesting or fiddling with politics will save our souls . . . Our only hope is in getting the facts of our results before the people. This might help to convince everybody that the next war would be fatal. For this purpose, actual combat-use might even be the best thing."[80]

Suppressing moral opposition to the bomb went hand in hand with enforcing the schedule for its production. The schedule itself took on the character of a transcendent principle, by reference to which political protest could be branded as illegitimate. For Oppenheimer, moral and political discussion was quite simply a *waste of time.* He enforced a disciplined focus on the narrowly technical problems of bomb construction. Metallurgist Ed Hammel recalled, "Los Alamos . . . was by '44 given the schedule of the production plants and associated pressure beneath, and we . . . were absolutely forbidden to get involved in any of these [political] things . . . That was canned as soon as Oppenheimer heard about it . . . Not only [was it] forbidden, but it was coming out of the program time, and that was unacceptable."[81]

The schedule itself helped to produce among the scientists a narrow orientation toward the merely technical problems of producing the bomb, as their focus on building the "gadget" crowded out any moral qualms. Wilson pointed to the scientists' sense of urgency as a reason for the overall lack of political discussion during the war: "At Los Alamos, we worked frantically so that a weapon could be ready at the earliest moment. Once caught up in such a mass effort, one did not debate at every moment, Hamlet-fashion, its moral basis."[82] V-E Day—May 8, 1945—did not lead Wilson to reexamine his commitment to the endeavor, though he later regretted that he had not left the project. "The thought never occurred to me," he said. "Nor, to my knowledge, did any of my friends raise any such question on that occasion." He was pulled along by the dynamism of the project, which structured his experience of time:

> Perhaps events were moving just too incredibly fast. We were at the climax of the project—just on the verge of exploding the test bomb in the desert. Every faculty, every thought, every effort was directed toward making that a success . . . Things and events were happening on a scale of weeks: the death of Roosevelt [April 12], the fall of Germany [surrender on May 7], the 100-ton TNT test of May 7, the bomb test of July 16, each seemed to follow on the heels of the other. A person cannot react that fast.[83]

Physicist Bernard Feld also emphasized the controlling pace of work, which prevented a reevaluation of the goals of the project in the wake of Germany's defeat: "Nobody stopped and said, 'We are not at war with the Germans any longer, do we have to stop and think?' We were caught up in this activity, which was all consuming. Nobody worked less than 15, 16, 17 hours a day. There was nothing else in your life, but this passion to get it done." The result was a "kind of tunnel vision."[84]

The laboratory took on the character of a total institution or a superorganism, one that claimed the scientists' whole lives. From the perspective of one scientist's wife, "Los Alamos was like a giant ant hill. The atom bomb was its queen and the Tech Area was her nest"—and "the Queen's demands for nourishment were unceasing."[85] In a letter to Oppenheimer, Szilard astutely suggested that the intensity of their labor had itself shaped the Los Alamos scientists' attitudes: "I expect you who have been so strenuously working at the site on getting these devices ready will naturally lean towards wanting that they should be used."[86]

It was in the period after the defeat of Germany that the Cowpuncher Committee was freezing the weapons' designs and enforcing a strict regime of scheduling. So it is not surprising that, as Segrè said, "the efforts to assemble the atomic weapon were redoubled during the late spring and early summer." Oppenheimer said that the project's "tempo" increased after the end of the war in Europe: "We were still more frantic to have the job done . . . We wanted to have it done before the war was over . . . I don't think there was any time where we worked harder at the speedup than in the period between the German surrender and the actual use of the atomic bomb."[87]

THE SCIENTIST IN THE TECHNOLOGICAL SYSTEM

If the Los Alamos scientists did not question the rightness of their mission, it was in large part because of the way in which their horizons were enclosed by the schedule. Groves had hoped that compartmentalization would make the scientists "stick to their knitting." The schedule proved far more effective in achieving this. And while Oppenheimer had opposed compartmentalization, he played a central role in supporting and lending legitimacy to the system of social discipline expressed in the project's schedule and institutional momentum. Through the schedule, everyday life at Los Alamos was woven into and structured by the vast technological and political-economic system of the Manhattan Project. Binding Los Alamos into this system meant the elimination of uncertainties in the design, manufacture, and human and moral dimensions of the bomb. Engineering the atomic system involved engineering the project's participants, molding them into a dedicated and disciplined workforce, devoted to the singular goal of bomb production and marching to the beat of a single drum. The Los Alamos scientists, in their fervor to get the bomb built *on time,* became the relatively unquestioning implementers of policy decisions made at higher levels of the organization.

Los Alamos was a closed world, and during the war, Oppenheimer was the scientists' "only contact with the world of Washington," at least the only one that they believed would represent their views and interests.[88] He paternalistically urged his colleagues to attend to their technical work and trust him to articulate their hopes and fears inside government. The bombing of Hiroshima and Nagasaki broke down the isolation of the Los Alamos community. The end of the war also temporarily interrupted the compulsive momentum of atomic weapons manufacture. After the initial euphoria of victory subsided, there was some time for somber reflection. With the war's end, many of the scientists emerging from the project, in the new "scientists' movement," would question and seek to break out from the limited role to which they had been assigned. They campaigned for ideas such as civilian and international control of atomic energy, as well as against secrecy. At the same time, however, many continued to respond to the allure of what Oppenheimer later called the "technically sweet" problems of weapons work, augmenting the sophistication, power, and numbers of the atomic weapons arsenal. Oppenheimer found himself in an increasingly tense and unstable position as he attempted to reconcile his role as weapons builder with his alter ego as academic "pure scientist" and as he mediated between the atomic scientists and the state.

CHAPTER SIX

Power and Vocation

"OPPIE'S GREATEST POEM"

The wartime mobilization of science for military ends led to an unprecedented incorporation of scientists into the apparatus of government, particularly as advisers to the new executive agencies established to manage atomic energy. This new insider role was exemplified by Oppenheimer, as were its accompanying tensions. Oppenheimer occupied a unique position from the end of World War II until the 1954 security hearings that excluded him from government. He was the personification of the new power of the scientists who emerged triumphant from creating the atomic bomb and who were widely credited with ending the war. Oppenheimer responded to the new expectations of him and rose to his new status. Friends who had not seen him since before the war were surprised at his confidence as a public figure and skill as an orator. Returning to Berkeley not long after the end of the war, classics scholar Harold Cherniss attended a large convocation to which Oppenheimer had been invited to address the student body. Oppenheimer spoke for about an hour to an audience of a few thousand gathered in the men's gymnasium—"and there was scarcely a whisper could be heard during all this time." According to Cherniss, "it was nothing that he said that held them spell-bound; it was this peculiar kind of magical influence that he could have." University of California president Robert Gordon Sproul had introduced the physicist by saying that this was "the Oppenheimer age."[1]

In 1946, *Time* magazine, saying that the atomic bomb had shattered the "ivory tower" and thrust scientists into the political fray, paid Oppenheimer the somewhat backhanded compliment of calling him the "most articulate of the new politicians."[2] Oppenheimer became the chief public articulator of the cultural and political meaning of the atomic bomb. His ornate prose promised a new philosophy for the age of the atom, combining popular hyperbole with a sense of gravitas. In expressing the meaning of the new atomic power, Oppenheimer fashioned his reaction at the Trinity atomic bomb test into an iconic moment. The story of his personal response to Trinity went through different permutations as Oppenheimer molded it as an oratorical device. In 1946, he told a university audience that at the test, "we thought of the legend of Prometheus, of that deep sense of guilt in man's new powers that reflects his recognition of evil, and his long knowledge of it."[3] The first publication of Oppenheimer's now-iconic reaction to the test was in a *Time* article on November 8, 1948, for which Oppenheimer was interviewed. The article stated, "Oppenheimer recalls that [at Trinity] two lines of the Bhagavad-Gita flashed through his mind: 'I am become death, the shatterer of worlds.'"[4]

The quotation was first given real prominence in Robert Jungk's best-selling 1958 book *Brighter Than a Thousand Suns*.[5] In a book published a year after that, William L. Laurence, the *New York Times* journalist brought in by General Groves to witness the Trinity test, claimed that Oppenheimer told him of this reaction at Los Alamos on the day of the test: "'At that moment,' I heard him say, 'there flashed into my mind a passage from the Bhagavad-Gita, the sacred book of the Hindus: "I am Become Death, the Shatterer of Worlds."'" Laurence wrote, "I shall never forget the shattering impact of those words." But in a *New York Times* article in late September 1945, Laurence had written, "To Prof. J. R. Oppenheimer of the University of California, who directed the work on the bomb, the effect, he told me, was 'terrifying' and 'not entirely undepressing.' After a pause he added: 'Lots of boys not grown up yet will owe their life to it.'"[6] In his 1947 book *Dawn over Zero*, Laurence still did not mention the Bhagavad Gita quotation. But he wrote of Oppenheimer, "This absent-minded scholar, who now finds outlet for his poetic vision through higher mathematics, turned out, in this very quiet and soft-spoken way, to be a veritable dynamo of action, animating the entire project with a vitality never seen in any laboratory. Los Alamos would go down in history as Oppie's greatest poem."[7]

The meaning of Los Alamos as a moment of transformation in the scientific vocation has indeed come to be framed by Oppenheimer's poetic invocation of the Hindu classic. The quotation "I am become Death" personalizes the destructive power of the atom and seems to call for a moral response. In this way, it encapsulates a humanistic formulation of the problems of the atomic age in relation to individual moral responsibility. If this was Oppenheimer's immediate response to atomic power, it suggests an immediate countering of technological power with liberal humanist morality. To many among the scientific community and the educated public, Oppenheimer seemed capable of bringing a civilized liberal humanism to bear on the unprecedented problems of nuclear weaponry. Fear of the atomic bomb as technology out of control was to some degree mitigated by the hope that this technological development was in the hands of a morally concerned elite.

It is impossible to know what in fact took place in Oppenheimer's mind on the morning of the Alamogordo explosion. But Frank Oppenheimer, who witnessed the test with his brother, said, "I wish I would remember what my brother said, but I can't—but I think we just said, 'It worked.' I think that's what we said, both of us, 'It worked.'"[8] The disjunction between these two responses—one moral and existential, the other coldly instrumental and technical (the two not necessarily psychologically mutually exclusive)—encapsulates the problem of Oppenheimer's persona and role at Los Alamos and in the postwar years. It mirrors the split between Oppenheimer's official function as a technical servant of the state and his public image as a humanistic spokesman for arms control and moral responsibility. It also mirrors Oppenheimer's dilemma in the face of competing conceptions of responsibility: to the state, to science itself, and to humanity.

Oppenheimer's postwar authority as scientific adviser and spokesman for the scientific community depended on his ability to embrace hopes for arms control and a peaceful postwar world while at the same time representing the utility of science as a source of military power. His authority was a careful balancing act, as he mediated between the scientific community and the state. In comparison with Oppenheimer, other scientists, notably Albert Einstein and Leo Szilard, were more consistent in their opposition to atomic weapons and the arms race, and this consistency gave them greater moral authority as spokesmen for scientific humanism. But unlike Oppenheimer, these figures were outsiders, without direct access to the closed circles where atomic weapons policies were made.

Oppenheimer was uniquely able to combine power with humanism and moral concern. He attempted to bring to the new role of the scientist as bomb builder and policy adviser the cultural authority that derived from the idea of "pure science" and from the Platonic image of the universality of scientific knowledge. In practice, this mediation between truth and power was a fragile political accomplishment, and one that he ultimately failed to maintain. The intensification of the Cold War limited his perspective and that of scientists generally. With the failure of negotiations for international control of atomic energy and the explosion of the first Soviet atomic bomb in 1949, and against the background of intensifying Cold War antagonism culminating in the Korean War, the American scientific community turned away from internationalism and toward the national-security state. This chapter examines the complex relationship between, on the one hand, Oppenheimer's cultural role in articulating the meaning of science and the place of the scientist in the modern world and, on the other hand, his political role as an adviser to the state.

INTERIM: OPPENHEIMER AND THE WARTIME SCIENTIFIC ESTABLISHMENT

Oppenheimer's position as an insider among the policy elite of the Manhattan Project was institutionalized in his membership on the Scientific Panel of the Interim Committee, set up in the spring of 1945 by Secretary of War Henry Stimson. The committee was to give advice on a wide range of matters relating to atomic weapons and atomic energy, and especially to advise on postwar nuclear policy.[9] The Scientific Panel—chaired by Oppenheimer, with Arthur Compton, Ernest Lawrence, and Enrico Fermi as the other members—was conceived as the voice of the scientific community in the corridors of power. However, the emphasis in creating the panel was less on giving the scientists input into decision making than on defusing potential opposition to the use of the bomb, such as had emerged at Chicago.[10]

That the advisory function of the panel was ritualistic was particularly clear in the case of its consideration of the question of the military use of the atomic bomb. In the assessment of historian Gar Alperovitz, the scientists "had virtually no impact on government decisions" during the war.[11] That the bomb would be used militarily had been an organizing assumption of the entire project, and as the project grew into a giant military-industrial system, it developed a powerful institutional momentum toward that end. Oppenheimer himself said that "the decision was implicit in the project. I don't know whether it could have been stopped."[12]

Oppenheimer and the Scientific Panel were able to consider the question of the use of the bomb only in the narrowest of terms. They acted as technical experts, providing information about the effects of the bomb.[13] The only alternative to military use that they were able to consider was the idea of a nonlethal demonstration, advocated by scientists at Chicago. This proposal, however, did not challenge the assumption that the atomic bomb was the key to ending the war. The demonstration idea was dismissed by Oppenheimer on technical grounds. He said that he could not think of how to make a non-lethal demonstration sufficiently spectacular that it would cause the Japanese to give up. This narrowly technical orientation was, in part, a result of the panel's insulation from knowledge of broader political, diplomatic, and military realities.[14] Oppenheimer later said, "We didn't know beans about the military situation in Japan. We didn't know whether they could be caused to surrender by other means or whether the invasion was really inevitable. But in the back of our minds was the notion that invasion was really inevitable because we had been told that."[15] The panel's extremely brief consideration of alternatives to military use of the bomb presupposed that an end to the war had to be brought about militarily rather than through diplomacy. Against this background, the demonstration idea was a nonstarter. Alperovitz wrote, "That the bomb would be used was essentially taken for granted when the Interim Committee did its main work," between May and July 1945. At the point during the war when scientists were included in the policy arena, they were in a position to be nothing more than a rubber stamp. The Scientific Panel merely provided an aura of rationality and of propriety and thereby helped to lend legitimacy to the bombings of Hiroshima and Nagasaki.[16]

There was one aspect in which the Scientific Panel did go beyond a narrowly technical role; this was on the question of international cooperation. At the panel's May 31 meeting, Oppenheimer presented his view that "Russia had always been very friendly to science and . . . that we might open up this subject with them in a tentative fashion and in most general terms without giving them any details of our productive effort."[17] In its June 16 report, the panel advised that before using the bomb, the United States should approach all its major allies, including the Soviet Union, with overtures regarding future cooperation.[18]

However, as the scientists got back to their technical work, their proposals for international cooperation were quietly, but decisively, shelved. The news of the atomic test profoundly affected Truman's dealings with the Soviets, but it did not lead toward greater cooperation, as the atomic scientists had hoped.[19] Indeed, the bomb removed the immediate necessity of such cooperation—Truman felt that he no longer needed the Soviet Union's help in the war against Japan. In contrast to the hopes of the Scientific Panel, Potsdam marked the beginning not of cooperation but of a policy of superpower confrontation, of which the bombings of Hiroshima and Nagasaki were an expression.[20] A few months after the bombings, Oppenheimer seemed to have briefly come close to recognizing this, but he expressed any concern only privately. In mid-October, he met with Secretary of Commerce Henry Wallace. According to Wallace's diary, the physicist "seemed to feel that the destruction of the entire human race was imminent . . . It seems that Secretary [of State] [James] Byrnes has felt that we could use the bomb as a pistol to get what we wanted in international diplomacy. Oppenheimer believes that that method will not work ... He thinks the mishandling of the situation at Potsdam has prepared the way for the eventual slaughter of tens of millions or perhaps hundreds of millions of innocent people."[21]

"WE STILL BELIEVE": OPPENHEIMER, ALAS, AND THE SCIENTISTS' MOVEMENT

Oppenheimer combined membership in the Manhattan Project's policy elite with paternalistic authority in relation to the laboratory scientists at Los Alamos. The laboratory's scientific rank and file, insulated from scientists at other sites by compartmentalization and without direct access to policy decisions, had little choice but to have faith in Oppenheimer as their representative. While the Los Alamos scientists harbored deep suspicion of General Groves and the project's military leadership, Oppenheimer encouraged his colleagues to defer to the judgment of the civilian political and scientific elites in the project's high command, from Vannevar Bush and James Conant to Secretary Stimson. Oppenheimer thus prevented the circulation at Los Alamos of Szilard's petition against use of the bomb and was able to persuade the scientists to forgo political meetings, subordinating their moral and political concerns to the institutional goal of developing the bomb. Oppenheimer gave legitimacy to the hierarchical structure and leadership of the Manhattan Project. The end of the war, however, disrupted this structure of authority and deference.

Peace interrupted the disciplined dedication to weaponeering that had so dominated the consciousness of the Los Alamos scientists and that had largely prevented the emergence of any significant critical reflection on the deeper meaning or morality of their work. Scientists' accounts of their reaction to Hiroshima emphasize how the initial euphoria of victory gave way to a stark realization of the horrors of the atomic age. Robert Wilson was one of the scientists most powerfully affected by Hiroshima: "The news of the tremendous suffering and damage and loss of lives ... was an epiphany that has changed my life ever since."[22] While most Manhattan Project scientists accepted the official justification of the bombings of Hiroshima and Nagasaki, many of them quickly became concerned that the full implications of the new weapon were not understood or were not being faced by the public and the country's political leaders. They were worried also about secrecy and military control, which had been necessary during the war but which they feared might prevent a proper public understanding of, and a long-term political solution to, the unprecedented global crisis presented by the atomic bomb.

Nevertheless, the idea that the Manhattan Project's scientists were collectively wracked with guilt over Hiroshima and Nagasaki is a misconception. Some were, but Oppenheimer's anguished confession that he had "blood on [his] hands," which he expressed to an unsympathetic Truman, has too often been taken as typical. It is usefully balanced by physicist Freeman Dyson's perception of the young scientists working at Cornell University in the late 1940s who were fresh out of the Manhattan Project. Dyson perceived that, "having no sense of tragedy, they also had no sense of guilt . . . They had come through the war without scars. Los Alamos had been for them a great lark. It left their innocence untouched."[23]

The overarching feeling among the scientific community at the end of the war was optimism and a strong sense of accomplishment and of their own potency in the creation of the atomic bomb. There was a powerful technological-utopian strain in the scientists' desire to find "a silver lining even in the destructive mushroom cloud of the atom."[24] Many of the scientists believed that the atomic bomb rendered conventional war obsolete and that it demanded a new approach to international relations. Oppenheimer himself had been greatly impressed with Niels Bohr's vision of atomic science as the keystone of a new internationalism. Bohr had spent his wartime exile from Nazi-occupied Denmark as an advocate for postwar arms control. He argued that the atomic bomb would be a weapon of unprecedented power and that its existence would demand a new approach to international peace and security. The only way to guarantee security, he argued, was through international control, to ensure that the science and technology be developed only for peaceful purposes after the end of the war.[25]

Central to Bohr's vision was the international fraternity of scientists that offered a model of peaceful cooperation across national divisions and an already existing set of relationships on the basis of which mutual trust could be promoted. In particular, he regarded norms of scientific openness and free communication as crucial for creating the kind of "open world" necessary for peaceful atomic cooperation.[26] Bohr failed to convince either Churchill or Roosevelt of the need for such cooperation, but he was more successful in proselytizing among his scientific colleagues. During the war, Bohr had visited Los Alamos as a consultant. Known to the scientists there as "Uncle

Nick," he was a patrician figure whose presence at the laboratory was felt to inject a sense of moral purpose into the work. His utopian vision for the role of atomic weapons in creating a peaceful world helped to maintain the dedication of the Los Alamos workforce to building the bomb. Oppenheimer later reflected that Bohr "made the enterprise which looked so macabre seem hopeful."[27]

Bohr's ideals resonated with the hopes of young scientists who, embracing their liberation from military strictures after the end of the war, formed what came to be called the "scientists' movement." Scientists' organizations sprang up at all the major Manhattan Project sites. Although it was at Chicago that the seeds of the scientists' movement had been sown, the first formal organization was the Association of Los Alamos Scientists (ALAS), established on August 30, 1945. Initially, scientists' political organizations were site-specific. As well as ALAS, there were the Atomic Scientists of Chicago, the Atomic Engineers of Oak Ridge, the Association of Oak Ridge Scientists at Clinton Laboratories, and the Association of Manhattan Project Scientists at Columbia University. These groups would later join together in the Federation of Atomic Scientists (which subsequently became the Federation of American Scientists).[28]

The ALAS scientists looked to Oppenheimer to represent their hopes. Oppenheimer was faced with the impossible task of reconciling the idealism of Bohr and the Los Alamos scientists with a narrow political pragmatism as a servant of the state. He counseled patience, urging his colleagues to have faith in the Truman administration, in the scientific leadership of Bush and Conant, and in himself as the scientists' representative in the corridors of power. However, as his commitments in Washington pulled him away from Los Alamos, he increasingly lost touch with the grass roots of his scientific constituency.

On September 7, 1945, ALAS completed a statement for public release, signed by almost all the civilian scientific employees of Los Alamos. The statement set out the position of the atomic scientists, describing the power of the bomb and its implications and arguing for the necessity of vesting control in an international organization. Victor Weisskopf wrote to Oppenheimer on behalf of ALAS, asking him to submit the statement to the Interim Committee so that it could be approved for release to the press. As Wilson put it, "Because we were still living on top of a mountain under the strictures of absolute secrecy, we turned it over, trustingly, to Oppenheimer to expedite its release."[29] Oppenheimer submitted the document as requested. When it was received by the War Department, however, Washington bureaucracy proceeded to block it.

Meanwhile, the ALAS members were becoming increasingly restless. Weisskopf again wrote to Oppenheimer asking whether, as a member of the Interim Committee, he would approve of their writing a "group statement, political only," for direct distribution to newspapers, radio commentators, and members of Congress. Oppenheimer replied that while he could not argue against individual letter-writing, he was very strongly opposed to any group statement or collective action. He was worried that the release of such a statement "would be a breach of faith with the administration." And he threatened that if the scientists released the document, he would give up his work in Washington and return to California. Then he dropped the bombshell that the ALAS statement that he had passed along to the War Department had been classified by the department as a state paper. Oppenheimer tried to put forward the most positive interpretation of this development. ALAS chairman William A. Higinbotham's naive initial reaction to Oppenheimer's version of events was, "Yes! Our document has become a state paper and has aroused a lot of helpful disc[ussion] in the cabinet." But the ALAS membership quickly realized that the statement could not now be released and that Los Alamos no longer had control over it.[30]

Soon after these exchanges, at an IBM (International Business Machines) luncheon, General Groves made a speech disparaging scientists' authority to speak out on policy questions, and his comments were widely quoted in the press. John Manley wrote to Groves advising him that Manley and a few others were trying to prevent their colleagues at Los Alamos from "doing anything which might embarrass the Administration." This task, he said, was made considerably more difficult by Groves's statement, and he warned the general of the danger of a schism developing between the administration, the Army, and the scientists.[31]

Norris Bradbury, who served as acting director of Los Alamos during Oppenheimer's many trips to Washington and who was soon to succeed him as director, wrote to Oppenheimer about the demand by the staff at Los Alamos to send a protest to the Interim Committee. Characteristically, Oppenheimer attempted to smooth ruffled feathers and to prevent a confrontation. While he would not object to their writing to the committee, he told Bradbury, "the Interim Committee is largely defunct and I doubt whether the desired answer will be forthcoming." This was certainly news to the scientists at Los Alamos—another indication of their distance from power and their reliance on Oppenheimer as their representative in Washington. And Oppenheimer added, conveying the impression of intimacy with those in power, that "General Groves and I can both assure you that he spoke in New York, as at other times, as an individual and I can assure you that the quoted newspaper accounts do not appear to correspond with the views of any of the members of the Interim Committee, nor, I believe, to official War Department views."[32]

Oppenheimer had begun to one-sidedly present the official line to the scientists. By binding the scientific community to the state, he helped to undercut the scientists' capacity to engage with the public. This was evident as Oppenheimer turned to the matter of what he called "the famous memo"— that is, the now-classified ALAS statement. "It is my feeling," he wrote to Bradbury, "and the general feeling of all with whom I have talked, that public discussion of the issues involved is very much to be desired, but that it should follow rather than precede the President's statement of national policy which will be conveyed in his message to Congress." The presidential message was to be the announcement of a new bill on atomic energy. "We do not anticipate further great delays in this message," Oppenheimer said, his use of the first-person plural suggesting, reassuringly, that he himself was involved in the process.[33]

Oppenheimer emphasized his intimacy with the administration and the ultimate harmony of the scientists' goals with those of the government. His appeal to his constituency was successful. Higinbotham told an ALAS meeting, "We have one representative . . . that is, Oppic . . . We still believe and urge you to go along with Oppie and the administration." The meeting agreed not to issue a statement until after the president's speech and, without dissent, carried the motion "that Willy tell Oppie that we are strongly behind him." The general feeling was that "as long as Oppie was our voice, we leave everything up to him."[34]

But some were no longer willing to leave everything up to Oppenheimer; they came to believe that he was being used by the administration to keep the scientific community in line. Wilson wrote out his own version of the ALAS statement and mailed it, as an individual, to the *New York Times*. He reflected, "Mailing it was a serious violation of security. But it made the front page of the *Times,* and no one has ever questioned my right to send it. For me, it was a declaration of independence from our leaders at Los Alamos, not that I did not continue to admire and cherish them. But the lesson we learned early on was that the Best and the Brightest, if in a position of power, were frequently constrained by other considerations and were not necessarily to be relied upon."[35]

A further challenge to Oppenheimer's authority, however, was to come with the presidential announcement, upon which the scientists had been enjoined to wait before publicizing their views, and the introduction of the

May-Johnson bill to Congress on October 4.[36] Most worrying for the scientists was the extent to which the bill would hand control of atomic energy over to the military. The scientists had the impression that the Army was attempting to rush the bill through Congress without adequate debate or scrutiny. They feared that if the bill passed, it would mean the indefinite extension of the wartime regime of the Manhattan Project.[37]

On October 7, Oppenheimer brought from Washington a copy of the May-Johnson bill and discussed it with the ALAS executive committee. He urged the scientists to have faith in the Truman administration and not to criticize the government's position. He was able to overcome the scientists' initial adverse reaction, and by the end of the session the committee had voted unanimously to endorse the bill.[38] In its support for the bill, ALAS stood in contrast with the scientists' organizations of the other Manhattan Project labs. The influence and independence of the broader scientists' movement proved harder to contain.

After visiting Chicago from Los Alamos, Herbert Anderson, himself formerly a member of the Met Lab, wrote angrily to Higinbotham: "We had been asked by our representatives in Washington to withhold comment lest this cause undue controversy and delay the acceptance of the measure." And he said that "I must confess my confidence in our leaders Oppenheimer, Lawrence, Compton, and Fermi, all members of the Scientific Panel advising the Interim Committee and who enjoined us to have faith in them and not influence this legislation, is shaken."[39] The same day (October 11), Oppenheimer, Fermi, and Lawrence sent a telegram to Stimson's successor, Robert Patterson. The scientists wrote that they "strongly urge[d] the passage of the legislation," defending it as "the fruits of well-informed and experienced consideration."[40] A rift was becoming apparent between the scientific leadership who had formed the Interim Committee and the general opinion of the rank-and-file laboratory scientists. In the following week, the previously unquestioned leadership of the wartime elite was challenged, as scientists from the various organizations descended on Washington to express their views on the atomic energy legislation.

Before returning to Washington, Oppenheimer was given another opportunity to rally Los Alamos behind him. On October 16, he resigned as director of Los Alamos; at the ceremony held that day, he accepted, on behalf of the laboratory, the Army-Navy Award for Excellence and a Certificate of Appreciation from the secretary of war.[41] Faced with dissent and controversy over the impending domestic legislation, Oppenheimer repeatedly sought to downplay its significance, stressing instead the higher goal of international control of atomic energy as a vehicle of world peace.[42] This, he suggested, was the scientists' true purpose, since it arose from the very nature of the scientific community as an international fraternity. It was to these utopian themes that Oppenheimer appealed in his final speech as director of Los Alamos. "By our works," he told his colleagues, "we are committed, committed to a world united, before this common peril, in law, and in humanity."[43] As when he had spoken on the occasion of Roosevelt's death, Oppenheimer wove the laboratory's instrumental work into a messianic narrative of a moral mission. It was a performance delivered with true virtuosity, "his voice . . . pregnant with responsibility," and it reinforced his authority as "the man who guided the work and wove the threads together." As during the war, he again called forth solidarity in the pursuit of a communal goal and integrated Los Alamos behind his leadership. As resident Eleanor Jette put it, "That day he was us. He spoke to us, and for us."[44]

The following day, he was in Washington. But his role in speaking for the scientists was now more problematic. While repeating passionately the scientists' mantra that they needed freedom in pursuing their research, Oppenheimer was also there to give his support to the May-Johnson bill, widely opposed by the atomic scientists outside Los Alamos. Oppenheimer was asked for his opinions on the bill while testifying to a Senate subcommittee dealing with science legislation. Giving his views on the organization and funding of basic research, Oppenheimer appealed for the freedom of the scientific community from regulation, control, and formal accountability, even while unprecedented amounts of public money were to be directed its way. He made "a plea for not overorganizing the work of scientists, and for trusting, as we have in the past, their own judgment of what work is worth doing."[45]

In defense of this autonomy, Oppenheimer presented an image of the scientific community as self-regulating. He drew on a conservative discourse about science as traditional knowledge, embodied in what he called "a way of life." Opposed to progressive calls for the planning of science to meet goals of social welfare, Oppenheimer argued that such motivation could not be imposed externally. Rather, "it is only indirectly, through the complex mechanisms of education, taste, and value, that the need of society for science does get translated into the seed of the scientists." Oppenheimer presented scientific progress as the organic growth of a community, with its own traditions, values, and modes of social control. He invoked ideas of fellowship, community, and apprenticeship as an alternative to formal control: "The scientist does not work in a vacuum, though he sometimes talks as though he does. He needs freedom; that is not because he is an isolated individual, but only because he may be in a better position to plan his work than anyone else. Equally, and equally deeply, he needs a sense of community with his fellow men."[46]

This defense of scientific autonomy, however, stood in tension with Oppenheimer's support for the May-Johnson bill, which proposed far-reaching powers for a militarily oriented Atomic Energy Commission. When he found himself pressed on this point by Senator William Fulbright, what was most striking in Oppenheimer's response was just how little he had concretely to say about the atomic energy legislation: "The Johnson bill, I don't know much about."[47]

Oppenheimer's message to the senators and representatives was substantially the same as the advice he had given the scientists at Los Alamos: an injunction to trust the good men of the administration. He told the House Committee on Military Affairs the following day:

> The [May-Johnson] bill was drafted with the detailed supervision of Dr. Bush and Dr. Conant, with the knowledge and the agreement of the former Secretary of War, Mr. Stimson. I think that no one in the country carried a greater weight of responsibility for this project than Mr. Stimson. I think no men in positions of responsibility, who were scientists, took more responsibility or were more courageous or better informed in the general sense than Dr. Bush and Dr. Conant. I think if they liked the philosophy of this bill and urged this bill it is a very strong argument. I know that many scientists do not agree with me on this, but I am nevertheless convinced myself.[48]

Critics of the bill, including both scientists and politicians, were worried about the breadth of power that it would give to the Atomic Energy Commission. The commission's powers appeared dangerously undefined in the bill's wording. The scientists' organizations were, for example, concerned that the AEC's control over atomic energy would extend to the regulation of laboratory research within universities. Oppenheimer argued that such breadth and vagueness were a necessary response to the novelty and uncertainty of atomic energy.

Just as Oppenheimer's argument for the bill was predicated on trust in Stimson, Bush, and Conant, so his conception of the operation of the commission was predicated on trust in the future commissioners. More important than written provisions, to Oppenheimer, was the character of the commissioners themselves. So, while admitting that "if it is construed unwisely; that is, if it is executed unwisely, it could stop science in its tracks," he argued that "the whole philosophy of the bill is that it will be possible to find a commission that will execute these provisions wisely."[49]

Further, unlike many of his scientific colleagues, Oppenheimer was not overly perturbed by the fact that the May-Johnson bill permitted military officers to be appointed to powerful positions within the commission. He told the House committee, "I think it is a matter not what uniform a man wears but what kind of man he is." "I cannot," he said, "think of an administrator in whom I would have more confidence than General Marshall."[50]

During the war, Oppenheimer had been initiated into a small but powerful policy elite, which was subject to almost no outside scrutiny or accountability. His appeals for trust in this elite signaled to others that he himself was now fully a member. But Oppenheimer failed to understand that such personal trust came less easily and was much less appealing to members of Congress, who valued their powers of scrutiny and oversight, and to those scientists who had felt marginalized and disempowered by their exclusion from key decisions during the war. To those situated at a greater social distance from the Manhattan Project's inner circle, Oppenheimer's advertisement of his easy familiarity in this elite could trigger suspicion. As one scientist commented, "When he started referring to General Marshall as 'George,' we knew what a change had come over him."[51]

So Oppenheimer combined a defense of the autonomy of the scientific community with an equally elitist defense of broad powers for a scientific-military administrative elite. In both cases, his argument was for personal trust rather than formal accountability and control. Oppenheimer was performing a dual role—as spokesman both for the scientific community (pressing for research autonomy) and for the administrative elite (pressing for strong administrative power with minimal accountability). He struggled to reconcile the former with the latter, but he seemed unprepared to deal with the potential for conflict between these roles.

If Oppenheimer had not expected such conflict between his constituencies, it was because of his belief that he himself had been able to bridge them. His rhetoric, appealing to notions of fellowship, community, and transcendent purpose, tended to paper over such grubby questions of power and conflict in favor of an image of national solidarity and moral cohesion wisely presided over by a trusted elite, in which he was embedded.

While not denying that the war had generated important changes, Oppenheimer sought to recapture a status quo ante bellum. This was also an effort to define the meaning of his own role as a scientist. Oppenheimer's arguments for scientific autonomy amounted to an elitist defense of the academic establishment, an appeal to an image of academic "purity" in denial of the new interweaving of academia with the military-industrial complex. Despite the new complexity of the scientific role, Oppenheimer sought to extract a sphere that he painted as somehow inherently untouched by the war. Rather than science having been transformed by the war, this rhetoric portrayed science as having been put on hold, or deep-frozen, during this time; its pristine body could now, after the war, be revived.[52]

Oppenheimer argued that the bomb project had been "an enormous technological development . . . but it was not science, and its whole spirit was one of frantic exploitation of the known; it was not that of the sober, modest attempt to penetrate the unknown." It followed from this that the mission-directed and government-controlled wartime organization of science could in no way be adequate as a model for scientific institutions in times of peace. "This is," he stated, "a plea for leaving much of the scientific strength of the country in the universities and technical schools, the small institutions in which scientists have worked in the past and in which they will have the leisure and privacy to think those essential, dangerous thoughts which are the true substance of science."[53] A distinction between science and technology, and the definition of the war's legacy as merely technological, was essential to Oppenheimer's negotiation of, on the one hand, academic autonomy and, on the other, control of atomic energy for national security. The science of the atom would be autonomous. The technology of atomic energy would be controlled and directed by the government.

Such a demarcation was also implicated in Oppenheimer's presentation of self. Indeed, Oppenheimer distanced himself from the very topic addressed by the May-Johnson bill. His statement, he told the senators, would be "somewhat academic and corresponds to my position as professor of physics rather than to my position as a maker of bombs." When Oppenheimer testified before the House committee, specifically on the May-Johnson bill, he was asked, as a matter of course, what his qualifications were on the subject. He replied airily, "I have practically no qualifications, Mr. Chairman. I am a physicist who taught in California, in Berkeley and in Pasadena, before the war. In 1941 I became interested in the possibility of making atomic weapons, and since the inception of the laboratory at Los Alamos I have been its director. So I know a little bit about the making of bombs."[54] Although it was Oppenheimer's managerial role as director of a large-scale wartime weapons laboratory that was of most interest to the representatives and senators, he often distanced himself from what soon came to be called "big science." While pointing out the demand of experimental laboratories for those "large and expensive gadgets which physicists like to play with," Oppenheimer said, "You see my equipment [indicating pencil]. This is a rather luxurious specimen, so I am not a very good person to talk about it."[55] When it suited him, Oppenheimer would play the cloistered theoretician, unconcerned with matters of the world. Much of his public discussion of atomic energy after the war was characterized by this aloofness, as if to suggest that his sights were focused on contemplating matters deeper or higher—at any rate, less mundane. It was as though atomic energy were something of a nuisance: it had to be sorted out, but it was only a rather nasty and annoying interlude from the real preoccupations of the scientist.

Oppenheimer's claim to speak on behalf of the scientists was further troubled by the fact that he was now testifying alongside delegates from the various new scientists' organizations. At the morning session of the Senate subcommittee on October 17, Oppenheimer was joined by Howard J. Curtis from the Association of Oak Ridge Scientists at Clinton Laboratories. There were a number of key points of disagreement between Curtis's testimony and that of Oppenheimer. These led to a sharp exchange when Oppenheimer asked to comment on the testimony that Curtis had just given. Curtis had, for example, rejected Oppenheimer's distinction between science and technology, addressing this as "a misconception which has crept into the press recently." No such distinction could be maintained, Curtis said: "The two are so closely connected that it would be impossible to pick out any single fact and say 'this is a scientific fact, devoid of industrial applications,' and any attempt to do so seems ludicrous."[56]

Curtis concluded that no institutional separation could be maintained between free scientific research and secret, military-oriented bomb research: "If the so-called secret of the atomic bomb is to be kept in this country, then American science as we have known it, will cease to exist." Oppenheimer responded that he could "see no technical difficulty about keeping considerable parts of this secret without interfering in a major way with [research] . . . People have kept military things secret in the past. If they wish to, they can in the future."[57]

Curtis had stressed that the "only . . . solution to the secrecy problem" lay in the establishment of international control of atomic energy. Oppenheimer argued in response that the May-Johnson bill was a necessary stopgap that would allow the continuation of atomic research until the creation of such an international organization.[58]

Oppenheimer's authority in such situations depended on the management of potentially conflicting constituencies. Alice Kimball Smith admiringly referred to Oppenheimer as playing the role of the "very helpful elder statesman," aiding the debate by presenting both sides of the argument.[59] Certainly, this is the image at which Oppenheimer was aiming, one that would transcend the differences between the factions on whose support he relied. There seems to have been some difficulty at the time in figuring out exactly what position he was taking; Oppenheimer's Senate testimony was reported by one newspaper on the following day as an "oblique attack" on the May-Johnson bill.[60]

Oppenheimer's intentionally mixed performance was, however, disappointing to the scientists who had relied on him as their spokesman. His statement that the administration's atomic energy bill was something about which he knew little came as an unpleasant surprise to scientists at Los Alamos: they had previously been led to believe that he was intimately involved in its preparation, and they had been counting on his benevolent influence on its content and direction. Meeting ALAS members after his testimony, Oppenheimer faced—in contrast with his heroic send-off from Los Alamos—what one scientist described as "the coolest reception I have ever seen Oppie given by a group of scientists." At a meeting of the ALAS executive committee on October 25, Weisskopf suggested not only that the group's "future action" be "not based on [the] assumption that [the] administration is with us," but also that "Oppie's suggestions be studied more critically."[61]

The tide was turning against the May-Johnson bill. In addition to the intense lobbying by the atomic scientists, the bill began to face opposition within the administration, fueled by worries about the lack of clear political accountability of the future AEC. The administration put its support behind a competing bill put forward by Senator Brien McMahon of Connecticut. This bill provided for a civilian commission and did not include such heavy-handed security provisions and restrictions on information as had its predecessor.

However, the sense of victory was short-lived: the House managed to introduce so many changes to the bill that, as historian Lawrence Badash wrote, "the product resembled May-Johnson more than it did McMahon."[62] The Atomic Energy Act, signed by the president at the beginning of August 1946, proved to be a Pyrrhic victory for the scientists' movement. Though nominally civilian, the commission incorporated a powerful Military Liaison Committee, and its work, in practice, came to be oriented primarily toward military goals. Historian Michael Sherry observed that "the struggle over civilian control obscured how civilian elites matched the zeal of military officers in pursuing national security."[63]

Oppenheimer was damaged by his support for the May-Johnson bill, but he weathered the storm. His ability to come out relatively unscathed was due primarily to the fact that he had never identified himself wholly with the bill. Rather, he had always emphasized what he regarded as the higher goal of the international control of atomic energy. It was with this transcendent goal that Oppenheimer was most strongly associated by the scientists, as he articulated a vision of a utopian mission for the scientists arising both from the nature of scientific community and from their work on the atomic bomb. On November 2, 1945, Oppenheimer spoke to ALAS members packing the Los Alamos movie theater. His speech wove together their instrumental work on the bomb and a sense of transcendent mission. Oppenheimer's skill here was in connecting the theme of special scientific responsibility with a defense of the administration. As he did so, he sought to present himself as chief mediator of the relationship between science and the state.

In this speech, Oppenheimer set out a conception of scientific vocation that presented scientists' wartime work on the bomb as legitimate and that also suggested a way in which scientists were centrally implicated in, and responsible for, the postwar problems arising from the bomb. He portrayed the role of the scientist as embodying certain inherent values, and he claimed that the atomic bomb was not a breach but a fulfillment of these values. Whatever the individual motivations—whether fear, curiosity, or political principles— that led scientists to engage in building the atomic bomb, he argued that there was a deeper reason for their involvement: "When you come right down to it the reason that we did this job is because it was an organic necessity. If you are a scientist you cannot stop such a thing. If you are a scientist you believe that it is good to find out how the world works; that it is good to find out what the realities are; that it is good to turn over to mankind at large the greatest possible power to control the world and to deal with it according to its lights and its values."[64] Oppenheimer thereby presented science as a *calling*, valued for its own sake.

However, Oppenheimer also sought to present this pursuit of science for its own sake as connecting with more universal human values. He did so by representing the atomic bomb itself as a vehicle for wider human aspirations. While there had "always been good arguments" for overcoming war, and specifically for the organization of a world federation or United Nations organization, Oppenheimer argued that the atomic bomb provided a new urgency and a new opportunity for achieving such goals. Atomic weapons, he said, are a universal problem for humanity, "a peril which affect[s] everyone in the world . . . a completely common problem." Because of that, the bomb was "not only a great peril, but a great hope." In this way, Oppenheimer connected technological determinism with the historical agency of the scientists. And he wove together the roles of pure scientist, bomb builder, and moralist.[65]

In arguing for a special role for scientists, Oppenheimer represented the communal structure of the scientific vocation as carrying universal significance. He argued that atomic energy "is a new field, in which just the novelty and the special characteristics of the technical operations should enable one to establish a community of interest which might almost be regarded as a pilot plant for a new type of international relations." The internationalism of science, as a model for a new international order, would combine with the globalism of the nuclear threat to produce "a new spirit in international affairs."[66]

Oppenheimer balanced the potential radicalism of this line with his now-familiar appeal for trust in the powers that be. Secretary Stimson, he said, shared the scientists' "hope . . . that there would be a new world." He also praised President Truman: "Certainly you will notice, especially in the message to Congress, many indications of a sympathy with, and an understanding of, the views which this group holds."[67] Thus, Oppenheimer set out an image of a revolutionary or messianic role for the scientist in ushering in a new world order, but he tempered this with an appeal for trust in the political establishment. In so doing, he staked a claim to a personal role as conduit between the atomic scientists and the government, connecting the professional ethos of science with the power of the state.

TOWARD INTERNATIONAL CONTROL: SCIENTIFIC MESSIANISM AND TRIUMPHANT AMERICAN LIBERALISM

If Oppenheimer's stance on the domestic politics of atomic energy strained his relationship with his scientific constituency, the issue of international control maintained his status as the legitimate voice of scientists' hopes. Once more he performed a bridging function, this time through his central role in early 1946 on David Lilienthal's Board of Consultants to the State Department. Lilienthal, the former chief of the New Deal organizational and engineering feat the Tennessee Valley Authority, was tasked by Under Secretary of State Dean Acheson with formulating a plan for the international control of atomic energy. In his work on what became known as the Acheson-Lilienthal report, Oppenheimer welded the ideology of the scientists' movement to the policy structure of the U.S. government and channeled the aspirations of the scientists' movement into support for existing political structures.[68] Oppenheimer's influence is clear in the report's statement that the development of atomic energy "may contain seeds which will in time grow into that cooperation between nations which may bring an end to all war" and in the report's invocation of Bohr's ideal of the international Republic of Science: "There can be no international cooperation which does not presuppose an international community of knowledge."[69] Oppenheimer was gratified to receive a letter from Bohr endorsing the report: "In every word of it I find just the spirit which I think offers the best hopes for the development in which we all put our whole faith . . . From page to page I recognized your broad views and refined power of expression."[70] Through his work on the report, Oppenheimer helped to give American foreign policy the aura of scientific legitimacy and to paint this policy as a realization of utopian scientific modernism.[71]

From their first encounter, Lilienthal was particularly impressed with Oppenheimer and acted as his governmental patron from that time on. For good or ill, Oppenheimer came to be closely identified with Lilienthal's regime. Their initial meeting was in a Washington hotel room on January 22, 1946. Oppenheimer, Lilienthal recorded in his diary, "walked back and forth, making funny 'hugh' sounds between sentences or phrases as he paced the room, looking at the floor—a mannerism quite strange . . . I left liking him, greatly impressed with his flash of a mind, but rather disconcerted by the flow of words." Lilienthal's awe at Oppenheimer's intellectual powers soon grew. He described his meeting with Acheson, Conant, Bush, Groves, and Oppenheimer the next day as "one of the most memorable intellectual and emotional experiences of my life" and made particular reference to Oppenheimer again: "The scientist who more than anyone else was able at Los Alamos to find a way to turn the knowledge of nuclear forces into a weapon that shattered the whole world, as we knew it, at Hiroshima; an extraordinary personage (and as I learned today a really *great* teacher)."[72]

To Lilienthal, Oppenheimer embodied the scientific aspects of the atomic bomb project, whose secrets he was now learning for the first time. He was thrilled when, at a meeting of the advisory group, "Oppenheimer talked to us, without limitation (i.e. including some of the top secrets, chiefly scientific discoveries not 'released' about fundamentals)." A few days later, briefed further by Oppenheimer, Lilienthal wrote in his diary, "No fairy tale that I read in utter rapture and enchantment as a child, no spy mystery, no 'horror' story, can remotely compare with the scientific recital I listened to for six or seven hours today . . . I was told well, technically, dispassionately, but interspersed with stories of the decisions that had to be made, the utter simplicity and yet fantastic complexity of the peering into the laws of nature. That is the essence of this utterly bizarre and, literally, incredible business."[73] Lilienthal later told AEC attorney Herbert Marks of his intense admiration for Oppenheimer: "[It] is worth living a lifetime just to know that mankind has been able to produce such a being."[74]

Oppenheimer's command of esoteric knowledge was particularly attractive to Lilienthal, because of his own modernist-technocratic faith in information and expertise. Lilienthal believed that his committee had "an opportunity to analyze what is called a political problem in a scientific spirit . . . We started somewhat as a chemist might, tackling a technical problem: with the facts as he found them."[75]

Oppenheimer likewise urged the "injection of the spirit of the scientists into this problem of atomic weapons." He again drew on the notion of science as a universal culture, in the reach and significance of its knowledge and in the internationalism of scientific community: "Science, by its methods, its values, and the nature of the objectivity it seeks, is universally human."[76] Speaking to a crowd of fifteen hundred at Cornell University in early May 1946, Oppenheimer presented wartime Los Alamos itself as exemplifying the capacity of scientists of different nationalities to work together in a common cause.[77] The universality of scientific culture was, in a sense, realized through the universality of the atomic threat. The bomb was "a new mechanism for altering the political complexion of the world."[78] Oppenheimer wove the atomic bomb into a narrative of progress, the weapon's global destructiveness providing a vehicle for the achievement of universal human ideals. In the same month, he told an audience in Pittsburgh that the only solution to the problem of atomic weapons was an end to war in general, but that the existence of the bomb itself provided an unprecedented opportunity to achieve this: "The atomic bomb, most spectacular of proven weapons, the most inextricably intertwined with constructive developments and the least fettered by private or vested interests or by long national tradition is for these and other reasons the place to start."[79]

Oppenheimer thereby presented the bomb as a medium for transcendence, a realization of modernist dreams of breaching the constraints of history and tradition. The technological achievement of the bomb paved the way for a purely rational solution to international relations. In this way, Oppenheimer mobilized the technological-utopian rhetoric of the scientists' movement to justify the bombing of Hiroshima and Nagasaki. He told a subcommittee of the Senate Committee on Military Affairs in October 1945 that the intensity of the scientists' wartime work was fueled by this philosophy: "We [at Los Alamos] thought that since atomic weapons could be realized, they must be realized for the world to see because they were the best argument that science could make . . . for a more reasonable and a new idea of the relations between nations." Senator Fulbright responded, "In other words, that is one of the justifications for its use. It took the shock, we will say, of Hiroshima, to bring the world to a consciousness of what another war might mean, and it therefore gives the reason for seeing that there is no more war. I think in that sense it does justify its use, regardless of how regrettable it may have been in that particular instance." Oppenheimer replied, "That is in my opinion," and he added, "I know that my colleagues share these views."[80] As he presented the concrete horror of the atomic attacks as a necessary means to the abstract good of pax atomica, Oppenheimer sought to connect the instrumental scientific role of bomb builder with a broader role for the scientist as a representative of moral progress. Through his membership on the Board of Consultants, Oppenheimer could be at once technical adviser to a governmental-bureaucratic committee and spokesman for utopian-humanistic aspirations for a world freed from war.

Oppenheimer's hope that a panel of experts could achieve a purely technical solution to the problems of world order was to be dented by increasingly fraught political conflicts. The appointment by Secretary Byrnes of Bernard Baruch to put forward the Acheson-Lilienthal proposals at the United Nations was recognized by those who had worked on the original report as signaling the impending defeat of their hopes.[81] The seventy-six-year-old retired financier, whom Oppenheimer and Lilienthal referred to as "the Old Man," could not have stood in sharper contrast with the sort of youthful, forward-looking technical elite that Lilienthal saw embodied in Oppenheimer and the atomic scientists.[82]

Oppenheimer and Lilienthal's fears were confirmed when Baruch began to demand changes to their plan. Baruch and his coterie were not in sympathy with the philosophy of the Acheson-Lilienthal report. Herbert Swope, one of Baruch's assistants, called the report "a set of pious platitudes." Swope was particularly annoyed by the scientists' "hoity-toity" talk of the "sanctity and illimitability" of science.[83] Baruch took exception to what he saw as the arrogance of the scientists' view that they "were wiser or more noble than others when it came to dealing with the world's fate." To him, the fundamental questions at stake were political.[84]

Baruch specifically objected to the lack of provision in the Acheson-Lilienthal report for enforcement or sanctions in case of treaty violation, and this became the key difference between the Baruch Plan and the earlier report, which it superseded. Baruch insisted that any agreement include penalties for violation, a detailed schedule for transition to UN control of atomic energy, and, crucially, the removal of the Security Council veto over penalties for treaty violation. The State Department had argued that it was pointless to spell out penalties in this way. If a major nation violated the treaty, the UN would be unable to compel compliance without a war, and talk of penalties would invite rejection by the Soviets.[85]

Any remote possibility of Soviet acceptance of the plan was also safely put to rest by the U.S. atomic tests in the Pacific, carried out for the Navy's Operation Crossroads. The first of the two tests occurred at just over two weeks into the UN negotiations, with a B-29 dropping an atomic bomb on captured Japanese naval vessels off Bikini atoll. The Soviets interpreted the tests as a signal that America's real aim was to maintain its atomic monopoly and as an attempt to put pressure on the negotiations.[86] The tests also weakened the American public's sense of the urgency of arms control, normalizing the atomic bomb as merely another weapon, not an outstanding moral and political problem. Journalist Norman Cousins wrote that "after four bombs, the mystery dissolves into a pattern. By this time there is almost a standardization of catastrophe."[87] When Baruch resigned as the United States' representative in January 1947, the talks at the UN had already become a meaningless exchange of propaganda.[88]

In his journal entries, Lilienthal described Oppenheimer's belief in the importance of the Baruch Plan as a fork in the road between peace and war, and he noted the physicist's consequent despair at the demise of the talks. In July 1946, Oppenheimer had confided in Lilienthal his belief that if the talks ultimately failed, "this will be construed as a demonstration of Russia's warlike intentions. And this will fit perfectly into the planning of that growing number who want to put the country on a war footing, first psychologically, then actually. The Army directing the country's research; Red-baiting; treating all labor organizations, CIO first, as Communist and therefore traitorous etc." Lilienthal recorded in his diary, "He is really a tragic figure; with all his great attractiveness, brilliance of mind. As I left him he looked so sad: 'I am ready to go anywhere and do anything, but I am bankrupt of further ideas. And I find that physics and the teaching of physics, which is my life, now seems irrelevant.'" This, Lilienthal reported, "wrung my heart."[89]

However, this was a dramatic part that Oppenheimer had been prepared to enact. Failure of the UN negotiations had been most likely from the beginning. As early as April, Oppenheimer had told Baruch that he believed the measures called for by the Acheson-Lilienthal report were incompatible with the Soviet system of government. Oppenheimer essentially saw the plan as a way for the United States to do the right thing and give the Soviets a chance to cooperate, even though he strongly doubted that they would. But he apparently did not question why the United States would pursue such a plan. He stopped short of attempting to gain a critical understanding of the power-political interests or diplomatic gamesmanship behind the plan.[90]

In treating the problems of atomic energy as technical ones to which a rational solution could be applied, Oppenheimer and the Board of Consultants abstracted these problems from the asymmetrical nature of atomic power in the immediate aftermath of the war. An unwillingness to confront issues of power (particularly to analyze American power from a realist perspective) was inherent in Oppenheimer's positivistic-technical approach. This turning away from the realities of power politics was also involved in Oppenheimer's characteristic rhetorical appeals to notions of "fraternity" and "community" and his belief that through the bomb, these values could come to form the basis for a new "spirit" in international affairs. He could see power politics at work only in the Soviet rejection of the proposals. The Acheson-Lilienthal report and the Baruch Plan were, he publicly maintained, rational proposals put forward in a spirit of generosity. The desire to present American science, and by extension the American atomic bomb, as transcending the world of power politics strongly informed the scientists' understanding of the Baruch Plan. Their spokesmen reacted swiftly and angrily to any challenges to this understanding.

The most significant such challenge came from British physicist P. M. S. Blackett. Blackett was a pioneer of operational research (OR) during the war, and in 1941 he participated in the writing of the British government's MAUD Committee report on the possibility of building an atomic bomb. Between August 1945 and spring 1947, he served on the Attlee government's Advisory Committee on Atomic Energy. In 1948, he received the Nobel Prize for Physics for confirming the existence of the positron. His book *Fear, War, and the Bomb,* first published in 1948 and released in America a year later, attacked the sacred canons of the American scientists' movement one by one, beginning with the official interpretation of Hiroshima.[91]

The American scientific elite largely supported the official account presented by Henry Stimson: that the bombing of Hiroshima had ended the war, saving in the process between half a million and one million American and Japanese lives—a figure that was plucked from thin air and that survives as a powerful myth.[92] Blackett argued that on the contrary, the purpose of the atomic bombings had been not only to end the war in the Pacific but also, and even more importantly, to end the war on American terms by preempting the agreed-on date for the Soviets' entry into the war against Japan. He concluded that "the dropping of the atomic bombs was not so much the last military act of the second World War, as the first major operation of the cold diplomatic war with Russia now in progress."[93]

Blackett's analysis of Hiroshima could not have been more unwelcome to the U.S. government and American scientists.[94] Stimson's version—that the atomic bomb saved both American and Japanese lives—was conducive to the scientists' self-image as universalistic representatives of human, rather than national or sectional, interests. The central canon of the scientists' movement was that the atomic bomb was not merely another weapon, but a new, transcendent force leading to the end of war altogether. Blackett's analysis, on the contrary, suggested that the atomic bomb was born and first used precisely as an instrument of machtpolitik (power politics) and that it signaled not a new era of peace, but the beginning of the Cold War.[95]

The power-political motives behind the atomic bombings were obscured not only by Stimson's official justificatory rhetoric. Blackett thought that even when American scientists started to have doubts about the validity of the official account, they opted for blanket pessimism rather than realistic analysis. Because they found the idea that the bombs were used to "win a diplomatic victory" to be "too morally repugnant to be entertained, [their] only remaining resort is to maintain that such things just happen, and that they are the 'essence of total war.'"[96]

As an example of this pessimistic "essence of total war" thesis, Blackett quoted a statement in which Oppenheimer seemed to depart from the official U.S. account. Oppenheimer had written that "every American knows that if there is another major war, atomic weapons will be used . . . We know this because in the last war, the two nations which we like to think are the most enlightened and humane in the world—Great Britain and the United States—used atomic weapons against an enemy which was essentially defeated." The statement that Japan was "essentially defeated" before the bombs were used seemed to put the event in a different light than did the Stimson version. But to Blackett, even this statement of Oppenheimer's did not demonstrate a real understanding of the reasons for the atomic bombings. Instead, he interpreted what Oppenheimer had said as suggesting that there was no real reason for the bombings, that they simply followed from the pursuit of total war, and therefore that such bombs would be used again by any side possessing them. This was a view that created "an atmosphere of imminent world destruction . . . in which clear thinking was at a discount and emotion triumphant." Oppenheimer's statement, Blackett said, exemplified the kind of "belief that provides the breeding ground for hysteria."[97]

Blackett presented his own view as being "in decisive contrast" both to Stimson's official explanation and to Oppenheimer's pessimistic argument. For Blackett, it was clear that the bombs were dropped "for very real and compelling reasons—but diplomatic rather than military ones." He saw a diplomatic war that had, in a sense, been initiated with the atomic bombings and was now being continued at the arms control negotiating table.[98]

It followed from Blackett's analysis that it was impossible to take seriously the image of universalism and generosity surrounding the American proposals to the UN in 1946. The Baruch Plan, for Blackett, was thinly disguised atomic diplomacy. And Blackett focused criticism not only on the U.S. delegation, but also on the original Acheson-Lilienthal report (drafted by Oppenheimer), on which the American proposal was based. An "essential asymmetry as between America and Russia," Blackett noted, was "inherent in the early stages of the Plan." The Soviet Union, he said, was "keenly aware of the immediate danger to her military security and the long-range danger to her economic development underlying the idealistic phraseology of the Lilienthal Plan." Even though it drew on the language of idealistic internationalism, Blackett argued, "support for the Baruch Plan falls into place as a consistent part of the Anglo-American policy of 'containing' Communism at all possible points."[99]

Blackett maintained that the American scientific elite had constructed an ideology that both legitimated American policy and obscured the realities of the place of atomic weapons in global politics. The scientists formed part of the group of "idealists and liberals" who "sincerely believed the [Baruch] Plan to be so equitable and even generous that its rejection by the U.S.S.R. could only be attributed to their willful neglect of their own self-interest . . . Such people, on seeing the U.S.S.R. reject these 'generous' proposals, tended in many cases passionately to implore the Russian leaders to realize the danger in which they stood from American atomic bombs." The aura of impartiality, universalism, and objectivity that the atomic scientists constructed for themselves masked the extent to which they were merely supporting American interests in the emerging Cold War.[100]

Oppenheimer's close colleague Isidor Rabi led the reaction against Blackett in the American scientific community. Dismissing Blackett's view that Hiroshima was the first act of the Cold War, Rabi went on to defend the American position at the UN. The Baruch Plan, Rabi said, was a "great and generous gesture. It was an offer to surrender our greatest weapon of military power in the interest of the security of all nations." The only reason for the Soviets' position was "the Original Sin of Communism, the intrinsic inability of a totalitarian state to withstand impartial inspection from outside," as well as the Kremlin's sheer "ignorance" and "cussedness."[101]

Oppenheimer did not himself respond publicly to Blackett.[102] But he never developed the type of radical critique of American atomic diplomacy that the

British physicist and socialist developed. While Oppenheimer saw Baruch's punitive approach as dooming the UN negotiations and was also dismayed by the decision to go ahead with the Bikini tests, his critique was limited by the political and intellectual constraints that he accepted as a scientific-governmental insider.[103] Furthermore, his perspective on the arms control negotiations was framed overall by his view that the "open world" that would be required for success on this front was incompatible with the Soviet system. As he focused his blame for the failure of the Baruch Plan on Soviet obstructionism, Oppenheimer's internationalist liberalism began its mutation into Cold War chauvinism.

Oppenheimer was now enlisted in the program, announced by Truman in the spring of 1947, of containing Communism. The physicist was present at the Harvard commencement on June 5, 1947, when his hero, General George C. Marshall—now secretary of state—announced his economic plan to bolster the Western European democracies against the Communist threat. Historian Frances Stonor Saunders observed, "It was no coincidence that [Marshall] had decided to deliver his speech here [at Harvard], rather than on some formal government podium. For these were the men assigned to realize America's 'manifest destiny,' the elite charged with organizing the world around values which the Communist darkness threatened to obscure."[104] This was a mission, with its connotations of American benevolence and idealism, to which Oppenheimer felt privileged to subscribe.

In September 1947, speaking to an audience of military officers as well as officials in the Foreign Service and the State Department, Oppenheimer painted a picture of Western reasonableness struggling against Soviet intransigence. He blamed the failure of the Baruch Plan on the fact that the "cornerstone of our proposal is an institution which requires candidness and great openness in regard to technical realities and policy." In a strong echo of diplomat George F. Kennan's anonymous July 1947 article in *Foreign Affairs,* "The Sources of Soviet Conduct," Oppenheimer asserted that the cooperative and open pattern of control embodied in the American UN proposals stood "in a very gross conflict to the present patterns of state power in Russia, namely the inevitability of conflict between Russia and the capitalist world."[105]

Oppenheimer contrasted the closed society of the Soviet Union with the political openness embodied both by American democracy and by the arms control structures of the Acheson-Lilienthal report, themselves modeled on the open structure of science. He thus came to portray the American republic and the Republic of Science as morally identical. Science depends, he argued, on the minimization of both secrecy and coercion, and those are ideals that "are very deep in our ethical as well as in our political traditions, and are recorded in earnest, eloquent simplicity in the words of those who founded this nation." Oppenheimer considered it a great puzzle of World War II that "the atomic bomb, born of a way of life, fostered throughout the centuries, in which the role of coercion was perhaps reduced more completely than in any other human activity, and which owed its whole success and its very existence to the possibility of open discussion and free inquiry, appeared in a strange paradox, at once a secret, and an unparalleled instrument of coercion." How could science have begot the atomic bomb? For Oppenheimer, this genealogy was nothing more than a bizarre "paradox." The bomb was spawned by science, but its nature was alien to science. Even when he described the atomic bomb as an instrument of power, he distanced both his profession and his nation from the bomb's violence. So, despite Hiroshima, Oppenheimer was able to assert that Americans are "stubbornly distrustful" of the use of power in foreign affairs and that "we seem to know, and seem to come back again and again to this knowledge, that the purposes of this country in the field of foreign policy cannot in any real or enduring way be achieved by coercion."[106]

The representation of the scientific community as an "open society," an image mobilized by scientists immediately after the war in appeals that state patronage be combined with professional autonomy, was now rhetorically attached to the Cold War dualism of Western enlightenment versus the darkness of the Eastern bloc. As Oppenheimer represented scientific reason as antithetical to Soviet Communism, he sought to place an image of science at the center of American national ideology. In so doing, he was solidifying his own position as representative of a civilian scientific elite in a strategic alliance with the national-security state.[107] By presenting science itself as an American value to be defended against the Soviets, Oppenheimer strove to bolster his and other scientists' authority within the polity. He suggested that the national-security state should sponsor scientists not only for their instrumental function as builders of the nation's atomic arsenal, but also for their ideological or legitimatory function in embodying and articulating what it was that that arsenal was in place to defend. Cultivating science would mean not only winning the technological and strategic war; Oppenheimer suggested that it was also, and perhaps more importantly, the key to winning the ideological Cold War. It was by presenting itself as the righteous defender of Western civilization, Oppenheimer suggested, that America could win allies in this ideological struggle: "We want the intellectuals of Europe to be friends of the United States."[108]

This ideological mobilization of science meant that paradoxically, as more and more of America's scientists embraced a narrowly instrumental militaristic role, and as their science was carried out in secret under the patronage of the military, this scientific elite increasingly proclaimed the openness of science as being at the heart of what separated America from its Communist enemy. There was also a pragmatic reason for this rhetoric. While scientists collectively embraced military funding and often subscribed to military goals, they were nervous of being controlled by the military. They valued their social status as professionals, a status that was in part connected to an image of "pure science" and to the degree of autonomy that this involved. Oppenheimer argued that this autonomy was necessary for the vitality of science: the scientific community "must not be sewed up so tightly that it is not a part of the living culture and development of the country." He advocated the combination of state patronage with support for "basic research" in a university setting. This might, in the short term, reduce the numbers working directly on atomic armament, but in the long run it was important in "cultivating a corps of people" with the knowledge and skills necessary to ensure "a heroic future" for America in atomic energy.[109]

Oppenheimer, therefore, sought to incorporate a reverence for "pure science" into American Cold War culture. As he embraced the Cold War, he also attempted to preserve within this new political context the combination of patronage and autonomy from political accountability that the scientific community had achieved after the war. America's civilian scientific elite hoped to fend off the bugbear of "military control" by adopting the military's goals as their own and by presenting themselves as the most qualified to advance these goals. In the late 1940s, the military was the main financial supporter of American science. And even after the establishment of the National Science Foundation (NSF) in 1950, Department of Defense support for scientific research dwarfed NSF budgets.[110] Even as they proclaimed their autonomy, America's scientific elite became increasingly subservient to the military goals of the Cold War and the arms race.[111]

REDEFINING THE SCIENTIFIC ROLE: WEAPONS AND "RESPONSIBILITY"

A new pessimism and self-doubt became evident in the *Bulletin of the Atomic Scientists* between 1947 and 1948. One article noted the paralysis of the scientists' movement in the wake of the failure of the Acheson-Lilienthal report and put it down to "the schematic and over-simplified one-world-or-none

reasoning," which led to the scientists' being "trapped by their own logic."[112]
The scientists had argued that Hiroshima was a point of decision, presenting
either an opportunity for an international regime leading to the elimination
of war, or a path to an atomic arms race and atomic brinkmanship, leading
most probably to a third world war. The failure of the goals of the Acheson-
Lilienthal report signaled clearly that the first route had been missed (had it
ever really existed) and that the United States was heading inexorably along
the other path. As Oppenheimer put it, "the jig was up." After the excursion
into politics, it was now time for "getting back to [scientific and technical]
work."[113]

Oppenheimer's advice was that the scientists should accommodate themselves to this new reality. While they should not give up the hope for arms
control as an ultimate goal, he argued that they could no longer advocate or
work toward it in the same way that they once had. The atomic scientists,
he said, must recognize that their role was no longer that of "the prophets of
doom coming out of the desert, but rather that of a group of specialized and,
in their way, competent, men who must be sensitive to all avenues of approach
which are hopeful and who are after all intellectuals and not politicians."[114]
The role of the intellectual, in Oppenheimer's account, was a stringently
restricted one. Oppenheimer's distinguishing of the "intellectual" from the
"politician" now mandated a retreat from the polis. He appealed to an image
of a state of intellectual purity that, though presently lost, could in principle
be recaptured.

Oppenheimer took the opportunity of a public lecture at MIT, in late November 1947, to outline his new understanding of the responsibilities of scientists. The speech, titled "Physics in the Contemporary World," marked
an important turning point in his public rhetoric. This new conception of
the scientist's role lacked any of the utopianism that had characterized his
speech to ALAS two years before. Scientists, he now said, are not qualified
to solve humanity's problems: "The study of physics, and I think my colleagues in the other sciences will let me speak for them too, does not make
philosopher-kings. It has not, until now, made kings. It almost never makes
fit philosophers." Oppenheimer spoke of "how much the applications of
science . . . have cast in doubt that traditional optimism, that confidence in
progress, which have characterized Western culture since the Renaissance."
The scientist could offer no general route to salvation. The world had proved
itself too corrupt to be remade in the scientific image. And he represented the
scientific engagement with this world, in war and the atomic bomb, as a fall
from grace. In facing the reality that they could offer no universal salvation,

scientists should, Oppenheimer suggested, focus on their own personal salvation. In other words, they should attend to the special duties of their vocation: "The true responsibility of the scientist, as we all know, is to the integrity and vigor of his science. And . . . [scientists] have a responsibility for the communication of the truths they have found . . . That we should see in this any insurance that the fruits of science will be used for man's benefit, or denied to man when they make for his distress or destruction, would be a tragic naïveté."[115] The "responsibilities" of the scientist were, Oppenheimer argued, purely vocational and strictly delimited. In performing these duties, the scientist fulfilled a personal calling. But that should not be mistaken for offering any comfort to humankind as a whole.

It was an appeal to an image of lost innocence. For Oppenheimer, the embodiment of this innocence was the Göttingen mathematician David Hilbert. To the question of the relationship between science and technology, Hilbert had answered, "*Sie haben ja gar nichts mit einander zu tun.* They have nothing whatever to do with one another."[116] While Oppenheimer recognized that it was no longer possible to accept such a nonchalant dismissal of the question, there was something in Hilbert's attitude that he wanted to recapture, reformulate, and revitalize. It could not now be denied that there was a powerful relationship between science and technology. For example, the demand of modern societies for technology was, Oppenheimer knew, a central reason for the social support of science. But in his view, it was crucial to recognize the essential differences between the two enterprises. For "no scientist, no matter how aware he may be of these fruits of his science, cultivates his work, or refrains from it," merely because of the technological and social benefits or problems that might result from the scientific exploration. The individual scientist's "compelling motive" for his work was quite different from any "social justification" for that labor. Oppenheimer's argument was that even today, the individual motive should differ little between the contemporary scientist and Hilbert. The purity of soul of this mathematician could be recaptured through an understanding of science as a calling.[117]

In order to recapture this spirit, however, scientists had to confront the recent history of the war. A recognition of sin was the first step on the path to redemption. Hence Oppenheimer's famous admission and admonition that "in some sort of crude sense, which no vulgarity, no humor, no overstatement can quite extinguish, the physicists have known sin; and this is a knowledge which they cannot lose." This was, of course, a reference to Hiroshima and Nagasaki. But while Oppenheimer admitted that the bomb was a product of physics, he also insisted that there was a sense in which they should be recognized as separate. Crucially, in Oppenheimer's rendition, bomb building was not physics. The physicists were acting as engineers, merely applying physics. In "the last world war . . . the demands of military technology . . . distracted the physicists from their normal occupations." The war created "a great gap in physical science," from which science had only begun to recover.[118]

When Oppenheimer spoke of "sin," he had in mind this abandonment, however necessary, of physics as a vocation. *Time* magazine expressed what was the most common interpretation of Oppenheimer's words "the physicists have known sin" when it said, "As if to expiate this sense of sin Oppenheimer threw himself into the campaign for international atomic regulation."[119] While Oppenheimer allowed this ambiguity, that foray into politics was, in his characterization of sin, as much part of the sin as was the building of the bomb. Certainly the physicists could not have done otherwise than give their aid in time of war, and Oppenheimer drew a parallel between scientists working at Los Alamos and physicists in Europe who had joined the Resistance. (He did not mention the Nazi bomb project under Werner Heisenberg.)[120] But the price that they paid in leaving their laboratories and desks for the world was a loss of purity. This purity could be recaptured by a revitalized understanding of the meaning of physics as a vocation. Such an understanding would mean attending to their vocational duties, but also recognizing the limits of these responsibilities.

Paradoxically, however, this idea of recapturing a vocational purity could legitimate the instrumental role of the scientist in the service of the national-security state. Such an ethos has one meaning in a situation in which the institutional and social basis of scientific research is independent and insulated from institutions responsible for industrial and military technology. But this appeal to an idea of "purity" had quite another meaning and set of implications after World War II, when the institutional separation of academia from industry and from the military had been considerably eroded, when university research was funded by the Office of Naval Research and other military bodies, and when a sophisticated research and development organization was in place to rapidly convert scientific findings into technological applications. It was in this radically new setting that Oppenheimer sought to invoke a traditional notion of "pure science," retreat from the world, and a narrow framework of vocational duties. However, in this new context, a formulation that divorced the value of science from questions of the ends to which it would be applied had the effect of legitimating a narrowly instrumental function for the scientist.

Oppenheimer's call for a retreat to purity was in fact a capitulation to the demands of the national-security state.

THE ATOMIC "EXPLOSIVES" COMMISSION

Oppenheimer's increasingly restricted conception of scientific responsibility meshed with the narrowing of political possibilities as the Cold War intensified. It was as chairman of the AEC's General Advisory Committee (GAC), the nation's most powerful scientific advisory body, that Oppenheimer responded in practice to the new constraints and attempted to act on his restricted conception of the scientific role. He was elected to the chairmanship at the GAC's first meeting in January 1947 and held that position until the summer of 1952. In June 1947, Oppenheimer told the commission that since scientists were years away from achieving the technology for civilian nuclear power plants, there was currently very little for the AEC to concern itself with but weapons: "energy" would mean "explosives."[121] This was "quite a blow" to Lilienthal's hopes of developing an open and civilian-oriented atomic energy program. It signaled that the AEC would not easily be able to break its dependence on the military, nor escape from the shroud of secrecy. In late 1947 Oppenheimer said, "Without debate—I suppose with some melancholy—we concluded that the principal job of the Commission was to provide atomic weapons and good atomic weapons and many atomic weapons." Asserting the GAC's commitment to this task, Oppenheimer dismissed the internationalist hopes of the scientists' movement as mere nostalgia for an irretrievable non-nuclear world: "It must be recognized that within our hearts we have been hoping that the world will be the world it was ten years ago. This is no longer possible."[122]

Oppenheimer now responded with outright skepticism to some scientists' ongoing efforts for arms control. Turning down a request from physicist Harrison S. Brown that he attend a conference of the world-government-oriented Emergency Committee of the Atomic Scientists, Oppenheimer said that the proposed date conflicted with his commitments at Berkeley and Pasadena. He then added dryly:

> It will of course be clear to you that in matters potentially so important for our common hopes, no engagement, no matter how firm and how urgent, can be allowed to stand in the way. You will therefore inevitably interpret my reluctance to cancel my California commitments as a certain expression of doubt as to the helpfulness of the proposed conference. I think it is only right to make that expression explicit.[123]

Oppenheimer could barely conceal his contempt for what he saw as the self-indulgent and naive idealism of this sort of political organizing: it was time to give up political distractions from the pure vocation of the scientist.

Nevertheless, he still clung to the belief that even if international arms control was not possible in the foreseeable future, it was the only way forward in the long run and the only ultimate source of hope for the future. Therefore, he believed, the strategy of containment should not include measures that would tend to close off the possibility of one day choosing a more peaceful direction. The contrasting position, represented by Edward Teller, was that there was no foreseeable alternative to containment through nuclear strength. This difference was at the heart of the debate over the hydrogen bomb from 1949 into the early 1950s.[124]

On August 29, 1949, the Soviet Union carried out its first test of an atomic weapon. The test was detected by a U.S. Air Force B-29 flying near Japan, which picked up unusually high levels of radioactivity. On September 23, Truman announced the end of America's nuclear monopoly. The GAC was burdened with the heavy responsibility of deciding the U.S. response. With Oppenheimer on the GAC were the wartime science administrator and Harvard president James B. Conant, Caltech president Lee DuBridge, physicists Enrico Fermi and Isidor Rabi, metallurgist Cyril Smith, Bell Telephone Laboratories president Oliver E. Buckley, and the chief engineer and vice president of United Fruit Company, Hartley Rowe. Another member, Glenn Seaborg, a chemist and the co-discoverer of plutonium, was in Sweden on a lecture tour during the crucial meetings at which the GAC formulated its reaction.[125]

The Soviet bomb brought into sharp focus competing visions of American military strategy and defense. The Strategic Air Command (SAC) and a powerful lobby of scientists, including Ernest Lawrence, Wendell Latimer, Luis Alvarez, and Edward Teller, advocated strategic bombing and the search for ever more powerful weapons as the twin cornerstones of American military and defense policy. From 1942, Teller had been interested in the possibility of a "superbomb," the basic principle of which would be to use a regular atomic bomb to set off a fusion reaction in deuterium. The "Super," or hydrogen bomb, promised a weapon that could release one hundred to one thousand times more energy than existing fission weapons and could damage an area twenty to one hundred times larger.[126]

The Soviet bomb gave Teller and the H-bomb lobby the argument they needed to push through their program. Ernest Lawrence announced that there was "nothing to think over" and said that the H-bomb called for "the spirit of Groves." Lilienthal, on the other hand, was sickened by what he saw as the "bloodthirsty" attitude of these scientists.[127] The GAC members were likewise loath to take what they saw would be a decisive, and most probably irreversible, step toward nuclear escalation, yet they were split on how to address the question. The most adamant in his opposition to the H-bomb was Conant, and it is likely that this strengthened Oppenheimer's nerve to come out against the new weapon.[128] In his diary entry for October 29, 1949, Lilienthal recorded that Conant was "flatly against" the bomb and that Oppenheimer was "inclined that way." Both Rabi and Fermi saw the attempt to develop the weapon as inevitable, and Fermi saw it as a scientific duty. Lilienthal rebutted the idea that there was any such imperative: "I deny there is anything inevitable about political decisions."[129]

The GAC's report of October 30, 1949, took a stand against development of the H-bomb. Instead, the advisory committee advocated expanding the capacity to produce fission bombs, particularly tactical nuclear weapons. While forecasting that "an imaginative and concerted attack on the [H-bomb] problem has a better than even chance of producing the weapon within five years," the GAC opposed a crash program or "all-out effort" for its development: "We all hope that by one means or another, the development of these weapons can be avoided." They advocated making a "commitment not to develop the weapon," though they disagreed over whether this commitment should be unqualified or conditional on the course of action taken by the Soviets. The majority urged an unqualified commitment not to develop the weapon, while Rabi and Fermi wrote their own minority statement that made the more limited call for the United States not to "initiate" development of the bomb.

The GAC's opposition to the H-bomb was extremely cautious. Only in addenda did they deal with their moral and political objections to the weapon, separating these from the more dry and technical language of the body of the report. The majority addendum stated the fear that "a super bomb might become a weapon of genocide." And despite their more resigned sense that the weapon's development was probably inevitable, Rabi and Fermi reiterated this claim, adding that the Super was "an evil thing considered in any light."[130]

The GAC recommendation did not, however, end pressure for the H-bomb. The committee's report was countered by one from the Joint Chiefs of Staff, which declared that "the United States would be in an intolerable position if a possible enemy possessed the bomb and the United States did not." It was this military reasoning that was most persuasive to Truman, and on January 31, 1950, he announced the decision to pursue an H-bomb program. According to officials close to him, Truman had already made up his mind in favor of developing the weapon, on the basis that the Soviet Union only respected military strength.[131]

Less than a month after the GAC report on the hydrogen bomb, Lilienthal submitted to the president his resignation as AEC chairman. This was precipitated by his growing disenchantment with the militarism and secrecy that dominated the AEC's function and by his political isolation in his opposition to the H-bomb.[132] In the weeks after the GAC report, the *Rochester Democrat and Chronicle* reported that Oppenheimer and Lilienthal had "developed a noticeable coolness" and that "Oppenheimer is said to have bluntly advised Lilienthal that the best thing he could do for the future of the atomic program would be to resign as AEC chairman."[133]

President Truman, in response to Lilienthal's resignation letter, acknowledged that the AEC chairman had been "under tremendous pressure and often under destructive criticism." The president added, "Yours was the task to solve a problem which presented the strange anomaly of secrecy in a public undertaking."[134] The handling of the H-bomb issue was indeed framed by secrecy. In sharp contradistinction to the public campaign of the scientists' movement in 1945–47, the debate over the H-bomb was a closed affair within the policy elite; it took the publication of the transcript of the Oppenheimer hearings in 1954 to open the dynamics of this debate to public view.[135] The fact that the debate was shrouded in secrecy helped moral concerns to be subordinated to raison d'état and power politics.

After Truman announced his decision to pursue the hydrogen bomb, Lilienthal said that the AEC was now "nothing more than a major contractor to the Department of Defense."[136] Isolated from any broad constituency, and now without Lilienthal as a defender and ally, Oppenheimer adjusted himself to the new configuration of institutional forces. He stood down from his original outright opposition to the weapon, as did other scientists who had previously voiced moral opposition. But exactly what stance he took after the presidential directive was a key matter of controversy during the security hearings. It was later alleged that Oppenheimer continued to surreptitiously oppose the H-bomb, and even that his "insufficient enthusiasm" was responsible for delaying the program. Others presented a different view. According to Gordon Dean (the AEC chairman between 1950 and 1953), Oppenheimer "very actively" participated in discussions about and calculations for the new design. He was "enthusiastic now that you had something foreseeable . . . He was, I could say, almost thrilled that we had something here that looked as though it might work."[137] Whatever Oppenheimer's feelings about the

H-bomb, he no longer expressed direct opposition to development of the weapon. And he famously told the 1954 hearings that the Teller-Ulam design, developed in 1951, was "technically so sweet that you could not argue" about it.[138] It appears that Oppenheimer was ambivalent between, on the one hand, his moral objections to the Super and, on the other, both his fascination with the technical problems of its design and his likely unwillingness to stand in opposition to an executive decision.[139]

Oppenheimer resolved the dilemma by choosing a third way, throwing his weight behind the parallel development of tactical nuclear weapons. Through the early 1950s, he played an important role in supporting the Army's desire for a tactical nuclear capability, involving warheads for short-range rockets, missiles, and artillery. He became closely involved in Project Vista, a large-scale scientific and technical study of tactical nuclear weapons begun by Caltech in early 1951.[140]

Oppenheimer supported tactical nuclear weapons as an alternative to strategic bombing. Paradoxically, however, in so doing, he was engaged in the normalization of atomic warfare. His new position contrasted sharply with his statements in the aftermath of Hiroshima that atomic weapons called for a radical new way of thinking and a new world order, and could not be considered to be just any other weapon. In Project Vista, Oppenheimer was now deeply engaged in the problem of how to fit atomic weapons into the framework of conventional warfare, how to make them into usable military weapons. He motivated Project Vista with the question, "What contribution may one reasonably hope that the atom can make to our military power, the power for the prevention of war, the limitation of war, *and for the defeat of the enemy in the event that war does come?*"[141] In 1946, Oppenheimer had said that atomic weapons "are not police weapons . . . They are themselves a supreme expression of the concepts of total war." In 1951 he said, "It is clear that they [atomic weapons] can be used only as adjuncts in a military campaign which has some other components, and whose purpose is a military victory. They are not primarily weapons of totality or terror, but weapons used to give combat forces help that they would otherwise lack."[142] According to Rabi, Oppenheimer during this time was becoming increasingly "inclined toward a preventive war."[143]

An indicator of Oppenheimer's influence on military policy came when he flew to Europe in December 1951 for a meeting with General Dwight D. Eisenhower. The press reported that he was passing on information from nuclear tests in Nevada of small-scale tactical atomic bombs. The *Herald Tribune* said, "The development of atom bombs for use against troops in the field has been regarded in high American quarters as an Allied trump card in event of a Russian attempt to sweep westward to the English Channel. Dr. Oppenheimer's mission was understood to be a briefing of American generals on just what they could expect, so the information can be incorporated in their defense planning."[144] The introduction of tactical nuclear weapons to NATO, advocated by Project Vista, became the cornerstone of the Army's nuclear doctrine. At the same time, the development of tactical atomic weapons complemented rather than competed with the strategic nuclear program of General Curtis LeMay and SAC. In the judgment of historians, "the acquisition by NATO of a tactical nuclear capability did not, as it turned out, materially impede SAC's capabilities. Indeed, in some respects it promised to enhance them. Nor did it lessen reliance on strategic forces, as Oppenheimer and others who worked on the Vista study hoped it would."[145]

Freeman Dyson gave considerable critical attention to Oppenheimer's role as "scholar-soldier" in advocating the development of tactical nuclear weapons. Writing during the mid-1980s, Dyson reflected that "Oppenheimer's efforts to sell tactical nuclear weapons to the army succeeded all too well . . . The six thousand NATO tactical warheads now in Europe are an enduring monument to Oppenheimer's powers of persuasion."[146] Oppenheimer's advocacy of tactical nuclear weapons was, in part, an attempt to employ interservice rivalry between the Army and Air Force in the attempt to counter the H-bomb program and SAC's doctrine of strategic bombing. But it also went deeper, expressing an important feature of Oppenheimer's changing orientation as a scientist. Dyson argued that in the immediate aftermath of World War II, Oppenheimer briefly "transcended his role as an American weaponeer and became for a while an international statesman, a spokesman for world scientific community." However, with the failure of efforts for international arms control, the Soviet atomic test of 1949, and the outbreak of the Korean War in 1950, his moral and political "horizons" became increasingly "narrow." By the early 1950s, he was "again, as in the Los Alamos days, a good soldier committed to the service of his country's military strength."[147]

The ethic of scientific responsibility that Oppenheimer formulated in his writings and speeches, from 1947 onward, was close to a soldierly ethic of duty. Abandoning the utopian dreams of the early scientists' movement, Oppenheimer denied that science could be a vehicle for transforming worldly reality. Instead, he argued that scientists had to adapt themselves to that reality, attending to the demands of the day. Scientists were not prophets but specialized technical experts. And once again, Oppenheimer set about placing his technical expertise at the disposal of the military, in pursuit of tactical nuclear weapons.

Yet there is a fundamental ambivalence and ambiguity here. On what grounds was Oppenheimer choosing between tactical and strategic weapons? Was this decision purely a technical one? Or, as was later widely alleged, was Oppenheimer's stance against the hydrogen bomb based on a fundamental moral position, in which he set himself against state policy and military-technological momentum? Oppenheimer's conception of the vocational responsibility of the scientist "to his science" was of little use as a guide to how to address the complex issue of the hydrogen bomb, in which technical, military and strategic, and political considerations were tightly interwoven. Teller presented his single-minded pursuit of the H-bomb design as an exercise of his vocational duty.[148] For Teller, the same conception of scientific responsibility set out by Oppenheimer in his 1947 speech "Physics in the Contemporary World" made pursuit of the hydrogen bomb not only ethically acceptable, but a positive duty.

Was the hydrogen bomb, however, a logical development and extension of existing science, apart from the consideration of any substantive goals? The advocates of the hydrogen bomb presented it as a natural step in two related senses: first, as a scientific and technical step; and second, as a strategic step in the arms race. On this view, the natural trajectory of both weapons research and the arms race was simply toward bigger, more powerful bombs. In its recommendation against a crash program, the GAC sought to disrupt this image of natural progression. How, and whether, to respond to the Soviet bomb was a choice. The crux of the GAC's position, then, was that the "technological imperative" did not lead deterministically to the hydrogen bomb, but instead was contingent. It was therefore possible, and indeed necessary, to consider broader ends, whether military and strategic, political, or moral. In political terms, the GAC suggested that if the United States refrained from development of the H-bomb, that might encourage the Soviets to do the same. Faced with a choice, the committee argued, the government should forgo the destructive quantum leap of the H-bomb.

That advice, however, went beyond Oppenheimer's narrow conception of scientists' responsibilities as being to their science. In particular, the addition to the report of moral concerns about the genocidal nature of the weapon expressed precisely the sort of concern for humanity that Oppenheimer had earlier defined as superfluous to science. Although the moral and political arguments in the GAC report were published in appendixes, it is clear that the committee felt that its response to the H-bomb issue should not be limited to technical issues. Physicist Herbert York argued that moral and pragmatic objections to the Super were necessarily connected: "If either of these elements had been missing they would not have made the recommendation they did. If they had felt it was essential for national security they would have overcome (although probably with considerable regret) their ethical reservations. And if they had not been concerned about its excessive power they would not have so concerned themselves with the question of whether or not it was essential."[149]

Oppenheimer was faced with a complex moral and political decision of the kind for which a soldier's ethic of "duty" offered little guidance. Unlike a soldier, he was not locked into a chain of command; he was, rather, forced to negotiate a position within a shifting, conflict-ridden, and unstable political terrain. But his failure to formulate and articulate a consistent and coherent conception of what *substantive* moral values the scientist should preserve and defend meant that his action in relation to nuclear weapons ultimately collapsed into narrow pragmatism. He was unable to sustain a principled rejection of the H-bomb, and his antagonism to this weapon merely led him to expedient support of tactical weapons as alternatives. What was lacking, and what Oppenheimer's instrumental ethic of vocation prevented him from developing, was a consistent moral opposition to the militaristic appropriation of science and its application to destructive ends.

PLATE 1   The "long-haired" professor. Robert Oppenheimer (*left*) wearing jeans and a silver Santa Fe–style belt buckle while visiting his New Mexico summer ranch, Perro Caliente, with Ernest O. Lawrence in the early 1930s. Courtesy of American Institute of Physics, Emilio Segrè Visual Archives.

PLATE 2   Photograph for Oppenheimer's Los Alamos identification card. Courtesy of Los Alamos National Laboratory.

PLATE 3 Oppenheimer as director of the wartime Los Alamos Laboratory. He was "intellectually and even physically present at each significant step," and his "continuous and intense presence . . . produced a sense of direct participation" in the scientists working there. Photograph from *Bulletin of the Atomic Scientists;* courtesy of American Institute of Physics, Emilio Segrè Visual Archives.

PLATE 4 "Like a good host with his guests": Oppenheimer at a party in Fuller Lodge at Los Alamos, ca. 1945. Courtesy of Los Alamos National Laboratory.

PLATE 5 Los Alamos main gate. New arrivals at Los Alamos were fingerprinted and photographed and had to sign the Espionage Act. Courtesy of Los Alamos National Laboratory.

PLATE 6   A cartoon distributed at Manhattan Project sites during 1944. The purpose of the image, according to General Leslie R. Groves, was "to impress on all concerned the necessity of making decisions promptly and that a 'safety first' policy would insure defeat." Correspondence and Related Papers on the MED, 1942–70, Papers of General Leslie R. Groves, National Archives and Records Administration.

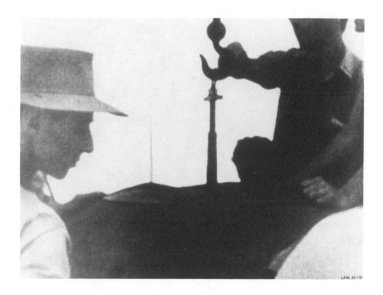

PLATE 7   During preparations for the Trinity atomic bomb test at Alamogordo, New Mexico, Oppenheimer is silhouetted against the bomb. Courtesy of Los Alamos National Laboratory.

PLATE 8   Oppenheimer and General Leslie R. Groves at the Trinity site. In the foreground are the remains of the tower on which the atomic bomb was mounted. Courtesy of Los Alamos National Laboratory; photograph by United Press International.

PLATE 9 "That day he was us": Oppenheimer (*left*) accepting the Army-Navy
Award for Excellence on behalf of Los Alamos Laboratory, October 16, 1945.
To Oppenheimer's left are General Leslie R. Groves, University of California
president Robert G. Sproul, and Commodore William S. "Deak" Parsons. Courtesy
of J. Robert Oppenheimer Memorial Committee, Los Alamos National Laboratory.

PLATE 10 "Father of the atomic bomb": Oppenheimer (*right*) with Henry D.
Smyth and General Kenneth D. Nichols. Courtesy of American Institute of Physics,
Emilio Segrè Visual Archives; photograph by United Press International.

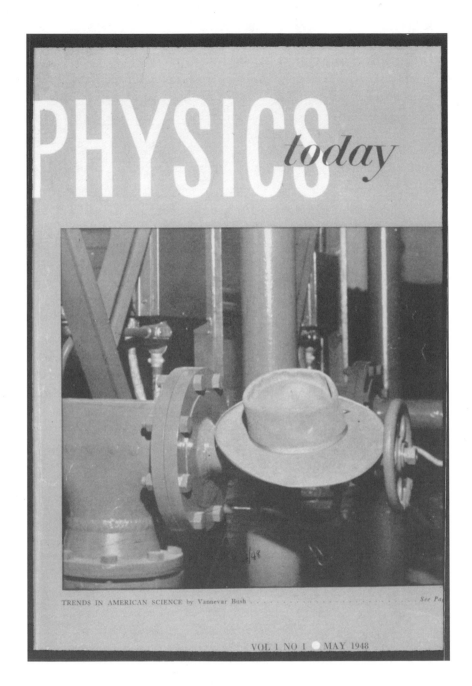

PLATE 11 The cover of the first issue of *Physics Today* (May 1948).
Oppenheimer's trademark porkpie hat rests against cyclotron piping. Courtesy of
J. Robert Oppenheimer Memorial Committee, Los Alamos National Laboratory.

PLATE 12 Oppenheimer with Albert Einstein at the Princeton Institute for Advanced Study. To the physicists, "Einstein was God and Oppie was His only begotten Son." International Communication Agency, United States Information Service; courtesy of American Institute of Physics, Emilio Segrè Visual Archives.

PLATE 13 . Oppenheimer waits to testify before a closed session of the House Un-American Activities Committee, June 7, 1949. Library of Congress; photograph by ACME Photo.

PLATE 14 "Now This!" Cartoonist Hy Rosen's depiction in the right-wing *New York Journal-American* (April 17, 1954) of the explosive impact of the Oppenheimer case.

PLATE 15 Herbert Block's interpretation of the message of the Atomic Energy Commission's Personnel Security Board in the Oppenheimer case. The cartoonist is referring to the board's criticism of Oppenheimer's lack of "enthusiasm" for the hydrogen bomb program. Herbert Block, *Herblock's Here and Now* (New York: Simon and Schuster, 1955), 23; originally published in the *Washington Post,* June 3, 1954.

PLATE 16   The security hearings took a physical toll on Oppenheimer, exaggerating his already thin and angular features. As one commentator put it, the hearings left him "a thin, gray, shrunken ghost." Courtesy of American Institute of Physics, Emilio Segrè Visual Archives.

PLATE 17   Oppenheimer in front of his inherited van Gogh painting, *Enclosed Field with Rising Sun.* Library of Congress; photograph by John Vachon, in *Look* magazine, April 1, 1958; courtesy of Ann Vachon.

PLATE 18   The tragic intellect. Oppenheimer during a return visit to Los Alamos in 1964, where he delivered a talk on Niels Bohr. Los Alamos National Laboratory; courtesy of American Institute of Physics, Emilio Segrè Visual Archives.

PLATE 19 Oppenheimer speaking at an American Physical Society press conference, January 27, 1966. Kitty Oppenheimer is sitting behind him. Courtesy of American Institute of Physics, Emilio Segrè Visual Archives; photograph by Mitchell Valentine.

PLATE 20 "I wonder of all the Oppenheimers I have seen, which is the real one?" Contact sheet with pencil marks by photographer John Vachon, selecting photographs for *Look* magazine, 1958. Library of Congress; courtesy of Ann Vachon.

CHAPTER SEVEN

"I Was an Idiot"

POWER AND INSECURITY

Occupying a nodal position within the complex scientific advisory apparatus that grew up within the U.S. government after World War II, Oppenheimer embodied what appeared to be massive de facto power accruing to the new advisors and consultants. It was a widespread belief that he "could block [policies] merely by expressing his dislike."[1] What this influence involved, however, was unclear, undefined by any fixed mandate or position. In addition to his formal positions, Oppenheimer exercised what one FBI agent called "considerable behind-the-scenes influence" over atomic energy policy, and his enemies regarded his influence as specially ubiquitous.[2] More than any other figure, Oppenheimer symbolized the idea, and the fear, that scientists might be "on top" rather than merely "on tap."[3]

Oppenheimer combined institutional influence within the state with public authority associated with his qualities of general humanistic cultivation and moral sensitivity. He frequently employed overtly moral, and sometimes even religious, language to talk about the implications of the atomic bomb and the role of the scientist. Even though Oppenheimer increasingly articulated a highly restricted conception of scientists' vocational responsibilities, he was publicly associated with a claim to broad cultural and moral authority on the part of scientists.

In the immediate aftermath of the war, this combination of special expertise and broad humanistic authority meshed with the symbolism of atomic energy as a transformative power ushering in a new era of peace and liberal modernity. However, the routinization of the bomb into the military arsenal, the increasing paranoia and secrecy of the Cold War, and the bureaucratization of the scientific role within the AEC and the national-security state made Oppenheimer's charismatic persona more problematic.

Oppenheimer's support from the scientific community and the organized scientists' movement had allowed him some degree of autonomy, even as he operated within bureaucratic organizations such as the AEC. But as the scientists' movement atrophied, as the atomic scientists increasingly divided into those thoroughly incorporated into technical and policy functions within the state and those who remained outsiders, Oppenheimer was less and less able to sustain an autonomous role in relation to the state and its agencies. As he accommodated himself to the military and the security state, he was vulnerable to the regimes of these institutions. And as his authority was increasingly defined by his advisory offices, rather than by his charisma in the eyes of the scientific community or the public, he became more and more subject to the institutional constraints of bureaucratic office and to the state's disciplinary apparatus.

This institutional dependence was, in fact, an element of Oppenheimer's authority from the beginning of the war. As Oppenheimer was emerging as the charismatic leader of Los Alamos, a quite different identity was being negotiated and constructed out of view of his scientific colleagues, in FBI and Army security investigations of, and interviews with, the physicist. Oppenheimer's insecurity in these interrogations contrasted starkly with the "effortless superiority" that he was seen to display in his public role as laboratory director. Even at the height of his accomplishments at Los Alamos and afterward, his position was in fact highly vulnerable. Behind the public face and out of view was a "backstage" identity—an ossified documentary identity, or "file person."[4]

Oppenheimer's willingness and ability to adapt to the situation and to others' expectations was crucial to his success at Los Alamos and afterward. This changeling quality was essential to, and was reinforced by, Oppenheimer's role in mediating between the different groups and interests involved in the alliance between science and the national-security state. His inconsistency—what has been seen as his lack of a unitary self or identity—was crucial to, and to a large extent an outcome of, his mediation between different fractions and interests in the technoscientific state. Yet even as Oppenheimer adapted, changed, and shifted, his past was collected in petrified form in files and dossiers. These records amounted to a documentary biography, one that could at any time be employed to challenge Oppenheimer's current presentation of self and to undermine his power.[5]

This chapter examines the institutional, social, and discursive dynamics involved in the destruction of Oppenheimer's authority and role as a state adviser. I argue that the interrogation and judgment of Oppenheimer was simultaneously an appraisal of the role of the scientist, and the relationship between science and the state, as these had developed from World War II.

BACKGROUND TO THE CASE: "UN-AMERICAN ACTIVITIES"

Since the beginning of his involvement in the war effort, Oppenheimer had been under scrutiny by the FBI, and by the time of the hearings in 1954, his FBI file was thirty inches thick.[6] Oppenheimer was also being investigated during the war by the Army's G-2 intelligence division, which dealt with Manhattan Project security. It was in interviews with security officers that Oppenheimer had invented the tortuous version of the "Chevalier incident" that would dog both men for years to come.

During the war, Chevalier had been shaken by the fact that when he had gone to New York to work for the Office of War Information, he waited half a year for security clearance and then was denied. Unaware of his friend's role in his difficulties, he wrote to Oppenheimer during that time, "All my foundations seem to have been knocked out from under me, and I am alone dangling in space, with no ties, no hope, no future."[7]

In 1946, FBI director J. Edgar Hoover put pressure on General Groves to provide the agency with the details of his wartime conversation in which Oppenheimer named Chevalier. But Groves told him, "I still feel that it would not be desirable for me to furnish you any details concerning the information reported to me by Oppenheimer, as it would endanger our relationship which must, in the best interest of the United States, be continued in its present state."[8] Between June and September 1946, the FBI interviewed both Chevalier and Oppenheimer, as well as the alleged contact, George Eltenton.[9] In August 1947, the AEC granted Oppenheimer a new security clearance for his work as GAC chairman, which he had begun in January of that year.[10] Nevertheless, the Chevalier incident, and the inconsistencies between his different accounts of it, would remain an unsettling presence beneath his new position of power.

Oppenheimer was under near-constant surveillance during this period. From when he left Los Alamos in October 1945 until the summer of 1947, Oppenheimer divided his time between appointments in Washington, D.C., and teaching, which he resumed at Caltech and Berkeley. In July 1947, he moved his family from Berkeley to Princeton, where he had accepted the position of director of the Institute for Advanced Study (again alongside his Washington commitments, particularly for the GAC). It appears that the Oppenheimers were aware during their postwar interlude in Berkeley that their phone was tapped. During one monitored phone conversation between Oppenheimer and Kitty in 1946, there was a clicking sound. Oppenheimer asked, "Are you still there? I wonder who's listening to us." Kitty replied ("lackadaisically," according to the report), "The FBI dear." Ralph Lapp recalled an incident at Oppenheimer's Princeton office about a year after Oppenheimer had taken up his appointment there: Lapp had begun to speak when Oppenheimer warned him, "Even the walls have ears."[11]

Oppenheimer's brother was also being hounded by the FBI, and information from Frank's FBI report found its way into the press. On July 12, 1947, the *Washington Times-Herald* published a front-page story branding Frank Oppenheimer as a former "card-carrying member of the Communist Party." The *Times-Herald* article carried the italicized disclaimer that the report on Frank *"in no way reflects on the loyalty or ability of his brother, Dr. J. Robert Oppenheimer."*[12]

However, only a few months later, the chairman of the House Un-American Activities Committee (HUAC), J. Parnell Thomas, was seeking to branch out from the investigation of Hollywood and decided to look at atomic espionage. He told Hoover he wanted to include "the tie-up between the Oppenheimers and the Soviet embassy." At the end of October 1947, a HUAC investigator, former FBI agent Louis J. Russell, testified before the committee about the Chevalier incident, and his remarks were widely repeated in the press. Chevalier was now accused of approaching not three scientists, but one— Oppenheimer. At this time, these revelations did not significantly damage Oppenheimer. Russell had said that the approach to Oppenheimer had been unsuccessful and that Oppenheimer had called Eltenton's activities "treasonable." Oppenheimer released a statement that he wished to "withhold comment, either confirmation or denial." For the time being, that seemed to dampen the flames; the story faded. Chevalier was subpoenaed to appear before California's Joint Fact-Finding Committee on Un-American Activities, known as the Tenney Committee. But he was dismissed without being asked about his wartime conversation with Oppenheimer.[13]

HUAC's growing confidence in confronting the scientific community was signaled by its attack on Edward Condon—briefly Oppenheimer's associate director at Los Alamos and, since late 1945, director of the National Bureau of Standards.[14] On March 1, 1948, after a drawn-out campaign of innuendo in the press, Thomas released a report that accused Condon of being "one of the weakest links in our atomic security."[15] The report was thin on evidence, dwelling on Condon's liberal political views and his membership in the Soviet-American Science Society (which HUAC alleged was a Communist front organization). The scientific community rallied to Condon's defense, with statements from Nobelists Albert Einstein and Harold Urey, the American Physical Society, and the American Association for the Advancement of Science, among others.

Oppenheimer was asked by Urey to sponsor an Emergency Committee of Atomic Scientists dinner in honor of Condon and in protest against the HUAC attack. Proceeds would go to the *Bulletin of the Atomic Scientists*. Oppenheimer, however, had become markedly cautious about putting his name to anything that could be construed as oppositional politics. He declined the offer, explaining in a telegram, "Have participated in and welcomed past expressions of confidence in Condon and am prepared to welcome future expressions." While he was "also deeply concerned at [the] general situation which made [the] attack on Condon possible," he added, "Believe, however, that these matters should all be kept separate."[16] It seems that Oppenheimer was wary of framing support for Condon within a general condemnation of anti-Communist politics and of connecting such support with the politics of arms control as represented by the *Bulletin* and the Emergency Committee.

The following month, spurred by the attack on Condon, the National Academy of Sciences (NAS) addressed, for the first time, the civil liberties of scientists. But the organization equivocated over what sort of response to make, some members worrying about compromising their "political neutrality" by entering the fray. At an NAS business session on April 27, Oppenheimer announced that most American scientists "will regard it as very odd, they will regard it as incomprehensible, that we should let pass an outrage of this kind and remain silent." It was necessary for the NAS to take a stand on "what every one of your colleagues—not every one, but an overwhelming majority—regards as a clear evil."[17] However, the statement that the academy eventually released was bland and not widely reported in the press.[18] Condon commented to chemist Martin Kamen, "It is amazing to me how some of the older scientists seem to be so completely lacking in perception of what is going on." Kamen replied, "I agree with you regarding the pusillanimity of the Academy hierarchy. It may be that these top scientists are suckers. I am not inclined to be so charitable."[19]

In November 1948, the NAS set up a Committee on Civil Liberties, chaired by James Conant and also including Oppenheimer and Oliver E. Buckley, the president of Bell Telephone Laboratories.[20] The committee's report, released in early February of the following year, was tortuously worded and hardly qualified as a ringing defense of civil liberty. The report concluded, "We believe that by a more careful separation of its punitive from its preventive aspects, the problem of security in government service will be rendered, not easy surely, but far more manageable."[21] The NAS sought to rein in what it saw as the more irresponsible uses of the security apparatus. It did not challenge the logic or legitimacy of the security state. Oppenheimer admitted to NAS president Alfred N. Richards that the report was "not very hot stuff." The institutional dependence of the scientific elite on the national-security state fostered political timidity in the face of Cold War anti-Communism.[22] In the case of Oppenheimer, this reticence was compounded by what he must have felt to be his own vulnerability to such attacks.

In September 1948, HUAC began to call witnesses in hopes of unearthing wartime Communist atomic espionage at Berkeley. Witnesses included Kamen, Communist Party organizer Steve Nelson, and Joseph Weinberg, a former student of Oppenheimer's.[23] At the end of the month, the committee released a preliminary report on its investigations, "Soviet Espionage Activities in Connection with the Atomic Bomb." The report claimed that during the war, there had been a concerted Communist campaign to channel secret information about the atomic bomb project to the Soviets. It accused former Chicago Metallurgical Laboratory scientists Clarence Hiskey and John H. Chapin of being involved in an espionage conspiracy against the Manhattan Project, and it claimed that Kamen had passed secrets about the work of the Berkeley Radiation Laboratory to representatives of the Soviet consulate in San Francisco. In addition, the committee cited the case of a "Scientist X" at the Berkeley Rad Lab who, the report claimed, had given information about the project—including a secret "formula"—to Nelson. Though not identified in this report, "Scientist X" was named by HUAC the following year as Weinberg. The report also mentioned Oppenheimer in connection with the Chevalier incident. Although the committee noted that Oppenheimer had "refused to cooperate" in the scheme, the fact that the report placed him amid the "interlocking associations" of a Communist network was, at the least, highly discomforting.[24]

In 1949, Oppenheimer's former students David Bohm and Giovanni Rossi Lomanitz were subpoenaed to testify before the committee. Both pleaded the Fifth Amendment when asked if they had been members of the Communist Party and whether they had known Nelson.[25] Oppenheimer regarded taking the Fifth as dishonorable, and when he himself appeared before the committee, on June 7, he adopted a different strategy. He had been called for "assistance" in the investigation of the Rad Lab and was told that he himself was not under suspicion. AEC general counsel Joseph Volpe, who sat beside Oppenheimer at the witness table, remembered, "Robert seemed to have made up his mind to charm these Congressmen out of their seats." The strategy apparently succeeded. At the end of the testimony, each of the six members of the committee came over to shake his hand. Congressman Richard Nixon later described him as having been "a cooperative witness."[26]

Exactly how cooperative Oppenheimer had been became public the following week when a Rochester, New York, newspaper published details of the testimony he gave regarding a former student, Bernard Peters. Oppenheimer had described Peters as "a dangerous man and quite Red." He said that Peters was an ultraleftist who had viewed the American Communist Party as not sufficiently dedicated to violent revolution. Incredibly, he even referred to Peters's escape "by guile" from a Nazi concentration camp as evidence of a suspicious tendency toward "direct action."[27] This betrayal of his own student deeply disturbed a number of Oppenheimer's scientific colleagues, including Condon, Hans Bethe, and Victor Weisskopf, all of whom wrote to Oppenheimer expressing their disappointment. Weisskopf told Oppenheimer that "we are all losing something that is irreparable. Namely confidence in *you.*" He begged Oppenheimer to "*set this record straight* . . . even if you have to pay for it by losing reputation somewhere else."[28]

Oppenheimer did write a letter to another newspaper, the *Rochester Democrat and Chronicle,* saying that his comments had been "misconstrued and . . . abused" and regretting that anything he said might "damage Dr. Peters and threaten his distinguished future career as a scientist." Oppenheimer added a point of principle: "Political opinion, no matter how radical or how freely expressed, does not disqualify a scientist for a high career in science." But this seemed to many of his colleagues to be too little, too late, and Oppenheimer's post hoc defense of the separation of science from politics smacked of hypocrisy.[29] Condon remained especially angry about this incident. When he had first learned of the HUAC testimony, he wrote to Oppenheimer, "One is tempted to feel that you are so foolish as to think you can buy immunity for yourself by turning informer. I hope that this is not true. You know very well that once these people decide to go into your own dossier and make it public that it will make the 'revelations' that have been made so far look pretty tame."[30]

Many years later, Condon told an interviewer that Oppenheimer's testimony "was the dirtiest two-faced thing I ever saw." "The thing that horrified me most," Condon said, "was, he, a Jewish boy, so soon after the six million had been cremated, and this was his personal protégé, also a Jewish boy—he said to this scoundrelly committee, 'I'm not sure how far I would trust Peters, because he resorted to guile in escaping from Dachau.'"[31] When he read of Oppenheimer's testimony, Condon wrote to his wife, "Oppie is really becoming unbalanced." He added, "If he cracks up it will certainly be a great tragedy. I only hope that he does not drag down too many others with him."[32]

Oppenheimer's conciliatory stance toward HUAC contrasted with other occasions on which he was less politic, and when flashes of arrogance helped to make him powerful enemies. On June 13, 1949, appearing again before Congress—this time in front of the Joint Committee on Atomic Energy—Oppenheimer criticized AEC commissioner Admiral Lewis Strauss's grounds for opposing shipment of radioisotopes to foreign countries. Oppenheimer assessed radioisotopes as being no more worth protecting from proliferation than a bottle of beer or a shovel. Strauss never forgave Oppenheimer for this public ridicule.[33]

The following day, Frank Oppenheimer and his wife, Jackie, appeared before HUAC to answer questions about Communist activities at the Rad Lab. In 1947, when questioned by a journalist, Frank had denied that he had been a party member. He had also written a letter to his employer, the University of Minnesota, denying this allegation. Now, testifying before the congressional committee, Frank admitted that he had been a member of the Communist Party but claimed he had left "long before" he began war research. Less than an hour after his appearance before the committee, he heard from news reporters that he had been fired by his university. From then until the early 1960s, Frank was on a blacklist and was unable to gain academic employment.[34]

Chevalier, finding himself blocked for promotion at the University of California, had resigned his academic post.[35] He also attributed the deterioration and eventual breakup of his marriage to the disintegration of his career. He was interested in Oppenheimer's HUAC testimony after reading reports that Oppenheimer had publicly vindicated him. He thought that if he could have an account of the testimony, it would help him in applications to find a new academic position. Oppenheimer replied, "I said that you had told me of a discussion of providing technical information to the USSR which disturbed you considerably, and which you thought I ought to know about." He added, "As you know, I have been deeply disturbed by the threat to your career which these ugly stories could constitute. If I can help you in that, you may call on me." In an attached letter, he told his old friend, "The reason for the formality of this letter which I enclose is that it seemed to me likely that you might find some occasion to use it, and that it ought to be couched in fairly formal terms. I do not feel at all formal about the troubles you are having; and I wish that I might think of an easy solution." Denied a U.S. passport, Chevalier made use of his dual nationality and moved to Paris.[36]

In early 1950, coinciding with Truman's announcement of the go-ahead for the hydrogen bomb, the country was rocked by revelations of Soviet espionage. Alger Hiss, a former special assistant to the secretary of state, had been accused of passing secret information to the Soviets; on January 21, he was convicted of perjury when he denied being involved with espionage. Hiss seemed the epitome of the East Coast liberal establishment—an upper-class diplomat, a graduate of Johns Hopkins University and Harvard Law School. His conviction fueled far-fetched allegations of a Communist conspiracy within the highest rungs of the American establishment, and it also bolstered the confidence of HUAC in taking on such elite figures.[37] Then, on January 24, a few days before Truman's H-bomb decision, Klaus Fuchs, working at Harwell nuclear laboratory in Britain, confessed to having spied for the Soviets while working in the Theoretical Division at Los Alamos. When he read in the papers about the Fuchs case, former Manhattan Project security officer Colonel Boris Pash immediately "felt that he would next be reading about Dr. Oppenheimer's involvement in such activities."[38]

Then, in May, Sylvia Crouch testified before the California Senate Committee on Un-American Activities in Oakland that in late July of 1941, Oppenheimer had hosted a "session of a top-drawer Communist group" at his Berkeley home.[39] Paul and Sylvia Crouch, though later discredited, were at this time prized ex-Communist informants for the Department of Justice and the array of legislative committees involved in the anti-Communist campaign.[40] The clear implication of the testimony was that Oppenheimer had been a party member, and a very senior one at that. Oppenheimer promptly issued a statement denying the allegation and asserting, "I have never been a member of the Communist Party."[41] The papers responded with splashy headlines such as "A-Scientist Oppenheimer at Red Meet, Probers Told," "Says A-Wizard Met with Reds," "Red Huddle Denied by Oppenheimer," and "Commy Meetings in Home Denied by Oppenheimer."[42]

Crouch's allegation, though unsettling, was not fatal. Powerful figures rallied to Oppenheimer's support. Nixon, who had been so impressed by Oppenheimer's performance before HUAC, took time during a campaign speech to assert his faith in the physicist: "I am convinced that Dr. Oppenheimer has been and is a completely loyal American and, further, one to whom the people of the United States owe a great deal of gratitude for his tireless and magnificent job in atomic research." General Groves wrote Oppenheimer a letter for use if he needed it, affirming his confidence in the physicist.[43] Such statements of support generated sympathetic treatment for Oppenheimer in the editorial pages, and there was frequent criticism of the California committee for being hysterical and overly credulous of informers. Some even carried paranoia full circle by suggesting that the Crouches could themselves be carrying out a Communist plot to sow distrust and discord in America's institutions.[44] Oppenheimer was able to ride out the storm. But Lilienthal's reassurance to Oppenheimer that the Oakland affair was "like a puff of wind against the Gibraltar of your great standing in American life" was perhaps an overstatement. For with each new public allegation, that standing was, little by little, eroded. The *Oakland Tribune* was already writing of "the Robert Oppenheimer case" and predicted that "there is more to come." A few days later, the *San Francisco Chronicle* described Oppenheimer as "the brightest star in the firmament" of science—but said that there was now "mud on the star." The controversy not only tarnished Oppenheimer's public image; it also took a personal toll. Toward the end of May, Oppenheimer wrote to Robert Bacher about what he called "the California doings": "I took it all very badly and feel now like a man slowly convalescing from a serious illness."[45]

Sylvia Crouch's allegations continued to hang around Oppenheimer's neck. During her Oakland testimony, she had claimed that Weinberg was one of those present at the 1941 gathering at the Oppenheimer home. This allegation was marshaled by the prosecution when, in 1952, Weinberg faced a grand jury, charged with perjury regarding his denials to HUAC two years earlier that he had been a Communist or had attended Communist meetings. As a result, the story of Oppenheimer hosting a closed Communist meeting was again in the headlines.[46]

In the spring of 1952, the assistant U.S. attorney prosecuting the Weinberg case requested a meeting with Oppenheimer and his lawyers. There Oppenheimer was brought face to face with Paul Crouch, who claimed to have accompanied his wife to the 1941 gathering at the Oppenheimer home and to have addressed the group about the new party line after the outbreak of war between Germany and the Soviet Union. Oppenheimer said he had never met the man before and proceeded to directly challenge Crouch on details of his story. For example, Crouch claimed that he and his wife entered the living room through the house's back entrance. Oppenheimer asked, "Up some steps at the back?" When Crouch assented, Oppenheimer stated, "There were no steps." It was not difficult to pick holes in Crouch's testimony. When questioned about the physicist's appearance, he said that Oppenheimer had

had the same crew cut that he had now. Oppenheimer and his attorneys pointed out that his hair had been long and tangled.[47] Oppenheimer was sure that he had been in New Mexico at the time the gathering at his house was supposed to have taken place. He and his lawyers set about the arduous task of trying to document his whereabouts on that date, ten years ago, by digging up records and interviewing old friends in New Mexico and Berkeley.[48] Weinberg was eventually acquitted on all counts in March 1953, though the judge expressed his view that, from the evidence, the security situation at wartime Berkeley had been "shocking."[49]

Oppenheimer was able to establish quite securely that he had been in New Mexico when the Crouches placed him as host of the meeting. But their allegations had been causing trouble for Oppenheimer for several years, and in the 1954 hearings he would again be called on to rebut them ("Transcript," 16, 216–18). Such attacks and allegations concerning Oppenheimer's prewar politics, though bolstered by the climate of McCarthyism, would most likely have continued to be fended off, without impact, were it not for the enemies in the military and the scientific community that Oppenheimer had made through his initial opposition to the Super. It was this, more than anything else, that led to the events of 1954. Herbert York wrote that "despite elaborate efforts to assert the contrary, the real stimulus for the removal of Oppenheimer's clearance was his opposition to the program for superbomb development."[50]

Teller, in particular, played a crucial behind-the-scenes role in manufacturing a case that Oppenheimer had intentionally hindered the development of the H-bomb. For example, a May 1952 FBI report noted that "Dr. Teller . . . believes that the H-bomb would have been a reality at least a year ago if it had not been for Oppenheimer's opposition . . . Oppenheimer either delayed or hindered the development of the H-bomb from 1945 to 1950 by opposing it on moral grounds . . . [and subsequently] opposed it on the basis that it was not feasible." Teller's account suggested that Oppenheimer would use any argument available to oppose the H-bomb. While Teller said that he did not believe Oppenheimer's motives were "subversive," that was strongly implied by the direction of his commentary.[51] He also mused that Oppenheimer was "a very complicated person . . . [who] in his youth was troubled with some sort of physical or mental attacks which may have permanently affected him. He also had great ambitions in science and realizes that he is not as great a physicist as he would like to be." At the same time, Teller was highly sensitive about being exposed as having informed on Oppenheimer, and he requested that those psychological tidbits be excluded from the report. Teller thought it would be obvious that he was the source, and he was worried that such exposure "could prove very embarrassing to him personally."[52] Teller's caution is made understandable by the comment of a close friend of his on Oppenheimer's perceived status in the physics community: "Einstein was God and Oppie was His only begotten Son."[53]

The campaign against him by pro-H-bomb scientists (clustered around the Berkeley Radiation Laboratory), senior Air Force figures, and AEC administrators had taken its toll on Oppenheimer. When his term as GAC chairman came to an end in the summer of 1952, Oppenheimer stepped down, as did his colleagues Conant and DuBridge.[54] All three men had decided to jump before they were pushed. Admiral Sidney W. Souers, a special consultant to President Truman, told Hoover in July that the administration was "making a clean sweep" of the GAC.[55] Just before his term came to an end, Oppenheimer expressed his weariness in a private letter to his brother: "By August, my six years on the Gen. Advisory Committee are over; they have seemed long."[56] But Oppenheimer's removal from the committee did not appease his enemies in the AEC, the military, and the administration; they wanted to see him barred completely from government work.[57]

If Oppenheimer had been increasingly frozen out of the Truman administration, he faced an even colder shoulder following Eisenhower's inauguration in late January 1953. When Lewis Strauss was appointed AEC chairman (to become effective July 1 of that year), he told Eisenhower that he "could not do the job . . . if Oppenheimer was connected in any way with the program."[58] Strauss, in close contact with Hoover, carried out an unrelenting campaign to undermine the physicist.[59] Oppenheimer had already begun to be excluded informally. In 1951, the Air Force stopped using him as a consultant. Secretary of Defense Charles E. Wilson simply abolished the department's Research and Development Board, of which Oppenheimer was a member; he bragged at a press conference in April 1954 that "we dropped the whole board. That was a real smooth way of doing that one as far as the Defense Department was concerned." In 1952, Oppenheimer's consulting work for the AEC totaled only two days, and in the following year only four days.[60]

The campaign against Oppenheimer was brought into public view by an anonymous article in *Fortune* magazine in May 1953, "The Hidden Struggle for the H-Bomb." The article alleged that Oppenheimer was the leading figure among a group of scientists opposing SAC and the H-bomb and accused him of trying to stop testing of the H-bomb in 1952. The article also placed Oppenheimer at the head of a shadowy group allegedly calling itself ZORC— "Z" for MIT physicist Jerrold Zacharias, "O" for Oppenheimer, "R" for Rabi, and "C" for Charles Lauritsen, who were said to be trying to undercut the

Air Force's "deterrent-retaliatory" strategy. The writer strongly suggested that these scientists' proposals for air defense (as an alternative to deterrence) were designed primarily as a way to hinder the Air Force program. The article finished by asserting that "there was a serious question of propriety of scientists' trying to settle such grave national issues alone, inasmuch as they bear no responsibility for the successful execution of war plans."[61] The author was Air Force reservist Charles J. V. Murphy, who thought that Oppenheimer was out to "denuclearize" American policy. During a recent tour of duty in the Air Force's top echelons, Murphy had had access to secret documents and private conversations regarding Oppenheimer and atomic policy.[62] The *Fortune* article marked Oppenheimer as fair game for press criticism. In September, a *Time* article eulogizing Strauss accused Oppenheimer and Lilienthal of indulging in the "paralyzing combination" of "hand wringing and baseless hope" in their opposition to the H-bomb.[63]

Perhaps most galling of all to the H-bomb lobby was a critique that Oppenheimer published in the July 1953 issue of *Foreign Affairs*. One foothold in Washington that Oppenheimer had maintained after stepping down from the GAC was on the State Department's Panel of Consultants on Disarmament, set up by Secretary Dean Acheson. The group made a series of radical and forward-looking proposals on limiting the arms race, including a moratorium on H-bomb testing. Among the ideas they forwarded to the incoming Eisenhower administration was the promotion of "candor" rather than secrecy in nuclear policy. Frustrated at the lack of governmental attention the proposals were receiving, Oppenheimer took the panel's idea to an elite public audience in a February talk at the Council on Foreign Relations and in the resulting article for *Foreign Affairs*. Here, he criticized overreliance on nuclear weapons and the "rigidity" of the administration's strategic ideology. Noting the futility of an inexorable arms buildup, Oppenheimer argued that "our twenty-thousandth bomb . . . will not in any deep strategic sense offset their two-thousandth." The arms race, he said, was leading to "a state of affairs in which two Great Powers will each be in a position to put an end to the civilization and life of the other, though not without risking his own." And Oppenheimer vividly likened this situation to "two scorpions in a bottle, each capable of killing the other, but only at the risk of his own life." He also criticized the culture of secrecy that shielded the real state of the arms race from democratic scrutiny.[64]

The publication of Oppenheimer's *Foreign Affairs* article suggested that the policy of easing him out of governmental power was allowing him to take a more publicly critical stance toward nuclear defense policy. Though he was increasingly cut off from insider influence, Oppenheimer's top-secret Q clearance gave him the authority to speak with insider knowledge on atomic weapons issues. The danger for the administration was that Oppenheimer would be an external critic with the authority of an insider. Strauss and his allies needed to find a way to destroy his credibility and authority entirely. At the same time, pressure was mounting from another source—Senator Joseph McCarthy. The *Fortune* article, with its suggestion of a shadowy "ZORC" conspiracy, had caught the McCarthy committee's attention.[65]

On May 12, 1953, McCarthy informed Hoover that his Senate Investigations Committee was planning to inquire into Oppenheimer. Hoover warned him that if he did so, he would have to prepare well, with "a great deal of preliminary spade work." It would be a symbolic battle: Oppenheimer was a "figure . . . around whom the scientists of the country have usually rallied." Hoover was particularly concerned that Oppenheimer should not "end up by becoming a martyr." Strauss also worried that McCarthy would bungle the investigation, agreeing with Hoover that "this was not a case which should be prematurely gone into solely for the purpose of headlines." A direct attack by McCarthy on the scientist would by extension implicate the administration. The desire to head off such a move was a key factor leading Eisenhower to endorse internal AEC action against Oppenheimer. All were aware, however, that action against Oppenheimer would be treated as an attack on the scientific community itself. In December, Hoover told the Secretary of Defense that "the scientists considered themselves sacrosanct and if anyone criticized them they all rose up in righteous indignation," adding that "my personal feeling was that they were no different from anyone else."[66]

Oppenheimer was the most senior scientist of the time to have his security clearance denied. He was also the most prominent and powerful individual ever to be brought down by such a proceeding. The hearings were regarded by many at the time as a watershed, having the potential to be either a great symbolic victory for the anti-Communists or an example of an overconfident security system finally overreaching itself.

THE CASE AGAINST OPPENHEIMER

The charges against Oppenheimer were initiated by William L. Borden, who had recently stepped down as executive director of the congressional Joint Committee on Atomic Energy. On November 7, 1953, Borden wrote to Hoover alleging that Oppenheimer was in all probability an "agent of the Soviet Union" ("Transcript," 837–38). The FBI forwarded the letter to the White House, and on December 3, President Eisenhower ordered that Oppenheimer's security clearance be withdrawn and that a "blank wall" be erected between the scientist and the nation's atomic secrets. Just before Christmas, Oppenheimer was given a letter from General Kenneth D. Nichols, General Groves's Manhattan Project deputy during the war and now the general manager of the AEC, informing the physicist that his clearance had been revoked.[67] Oppenheimer refused to resign as a consultant to the AEC in response to Nichols's charges. He told Strauss that resigning "would mean that I accept and concur in the view that I am not fit to serve this Government, that I have now served for some 12 years . . . This I cannot do." Seeking to clear his name, he exercised his right to appeal before an AEC Personnel Security Board (PSB).[68]

The hearings began on April 12, 1954. The PSB's task was to evaluate the charges and the evidence and to determine whether Oppenheimer's continued employment did pose a risk to the nation's security. The proceedings lasted almost a month, the board hearing the testimony of forty witnesses before wrapping up on May 6. The board sent its report to Nichols, who forwarded it to the AEC commissioners. On June 29, the commissioners announced their finding that Oppenheimer was indeed a security risk, and they upheld the withdrawal of his clearance. Oppenheimer was barred from access to government secrets. He could no longer be called upon to advise on matters of atomic energy, and his ties to the institutions of government were effectively severed.

Harold P. Green, the AEC legal officer who drafted the original letter of allegations sent by Nichols to Oppenheimer, subsequently criticized the ambiguous, quasi-legal character of the proceedings as inhabiting a gray area of "demi-jurisprudence." Formal rules, criteria, and procedures for security were relatively new in the late 1940s, replacing the more ad hoc and informal types of judgment made during the war. In consequence, there were "no established traditions . . . no publicly available precedents." The proceedings had "many of the trappings of the judicial process," but without the protections usually associated with law.[69]

The three-member board was chaired by Gordon Gray, president of the University of North Carolina, a former secretary of the Army under Truman, and a man firmly entrenched in the American establishment. Sitting alongside him were Dr. Ward Evans, a professor of chemistry at Northwestern University, and Thomas Morgan, retired chairman of the board of the defense contractor Sperry Gyroscope and a director of Lehman Brothers and Bankers Trust.[70]

As the hearings commenced, Gray reminded those present that this was "an inquiry and not in the nature of a trial. We shall approach our duties in that atmosphere and in that spirit" ("Transcript," 20). But as one commentator observed, the proceedings were "neither wholly an inquiry nor wholly a trial. The Board permitted them to fall into a muddled middle course."[71] The AEC's attorney, Roger Robb, was a highly successful criminal prosecutor, used to trying murder cases, and he pursued the Oppenheimer case in the same style. Robb was both behaving as a prosecuting attorney *and* acting on behalf of the PSB, a combination that has called into question the board's neutrality.[72]

According to Green, the hearings were "launched with the predetermined objective" of finding Oppenheimer a security risk. He remembered how Nichols had gloated that they were about to finally catch the "slippery sonuvabitch." There was considerable pressure for such an outcome from Strauss, as well as Hoover. Green alleged that hard-liners who would have "a predisposition to find against Oppenheimer" were selected for the board.[73] The three were assembled in Washington a week prior to the beginning of the hearings and were shown the files on Oppenheimer, including the FBI file— documents that were not made available to Oppenheimer and his attorneys.[74] During the last stages of the AEC's deliberations, Strauss told Eisenhower he was worried that other AEC commissioners—Henry D. Smyth, Eugene Zuckert, and Thomas A. Murray—would vote in favor of Oppenheimer. According to Green, Eisenhower urged Strauss to do all he could to ensure that they would find against Oppenheimer; and after the AEC's decision, Eisenhower wrote a letter of congratulation to Strauss on the result.[75]

Eisenhower was deeply worried that if the case did not go the right way, it would become a stick with which McCarthy could beat the administration and boost his own power. Despite Gray's pains to present the hearings as a neutral and routine administrative inquiry, the whole proceeding was haunted by the specter of McCarthy. Six days before it began, McCarthy stated on television that Communists in government were responsible for setting back the development of the H-bomb program for eighteen months. McCarthy said, "Our nation may well die" because of the delay, and he added, "I ask who caused it? Was it loyal Americans—or was it traitors in our government?"[76]

The initial secrecy surrounding the proceedings broke down. The AEC and Oppenheimer's attorneys came to regard a controlled release of information as preferable to sensationalized revelations from McCarthy. Also, during the AEC's deliberations, Zuckert left under a train-carriage seat a copy of key documents concerning the hearings. The following evening, the commissioners held an emergency meeting in which Strauss urged them to publish the transcript, this being preferable to possible unauthorized release if the documents had gone missing. Although the package was safely recovered, with no evidence of tampering, the publication went ahead. On June 15, before the commissioners delivered their judgment, copies of the entire transcript were passed to the press.[77]

Another peculiarity of the proceedings was the lack of clarity about the criteria against which Oppenheimer was being judged. The meaning of the label "security risk" was disturbingly ambiguous. Green noted the government's vacillation between two general standards of security. The first was the "Caesar's wife" approach—"Caesar's wife must be above suspicion." On this criterion, any item of information suggesting disloyalty, if found to be true, would be grounds for the denial of security clearance. The second was the "whole man" approach, according to which any "derogatory information" must be weighed against favorable factors, including the person's value to the nuclear program.[78] The tension between the two understandings of security was to prove critical during the hearings.

A further confusion about the meaning of "security risk" lay in the question of what information or events could be counted as relevant. Borden's letter to the FBI accused Oppenheimer of being an enemy agent. "The central problem," Borden wrote, "is assessing . . . whether he became an actual espionage and policy instrument of the Soviets . . . My opinion is that, more probably than not, the worst is in fact the truth." However, the evidence that Borden put forward to support this allegation was most indirect. It focused mainly on questions of association, particularly Oppenheimer's links with Communists in the 1930s and financial contributions to the party. As to the question of Oppenheimer's *actions,* Borden alleged only that Oppenheimer "was a sufficiently hardened Communist that he either volunteered to the Soviets or complied with a request for such information." This charge was based, however, on the premise that any Communist with the opportunity to do so would carry out espionage for the Soviets ("Transcript," 838).

As Borden framed his allegations, Oppenheimer did not need to be caught red-handed as a spy. Rather, given the fact that his advisory positions gave him ample opportunity to carry out espionage, all that needed to be shown was that Oppenheimer was a Communist (something that Borden regarded as clearly demonstrated by his associations). Even in Borden's original denunciation, then, the very clear allegation of espionage easily slid into an emphasis, for practical purposes, on more diffuse questions of association. And application of the "Caesar's wife" standard meant that, quite apart from any presumption of actual espionage, derogatory information (such as association with Communists) would be *in itself* sufficient to render a judgment of Oppenheimer as being a security risk.

Such matters of "character" formed the substance of Nichols's letter listing the charges Oppenheimer faced. The letter was a rambling hodgepodge of allegations: that Oppenheimer had belonged to Communist front organizations, that he had had Communist friends, that his brother Frank had been a member of the Communist Party, and that Oppenheimer was responsible for having Communists recruited to the Manhattan Project. Nichols's letter further expanded the question of security by casting doubt on Oppenheimer's "veracity, conduct and even your loyalty." Therefore, he said, "the Commission has no other recourse, in discharge of its obligations to protect the common defense and security, but to suspend your clearance" ("Transcript," 6).

Nichols's letter also gave central importance to the "Chevalier affair." Nichols highlighted the fact that Oppenheimer did not initially report the incident to the authorities and that only when compelled by Groves did he identify Chevalier ("Transcript," 5–6). The Chevalier affair, which was seen as suggesting Communist associations and a cavalier attitude toward security, formed the basis for the judgment that Oppenheimer's "defects of character" made him a security risk. This understanding of "security risk" was wholly separate from the original charge of espionage or from any actions of Oppenheimer's that might have been directly harmful to the United States. Rather, it was maintained that his character was such that he could not be trusted: it was a question of what he *might* do, rather than what he *had* done.

Another of Borden's allegations did suggest that Oppenheimer had actively harmed the security of the United States. This played an important role in the hearings and was emphasized in the findings of the PSB. Borden accused Oppenheimer not just of espionage, but of being a "policy instrument" of the Soviet Union, via his influence on American nuclear weapons policies. This accusation made Oppenheimer's views and his positions taken as a scientific adviser central to the case against him. Oppenheimer had used his "potent influence," Borden alleged, to hinder the nation's postwar development of atomic weapons and had "worked tirelessly . . . to retard the United States H-bomb program" ("Transcript," 838). Nichols charged that Oppenheimer had "opposed the development of the hydrogen bomb (1) on moral grounds, (2) by claiming that it was not feasible, (3) by claiming that there were insufficient facilities and scientific personnel to carry on the development, and (4) that it was not politically desirable." He also charged that Oppenheimer had continued to oppose the H-bomb project even after it became national policy. Like Borden, he highlighted Oppenheimer's influence within the scientific community and government: "You were instrumental in persuading other outstanding scientists not to work on the hydrogen-bomb project . . . [This opposition] has definitely slowed down its development" (6).

Interlocking in the case against Oppenheimer, therefore, were questions both of biography and character and of the proper role of the scientific adviser. Oppenheimer's motivations and loyalties were seen to reflect in important ways on the trustworthiness of his scientific advice. Intimate aspects of his biography were seen to be bound up with general questions of the role of the scientist and the relationship between science and the state.

IDENTITY AS AN OBJECT

Oppenheimer said that during the hearings, he had "very little sense of self."[79] The hearings made his identity a public object over which he was able to maintain relatively little discursive control. The writer André Malraux, for example, could not understand what he saw as Oppenheimer's passivity in relation to his accusers: "The trouble was, he accepted his accusers' terms from the beginning."[80] The disorienting quality of the proceedings for Oppenheimer, however, was due precisely to the fact that he had very little control over biographical information and hence over the definition of his self. The prosecution's strategy was to cast doubt on Oppenheimer's trustworthiness by disrupting any attempt to create a coherent biographical narrative.

Oppenheimer's memories and his accounts of himself could be shown to be inconsistent in relation to transcripts and memoranda produced from wartime and later security interviews. Collected as a file, these formed a documentary biography that developed a coherence and objective reality of its own, and in relation to which Oppenheimer's self-descriptions could be compared and judged. Robb, on behalf of the AEC, used this documentary record with great effect to shatter the coherence of Oppenheimer's self-presentation during the hearings.[81] Robb was aided by the fact that Oppenheimer and his counsel did not have access to key pieces of evidence. The FBI file, for example, though read by the security board, was not available to Oppenheimer and his attorneys, nor were many of the key classified documents used in the case.[82] When, early on in the proceedings, Oppenheimer's attorney Lloyd Garrison referred to a wartime telegram from Oppenheimer to Colonel James C. Marshall of the Corps of Engineers, both Robb and Gray challenged him as to what "a lawyer in private practice, is doing with parts of the files of the Manhattan Engineering District" ("Transcript," 123).

Control over the documentary record gave Robb authority in determining the real history. Oppenheimer very frequently deferred to Robb's description of events, accepting the documentary record as more reliable than his own memory.[83] Robb was greatly assisted by Oppenheimer's readiness to disavow his previous accounts of the Chevalier incident. Early in his testimony, Oppenheimer confessed to having "invented a cock-and-bull story" for Manhattan Project security officers in 1943 when he gave his original account of the incident. When asked why he had claimed that the intermediary had approached three people, he replied, "Because I was an idiot" ("Transcript," 137).

Robb subsequently forced Oppenheimer to acknowledge exactly those places where he had lied in the crucial wartime security interview. At this point in the hearings, Oppenheimer's attorneys had not yet been allowed to see this transcript, even though it was being used in cross-examination. When Robb confronted Oppenheimer with his account to Pash, in which he had described three people in the project being approached, Oppenheimer stated that "this whole thing was a pure fabrication except for the one name Eltenton" ("Transcript," 146). Robb asked, "Why did you go into such great circumstantial detail about this thing if you were telling a cock and bull story?" Oppenheimer responded, "I fear that this whole thing is a piece of idiocy . . . [It] seems wholly false to me" (149).

Oppenheimer still wanted to present himself as having *eventually* told Groves the whole story.[84] But the effect of Robb's cross-examination was to utterly discredit Oppenheimer's veracity. Robb put it to Oppenheimer that he had lied to Groves as well. He produced in support a telegram from Colonel Nichols to security officer Lieutenant Lyall Johnson informing Johnson of Chevalier, whose name Oppenheimer had given Groves the day before. The telegram read, "Oppenheimer states in his opinion Chevalier engaged in no further activity other than three original attempts." Robb drew Oppenheimer's attention to the circumstantial detail still present in this telegram, pointing out that Oppenheimer was "still talking about the three people."[85]

Toward the end of the hearings, Oppenheimer was questioned by Gray about why, if Oppenheimer had been trying to protect Chevalier, he had concocted a false story that "showed that Chevalier was deeply involved, that it was not just a casual conversation, that it would not under those circumstances just have been an innocent and meaningless contact, and that it was a criminal conspiracy" ("Transcript," 887). In his recommendations to the AEC, Nichols made clear the implication: "It is difficult to conclude that the detailed and circumstantial account given by Dr. Oppenheimer to Colonel Pash was false and that the story now told by Oppenheimer is an honest one . . . Is it reasonable to believe a man will deliberately tell a lie that seriously reflects upon himself and his friend, when he knows that the truth will show them both to be innocent?" In other words, the account Oppenheimer had offered during the hearings was more consistent with a motive to lie than the version he had given to Pash. According to Nichols, it was not until after the war, when Oppenheimer was interviewed by the FBI in 1946, that he began to tell the version of events presented at the hearings. And Nichols asserted that Oppenheimer had conferred with Chevalier about the latter's earlier FBI interview and had adapted his own account to be consistent. Nichols concluded, "From all these facts and circumstances, it is a fair inference that Dr. Oppenheimer's story to Colonel Pash and other Manhattan District officials was substantially true and that his later statement on the subject to the FBI, and his recent testimony before the Personnel Security Board, were false."[86]

In their final decision, the commissioners gave weight to Nichols's argument. In their findings "as to 'character,'" they wrote: "In the hearings recently concluded, Dr. Oppenheimer under oath swears that the story he told Colonel Pash was a 'whole fabrication and tissue of lies.' It is not clear today whether the account Dr. Oppenheimer gave to Colonel Pash in 1943 concerning the Chevalier incident or the story he told the Gray Board last month is the true version."[87] Oppenheimer's adversaries were able to claim that his duplicity was such that it was impossible to determine when he was telling the truth and when he was not.

The general sense that Oppenheimer was a man not to be trusted was all that was required to make a successful case for disqualification on the grounds of "defects of character."[88] This helps to explain why Edward Teller's testimony was so damaging. What was so significant in Teller's testimony was his expression of a feeling of vague unease about Oppenheimer's moral reliability: "I thoroughly disagreed with him in numerous issues and his actions frankly appeared to me confused and complicated. To this extent I feel that I would like to see the vital interests of this country in hands which I understand better, and therefore trust more" ("Transcript," 710). This attack on Oppenheimer's character meshed perfectly with Robb's strategy.

"WRITING A MAN'S LIFE": IDENTITY, SOLIDARITY, AND THE SYSTEM

The cases for and against Oppenheimer divided largely along the lines of the two approaches to evaluating a security risk: "Caesar's wife" versus the "whole man." Whereas Robb emphasized particular incidents that would count as black marks against Oppenheimer's character, the physicist's defense counted on recontextualizing such items within a general portrait of Oppenheimer as a man. This was Oppenheimer's strategy in his letter of reply to Nichols's charges: "The items of so-called derogatory information set forth in your letter cannot be fairly understood except in the context of my life and my work" ("Transcript," 7). The very intimate quality of his account of himself in that letter contrasted strongly with Nichols's dry and disjointed bureaucratic language. But the argument behind the "whole man" approach was put forward most forcefully by Oppenheimer's friend and colleague Isidor Rabi.

Robb attempted to co-opt Rabi by getting him to accept the primacy of the "facts" contained in the files, to which Rabi did not have access. Rabi, however, resisted this attempt by Robb to prioritize documentary "information" over his own personal experience with Oppenheimer: "I am in possession of a long experience with this man, . . . and there is a kind of seat of the pants feeling [on] which I myself lay great weight . . . I have seen his mind work. I have seen his sentiments develop." Rabi therefore rejected the idea that the file could speak for itself. "You have to take the whole story . . . That is what novels are about. There is a dramatic moment and the history of the man, what made him act, what he did, and what sort of person he was. That is what you are really doing here. You are writing a man's life" ("Transcript," 469–70).

As well as the tension between the two criteria for judging a security risk, the hearings centrally involved a related conflict: that between two conceptions of trust. In the first conception, trust was personal, founded on face-to-face interaction, and involved bonds of collegiality, friendship, and even family. In the second, it was impersonal and bureaucratic, operating through formal methods such as surveillance and record-keeping. The contrast between these two frameworks of trust was highlighted when Robb pressed Oppenheimer on why he had felt it safe to bring his brother to Los Alamos, knowing that Frank had been a Communist. Robb asked, "Tell us the test that you applied to acquire the confidence that you have spoken of?" Oppenheimer replied, "In the case of a brother you don't make tests, at least I didn't . . . I knew my brother." In Robb's hands, this became further evidence that Oppenheimer was not scrupulous in his application of security regulations: "I see," the lawyer said. "In other words, you felt that your brother was an exception" ("Transcript," 111).

Mathematician John von Neumann also highlighted the difference between bureaucratic and personal frameworks of trust, but in relation to collegial relationships. He told the board that they should take into account scientists' general inexperience during the war in dealing with "a universe we had not known before." What made this "Buck Rogers universe" new and particularly disturbing was "this peculiar problem of security." The problem lay in attempting to adjust to a way of thinking in which "people who looked alright might be conspirators and might be spies." Such things "do not enter one's experience in normal times" ("Transcript," 649–50). Particularly difficult was the requirement not to accept the facework commitments of others, not to base one's judgment of people on the fact that they "looked alright." In place of face-to-face forms of accountability, the security system substituted surveillance and documentation. The Oppenheimer hearings were, in a sense, the culmination of this process and of the conflicts that were bound up with it.

The AEC's official rhetoric demanded the subordination of the local, familiar, and personal to formal channels and bureaucratic procedures.[89] The need to subordinate personal inclinations to the supposedly transcendent requirements of the security system was a central trope in the hearings. The PSB's findings spoke of the requirement for "protection and support of the entire [security] system itself." A proper attitude toward security demanded, the board said, "a subordination of personal judgment as to the security status of an individual as against a professional judgment in the light of standards and procedures when they have been clearly established by the appropriate process. It must entail a wholehearted commitment to the preservation of the security system."[90] The ideology of security presented local loyalties to friends and colleagues as potentially at odds with loyalty to state and nation.

Oppenheimer's refusal to name Chevalier was taken as evidence that he would place his loyalty to friends, and his own judgment regarding their reliability, above his duties to the security system.[91] The PSB's report stated that Oppenheimer "repeatedly exercised an arrogance of his own judgment with respect to the loyalty and reliability of other citizens to an extent which has frustrated and at times impeded the workings of the system." "Loyalty to one's friends," the board said, "is one of the noblest qualities." However, "being loyal to one's friends above reasonable obligations to the country and to the security system . . . is not clearly consistent with the interests of security." As a result of this loyalty to friends, it was argued, Oppenheimer had "a tendency to be coerced, or at least influenced in conduct over a period of years." Oppenheimer's "susceptibility to influence" was referred to prominently as a reason for denying him clearance.[92]

This distrust of the local and the familiar extended to the collegial ties of the scientific community. The PSB expressed concern about what it called the "solidarity" of the scientific community:

The Board has been impressed, and in many ways heartened by the manner in which many scientists have sprung to the defense of one whom many felt was under unfair attack. This is important and encouraging when one is concerned with the vitality of our society. However, the Board feels constrained to express its concern that in this solidarity there have been attitudes so uncompromising in support of science in general, and Dr. Oppenheimer in particular, that some witnesses have, in our judgment, allowed their convictions to supersede what might reasonably have been their recollections.[93]

Adherence to the security system, therefore, required the formation of a particular model of identity and solidarity in which loyalty to the state was expected to trump all other commitments.[94]

CHARISMA AND THE HYDROGEN BOMB

Paradoxically, alongside the charge that Oppenheimer was unduly susceptible to being influenced by other people, it was also claimed during the hearings that he himself had an extraordinary capacity to influence others; that he was, so to speak, a scientific Svengali. Oppenheimer's perceived charisma was a significant theme running through both the testimony and the findings. Underlying this theme was the question of the role of the scientific adviser and the limits of scientific authority. Oppenheimer's charismatic power to persuade was portrayed as destabilizing the boundaries of scientific authority, blurring the "technical" into the "moral" and "political."

The PSB chairman, Gray, expressed the board's view of the importance of the issue of the H-bomb in the case against Oppenheimer. Oppenheimer's opposition to the H-bomb, it was alleged, departed from attention to the purely technical and was motivated by moral and political concerns in which Oppenheimer had no special authority ("Transcript," 250).

It was in defending himself against this charge that Oppenheimer made his famous remarks about the H-bomb problems being "technically sweet," a phrase he repeated twice. Early on in the proceedings, he speculated on how the GAC's response to the H-bomb question in 1949 would have been different if the technical problems of its development had then been solved to the degree that they were two years later:

It is my judgment in these things that when you see something that is technically sweet, you go ahead and do it and you argue about what to do about it only after you have had your technical success. That is the way it was with the atomic bomb. I do not think anybody opposed making it; there were some debates about what to do with it after it was made. I cannot very well imagine if we had known in late 1949 what we got to know by early 1951 that the tone of our report would have been the same. ("Transcript," 81)

Later on in the hearings, he said, "The program we had in 1949 was a tortured thing that you could well argue did not make a great deal of technical sense. It was therefore possible to argue also that you did not want it even if you could have it. The program in 1951 was technically so sweet that you could not argue about that" ("Transcript," 251). This statement was a useful rhetorical device, painting Oppenheimer qua scientist as a different emotional entity from his political or moral identity. He admitted that he was greatly troubled "when it became clear to me that we would tend to use any weapon we had." But he reconciled this with his role as a scientist by claiming that there was a "sharp distinction" between the questions of whether to build the bomb and whether to use it (250).

The scientist's concern, Oppenheimer admitted, should be focused on the technical feasibility of building the bomb, rather than the moral and political questions of its use. It was for that reason, he asserted, that the GAC separated such "general advice" touching on questions of morality and politics into the annexes of the 1949 report ("Transcript," 236). He admitted that he did bring some "freight . . . into the General Advisory Committee, and into the meetings that discussed the hydrogen bomb." This freight was that while he believed that nuclear weapons would "put an end to major total wars . . . the notion that this will have to come about by the employment of these weapons on a massive scale against civilizations and cities has always bothered me." But, he said, "I know of no case where I misrepresented or distorted the technical situation in reporting it to my superiors or those to whom I was bound to give advice and counsel" (87).

Despite Oppenheimer's assertion of the clarity of the boundaries between the technical and the political, in practice they were always ambiguous and uncertain. Anxiety about Oppenheimer's "influence" and powers of persuasion was closely bound up with the perception that these boundaries were permeable and hence that the powers of the adviser were pervasive, and Robb used this anxiety to his advantage. The political scientist Sanford Lakoff aptly described this strategy of Oppenheimer's accusers as the "imputation of serpentine charisma."[95] Physicist Luis Alvarez told the board that Oppenheimer was "one of the most persuasive men that has ever lived" ("Transcript," 803). Wendell Latimer, associate director of the Berkeley Radiation Laboratory and an advocate of the H-bomb, emphasized that Oppenheimer used his great "influence" in order to hinder the H-bomb program. As an example of Oppenheimer's personal power, Latimer said that during the war, General Groves had been very much under Oppenheimer's influence. Groves was "simply an administrator"; it was Oppenheimer who was doing "the thinking for the program." According to Latimer, Groves found Oppenheimer "overwhelming" and was "so dependent" on Oppenheimer's judgment "that I think it is reasonable to conclude that most of his ideas were coming from Dr. Oppenheimer" (663).

Over the years that he knew Oppenheimer, Latimer made what he called a "study" of the man, which he described as "most interesting" and even "amazing." "Unconsciously," he said, "I think one tries to put together the elements in a man that make him tick." Latimer dwelled on Oppenheimer's power as chairman of the GAC: he felt unable to understand how the GAC could, on rational grounds, warn against an H-bomb crash program. Searching for an explanation, he claimed, led him to hit upon Oppenheimer's magnetic influence. "I kept turning over in my mind how they could possibly come to such conclusions, and what was in Oppenheimer that gave him such tremendous power over these men." He described Oppenheimer's charismatic powers: "I have seen him sway audiences. It was just marvelous, the phraseology and the influence is just tremendous" ("Transcript," 663–64).

Even Conant was swayed, in Latimer's opinion. Himself a chemist, Conant, "in matters pertaining to theoretical physics, . . . trusted Dr. Oppenheimer completely." Oppenheimer's attorney asked Latimer "whether Dr. Conant's judgment in connection with the hydrogen bomb was based on a technical evaluation . . . a judgment as to the nuclear aspects of the problem, the scientific nuclear aspects of the problem." Latimer replied, "Those were the reasons which were given in the report. They were expressed in technical terms. I was by no means convinced that those were the real reasons behind the decision." These "technical reasons" sounded to Latimer "pretty phony." He admitted that they "would have been legitimate reasons if he [Conant] had been exercising his free judgment." However, Conant was "overwhelmed by his great confidence in Dr. Oppenheimer's judgment," and Latimer therefore doubted that "it was free judgment on his part" ("Transcript," 664 65).

Prima facie, the reasons given by the GAC against the H-bomb program (including the worry that it would divert manpower away from other needed programs) were technical, and therefore legitimate (although in Latimer's opinion, wrong). What rendered Conant's opinion illegitimate and his technical reasons "phony," in Latimer's view, was that it was possible to give a *causal,* quasi-sociological account of the origins of his beliefs, an account that Latimer framed in terms of Oppenheimer's charismatic influence.[96] The technical reasons that Conant and other members of the GAC gave for opposing the H-bomb were "phony" because they were pressed upon them by the powerful personal influence of Oppenheimer and were a front for Oppenheimer's political opposition to the H-bomb.

Oppenheimer's supporters also regarded him as charismatic, but they portrayed his charisma in a very different way. Caltech president Lee DuBridge said that, as chair of the GAC, Oppenheimer "was so naturally a leader of our group that it was impossible to imagine that he should not be in the chair ... He was the natural leader because we respected his intelligence, his judgment, his personal attitude toward the work of the Commission ... He was a natural and respected and at times a loved leader of that group."[97] DuBridge denied that this charisma was in any way insidious. On the contrary, he presented Oppenheimer as aiding the rational consensus of scientific opinion: "He encouraged a full and free and frank exchange of ideas throughout the full history of the Committee ... He never dominated nor suppressed contrary or different opinions" ("Transcript," 518). Rabi said that "it was always a miracle to the other members on the [GAC] how he could summarize three days of discussions and give the proper weight to the opinion of every member, the proper shade, and it rarely happened that some member would speak up and say, 'This isn't exactly what I meant.' It was a rather miraculous performance."[98] On this account, Oppenheimer's charismatic leadership brought out the consensus of the group and created an atmosphere in which others were able to present their views freely. Whereas Latimer presented Oppenheimer's charisma as anathema to reason, Rabi painted Oppenheimer as a disinterested vehicle of rational discourse.

The nature of Oppenheimer's personal influence was thus contested during the hearings. Centrally at stake was whether such charismatic personal authority and influence was compatible with the role of the scientist as adviser to government and servant of the state. To his opponents, Oppenheimer's charisma suggested that his powers were broad and undefined. It meant that his influence could breach the boundaries of the "technical" and that he could exercise "political" influence through his scientific advice, an influence that they believed to be incompatible with the interests of national security. His supporters, however, sought to dissociate charisma from worries about "influence" by portraying Oppenheimer as disinterested. While not denying that he had definite ideas of his own, they argued that what made him charismatic was precisely his willingness to suppress his own viewpoint for the purpose of rational debate and consensus. Rather than pushing any particular policy line, Oppenheimer, they said, helped to create a communicative environment in which a consensus could emerge freely and rationally. In contradistinction to what Lakoff called "serpentine charisma," they associated Oppenheimer with an ideal of communitarian scientific debate.

REASONS AND MOTIVES:
TRYING A MAN FOR HIS OPINIONS?

Was Oppenheimer's opposition to the H-bomb in itself sufficient to make him a risk to national security? Could the content of his scientific and policy advice legitimately be the subject of the case against him? Significant problems of legitimacy were involved in the pursuit of a case against Oppenheimer based on his stance toward the H-bomb. A number of senior scientists testifying before the board strongly condemned this aspect of the case.[99] The strongest critics were Conant and Vannevar Bush. Conant had himself been an outspoken critic of the H-bomb program, and he told the committee of his concern that the wording of Nichols's letter outlining the charges against Oppenheimer "would indicate that anybody who opposed the development of the hydrogen bomb was not eligible for employment on atomic energy work later." This, he argued, would be in conflict with American liberal democratic principles: "Such a position would be an impossible position to hold in this country" ("Transcript," 384). Bush presented the argument even more forcefully: "I think in fact the Republic is in danger ... I think this board or no board should ever sit on a question in this country of whether a man should serve his country or not because he expressed strong opinions." To prosecute a man for his opinions, he said, is "contrary to the American system."[100]

There was, however, an alternative construction of the relevance of the H-bomb issue to the case, one that was seen to sidestep such objections.[101] Bush himself, for example, stated that if the charges against Oppenheimer had been worded to say that "by improper motivation because this man had allegiance to another system than that of [this] country, he expressed these opinions in an attempt to block the program, then I would not have objected" ("Transcript," 567). While he was not endorsing that charge of improper motivation, Bush had to agree that his strong objections of principle would not apply to it. This was where allegations regarding Oppenheimer's associations and character became relevant to the assessment of his scientific advice. Conant recognized that the allegations regarding the H-bomb implicitly referred back to these prior charges: "It seems to me that . . . the implied indictment [is that] because of . . . Dr. Oppenheimer's association with alleged Communist sympathizers in the early days in his youth—that that somehow created a state of mind in Dr. Oppenheimer so that he opposed the development of the hydrogen bomb for what might be said reasons which were detrimental to the best interests of the United States, because they were interests of the Soviet Union which he in one way or another had at heart."[102]

On such a view, the issue of Oppenheimer's H-bomb advice, if it was to be relevant, had to be convincingly related to other aspects of the case against him, in particular the charges regarding his character, associations, and earlier history. This argument was in fact made by Oppenheimer's own attorney. Garrison announced early on in the hearings that he would present a *lay* case. The proper question, he told the board, was not whether Oppenheimer's scientific judgment was right or wrong but whether this was "an honest judgement." In other words, did Oppenheimer "do the best he could for his Government"? ("Transcript," 23). The critical issue was to be Oppenheimer's motivation. As Garrison put it after the proceedings had ended, "If Dr. Oppenheimer's motives were honorable, his recommendations were irrelevant."[103]

Garrison's strategy in shifting the question from reasons to motives suggests that he was confident of his ability to rebut the charges of disloyalty. However, the strategy was risky in that it allowed the issue of Oppenheimer's H-bomb advice to blur into, and be colored by, the more general doubts about his loyalty and character. In the hearings, questions of character were intimately related to the examination of the boundaries and legitimacy of Oppenheimer's scientific authority. Deciding the legitimacy of scientific advice became a question about the personal trustworthiness of the scientific adviser.

DISCIPLINING EXPERTS: THE BOARD'S FINDINGS

The recommendations of the Personnel Security Board explicitly cited Oppenheimer's H-bomb advice as a reason for withholding clearance: "We find his conduct in the hydrogen-bomb program sufficiently disturbing as to raise a doubt as to whether his future participation, if characterized by the same attitudes in a Government program relating to the national defense, would be clearly consistent with the best interests of security."[104] However, the type of objections raised by Bush and Conant, drawing as they did on powerful liberal democratic conceptions of political legitimacy, were never entirely dispelled. In framing the case against Oppenheimer, his accusers implicitly recognized the power of these challenges to the legitimacy of the proceedings, and the need to counter such objections. Gray and Morgan, in their findings (which formed the majority report of the PSB), felt compelled to dwell at length on these objections: "It is our conclusion that, whatever the motivation, the security interests of the United States were affected" by Oppenheimer's opposition, or, as they notoriously put it, his insufficient "enthusiasm" for the H-bomb ("Findings," 1017). In place of a distinction between loyal and disloyal motives, the board substituted a distinction between the technical and the moral, limiting the authority of the scientific adviser to the technical. It was this shift away from the question of the loyalty of Oppenheimer's motives that allowed the board to find—in their view, without contradiction—that Oppenheimer was both loyal and a security risk. But this move was arguably an expansion of the domain of motives now deemed illegitimate. Gray and Morgan regarded as potentially subversive any role for scientific advice beyond narrowly construed technical problems. It was Oppenheimer's concern with moral issues that made him a security risk. The implicit reply to Bush and Conant was, then, that the freedom of expert scientific opinion was to be respected, but that scientific opinion was to be regarded as "free" and "expert" only so long as it was separate from moral and political concerns.

The board members considered it their duty not only to make specific recommendations concerning Oppenheimer, but also to define the norms that would govern scientific advisers' sphere of authority. Their report stated that "one important consideration brought into focus by this case is the role of scientists as advisers in the formulation of Government policy." Bush invoked the democratic republic in defense of Oppenheimer's freedom of opinion. The board appealed to similar principles, but for the opposite reason:

> As a Nation we find it necessary to delegate temporary authority . . . to duly elected representatives and appointive officials as provided for by the Constitution and laws. For the most part, these representatives and officials are not capable of passing judgment on technical matters and, therefore, appropriately look to specialists for advice . . . These specialists have an exponential amplification of influence which is vastly greater than that of the individual citizen. ("Findings," 1015)

This power of experts was not only vast but also ill-defined and ambiguous. Gray and Morgan saw the PSB's task as establishing clear limits on this power. Their solution was to call for the bounding of the authority of experts within the realm of the "technical": "A question can properly be raised about advice of specialists relating to moral, military and political issues, under circumstances which lend such advice an undue and in some cases decisive weight. Caution must be expressed with respect to judgments which go beyond areas of special and particular competence" (1016).

For Gray and Morgan, the boundary dividing the technical from the moral was identical with a boundary between objectivity and emotion. They warned, "Those officials in Government who are responsible for the security of the country must be certain that the advice which they seriously seek appropriately reflects special competence on the one hand, and soundly based conviction on the other, uncolored and uninfluenced by considerations of an emotional character." Emotions, they demanded, must be regulated according to the national interest, and dominated by patriotism: "Emotional involvement in the current crisis, like all other things, must yield to the security of the nation." They mentioned, as an example of such "emotion," the feelings of guilt expressed by Oppenheimer and other scientists who had worked on the Manhattan Project. Gray and Morgan suggested that this guilt had clouded Oppenheimer's technical judgment. They concluded this section of the report with general recommendations for the political evaluation of expert advice: "In evaluating advice from a specialist which departs from the area of his speciality, Government officials charged with the military posture of our country must also be certain that underlying any advice is a genuine conviction that this country cannot in the interest of security have less than the strongest possible offensive capabilities in a time of national danger" ("Findings," 1016). The national interest was thus explicitly defined in militaristic terms. And patriotism was rhetorically united with objectivity and reason against subjectivity and emotion.

The board's findings with regard to Oppenheimer were presented as a simple application of these general principles. This provided a response to Bush: the board could claim that it did not "question Dr. Oppenheimer's right to the opinions he held" and that it was "willing to assume that they were motivated by deep moral conviction." The key criticism, however, was precisely that in the board's view, Oppenheimer's objections to the H-bomb were *moral.* As one commentator put it in a book published a year later, "They did not care what his moral scruples were. It was the fact that he had any at all which was derogatory and which made him a security risk."[105]

This exclusion of moral concerns was demanded both by science and by patriotism. Gray and Morgan wrote, "We are concerned . . . that he may have departed his role as scientific adviser to exercise highly persuasive influence in matters in which his convictions were not necessarily a reflection of technical judgment, and also not necessarily related to the protection of the strongest offensive military interests of the country" ("Findings," 1017–18). Here again was a militaristic definition of the national interest and the use of the national interest as a transcendent standard against which expertise was to be judged. This framework also casts light on Oppenheimer's "lack of enthusiasm" for the H-bomb program. "He did not show the enthusiastic support for the program which might have been expected of the chief atomic adviser to the Government . . . [and this] undoubtedly had an effect upon other scientists" (1017). The implication was that experts should demonstrate self-control in calling up only those emotions appropriate to the national interest. The expert's emotions were to be rigidly disciplined by an underlying patriotism. The emotional hardness of the hard scientist became, through these tropes, identical with the soldierly hardness of the patriot.[106]

JUSTIFICATION AND ACTION

The PSB's goal of closing the case with a neat statement of principle proved impossible to sustain in the face of competing political pressures. In particular, the charges regarding Oppenheimer's stance toward the H-bomb, from which the PSB's statement on the role of the adviser followed, became increasingly politically difficult. The instability of the PSB's position was signaled initially by the fact that the only scientist on the board dissented strongly from the majority opinion. Ward Evans wrote that there was "nothing wrong with [Oppenheimer's] character." On the matter of the H-bomb, Evans said that "he did not hinder the development of the H-bomb and there is absolutely nothing in his testimony to show that he did." And Evans rejected worries about Oppenheimer's powers of "influence": "If his opposition to the H-bomb caused any people not to work on it, it was because of his intellectual prominence and influence over scientific people and not because of any subversive tendencies." Most significantly, Evans expressed his concern about the effect that finding against Oppenheimer would have on American science: "His witnesses are a considerable segment of the scientific backbone of our Nation and they endorse him. I am worried about the effect an improper decision may have on the scientific development in our country." Fear of alienating the scientific community was the key factor that led to the AEC's equivocation over what place the issue of the H-bomb should have in the verdict against Oppenheimer.[107]

As the matter of J. Robert Oppenheimer passed from the PSB to the general manager to the AEC itself, the H-bomb allegations were subordinated to the more general issues of "character" raised by the Chevalier incident. The AEC commissioners' equivocation over what place the issue of the H-bomb should have in their verdict was due in part to fear of alienating the scientific elite. But probably more important was the fact that the premise—that there had been a delay in the production of the bomb—had been publicly denied by Eisenhower in response to attacks by McCarthy. The PSB's finding, since it implied such a delay, was in danger of embarrassing the administration.[108] Nichols now backed away from his inclusion of the H-bomb issue in the original charge. His prose was tortured as he attempted to drop the hot potato of the H-bomb allegations while at the same time defending his original decision to include them. It was necessary for the board to consider the H-bomb issue, he stated, "in order that the good faith of [Oppenheimer's] technical opinions might be determined." He conceded that although Oppenheimer had not shown appropriately "enthusiastic support" for the H-bomb program, no sinister motives were established. Instead, Nichols asserted that his chief concern was with the issue of character and associations.[109]

The majority report of the AEC was drafted by Lewis L. Strauss and signed by Eugene Zuckert and Joseph Campbell. Following Nichols's new position, they argued that they had not been swayed by the issue of Oppenheimer's H-bomb recommendations. However, Strauss brought to prominence another allegation not mentioned in the PSB's report at all. This was the charge that Oppenheimer had misled the PSB when he said that the GAC was "surprisingly unanimous" against the H-bomb. Glenn Seaborg, who had been in Sweden when the GAC met, had in a letter expressed opinions favorable to development of the H-bomb, and Strauss claimed that Oppenheimer had covered this up.[110] This allegation connected back to the issue of character, which was for Strauss, Zuckert, and Campbell the primary justification for finding Oppenheimer a security risk. The alleged lie regarding Seaborg's H-bomb opinion was listed in a section labeled "As to 'character,'" together with the Chevalier incident, as evidence of Oppenheimer's personal unreliability and "fundamental defects in his 'character'" ("Decision and Opinions," 1049).

Zuckert appended to the commission's statement his own concurring opinion. He also felt it important to state that he was not condemning Oppenheimer for his opinions. Oppenheimer's advice on the H-bomb did not provide grounds for finding him a security risk, since there was no evidence of improper motive ("Decision and Opinions," 1055). Zuckert said that he had "considered the evidence as a whole and no single fact was decisive." What concerned him was the accumulation of questionable incidents and associations. He wrote that "when I see such a combination of seriously disturbing actions and events as are present in this case, then I believe the risk to security passes acceptable bounds" (1052).

Campbell likewise stated that his decision was based not on the H-bomb issue, but on "character, loyalty, and associations." But he gave no further details as to the exact evidence that led to his view. Rather, he blandly stated that he regarded the PSB members as men of "honor and integrity, and that in their majority opinion Dr. Oppenheimer did not refute the serious charges which faced him" ("Decision and Opinions," 1057, 1058). Therefore, he said, it was his duty to uphold the recommendation of the PSB and the general manager to withhold clearance.

Commissioner Thomas Murray did not sign the majority report but instead drafted his own concurring opinion. He contradicted outright the PSB on the issue of H-bomb advice, stating, "Government cannot command a citizen's enthusiasm for any particular program or policy projected in the national interests. The citizen remains free to be enthusiastic or not at the impulse of his own inner convictions" ("Decision and Opinions," 1059). Yet despite dismissing the H-bomb allegations, Murray stated flatly that Oppenheimer was "disloyal." This conclusion depended on a definition of loyalty as strict adherence to security regulations. Oppenheimer was not "scrupulous in his fidelity to security regulations"; ergo, he was disloyal. Murray adopted the strict "Caesar's wife" criterion of security. Whether or not Oppenheimer had revealed secrets, his associations themselves constituted a breach of security rules. "No matter how high a man stands in the service of his country," Murray argued, "he still stands under the law. To permit a man in a position of the highest trust to set himself above any of the laws of security would be to invite the destruction of the whole security system" (1061).

The only AEC commissioner to support Oppenheimer's continued security clearance was the physicist Henry D. Smyth. He emphasized that the only question that mattered was whether Oppenheimer was likely to reveal secrets to the nation's enemies. In his view, nothing in the evidence presented suggested such a danger. This included the Chevalier incident, which the other commissioners thought so obviously damning. Smyth's definition of the problem also meant that Oppenheimer's H-bomb advice was irrelevant. He dealt with this matter only in order to address the question of whether Oppenheimer had suppressed Seaborg's opinion in the GAC report. Smyth doubted that Oppenheimer had intentionally suppressed the letter, and he added, contrary to Strauss, that Seaborg's letter did not express a formal conclusion in favor of the H-bomb ("Decision and Opinions," 1064). Moreover, Smyth radically disagreed with the other commissioners on the status of the security regulations. His comments sharply contradicted Murray's view, for example, that adherence to the system had to be perfect: "I would suggest that the [security] system itself is nothing to worship. It is a necessary means to an end . . . If a man protects the secrets he has in his hands and his head, he has shown essential regard for the security system." Smyth defended Oppenheimer, saying that the physicist's "further employment will continue to strengthen the United States" (1065).

Differences between the reports of the PSB and the AEC on the validity and significance of the H-bomb charges showed residual uncertainties over the normative framework by which science advisers were to be held accountable. The AEC was never going to succeed in formulating abstract standards demarcating a person's responsibilities as scientist, official, citizen, and human being. The justifications in the final report for the withdrawal of clearance were contradictory and confusing. However, while offering variant justifications, the majority on the PSB and among the AEC commissioners were agreed on the verdict: Oppenheimer's clearance was to be denied. The real message of the hearings was contained not in the report, but in the action taken.

The hearings were, above all, drama and ceremony, and at their center was the person of Oppenheimer. In the wake of the PSB's findings, the secretary of the American Physical Society, Karl Darrow, wrote to his colleagues Raymond T. Birge and Hans Bethe:

> I think that we can no longer handle this issue by just stating principles. The three members of the Board apparently had much the same principles, yet one of the three differed from the two others in the application of these principles. In my inclination to side with the minority of the Board, I find that I am not moved by dissent from the principles expressed by the Board, but by my feeling that Robert is perfectly safe and deserves the confidence of the nation. The crux of this matter is, that it is impossible to draw up principles which will pass some people and stop others, without stating very exactly what are the qualities of the people who should be passed; but this amounts to saying that these people are people like Robert, so that Robert himself becomes part of the definition of the principles.[111]

The disagreement between the two sides, Darrow argued, concerned not abstract principles, but Oppenheimer himself. The true meaning of the issues could not be detached from the man around whom the controversy swirled. Darrow's argument points to the symbolic quality of the denial of Oppenheimer's clearance. Struggles over the nature and extent of scientific authority and the relationship between science and liberal democracy converged on the personal role of Oppenheimer and the question of whether his example was to be emulated or repudiated.

A cartoon on the front page of the *New York Journal-American* portrayed the Oppenheimer case as an H-bomb exploding over Washington, D.C.[112] The "detonation" of this case was supposed to be controlled and contained, with access to the proceeding restricted and testimony kept confidential. But it quickly became an all-out battle for public opinion. Some prominent columnists, notably Joseph and Stewart Alsop, vigorously defended Oppenheimer. However, the transcript's revelations of Oppenheimer's confused and contradictory actions and testimony in relation to the Chevalier incident considerably dampened public support for the physicist.[113] In particular, newspapers emphasized the theme, stressed by the PSB and especially Commissioner Murray, of the supremacy of law over the individual person. Although there were numerous voices of dissent, the strongest tenor of the newspaper reaction was that the case exemplified the proper functioning of the security system. For many, the very fact that such a prominent figure as Oppenheimer could be proceeded against demonstrated the impartiality of the system. Efforts to excuse Oppenheimer were frequently cast as mere "special pleading."

The *New York Herald Tribune* provides a particularly interesting lens on opinion about the case. Its letters pages contained many statements in support of Oppenheimer, as well as plenty against him. Its opinion pages featured the Alsops' forceful defense of Oppenheimer in their column "Matter of Fact," as well as regular attacks on the physicist by right-wing journalist David Lawrence in his column "Today in Washington."[114] In its editorials, the *Herald Tribune* from the beginning tried to defend the legitimacy of the proceedings against the charge that they were merely a form of McCarthyism. It stated two days after the hearings began, "The encouraging aspect of the case is that the investigation should be in good hands and under sound procedures."[115] Following the PSB's report, the paper's editors asserted that the Gray Board

is a board of outstanding men and it has performed its arduous task with a seriousness, with a sense of responsibility and a feeling for the gravity of the issues involved which deserves the highest praise ... Given the limits set for them and the evidence presented, they could scarcely have avoided the conclusion that Dr. Oppenheimer falls within the category of a "security risk." Under the same laws others have been held security risks time and time again. Dr. Oppenheimer, despite his immense scientific contributions, stood—as under a government of laws he must stand—on an equal footing with those whose genius has been of a far lesser order.

"While the final determination remains to be made by the Atomic Energy Commission," the editors wrote on June 3, "the Gray report will stand as a vital document on its own merits."[116]

The *Herald Tribune* stated in an editorial on June 8 that the purpose of the government's security policies was to create "a number of tests which permit boards to establish, factually and objectively, whether an individual falls within the classification of a 'security risk.' Factually and objectively it has been determined that Dr. Oppenheimer does so fall."[117] When the AEC delivered its verdict, the editors asserted that "the matter can be expected to end here."[118]

Yet alongside such statements presenting the hearings as normal bureaucratic and legal procedure, there was also the argument that the pressing Communist threat made extraordinary measures necessary. The *Herald Tribune* editors wrote, "What weighs over everything else is the danger in which America finds itself. It is confronted by an enemy as implacable as resourceful, adopting every means of infiltration and subversion, taking advantage of the smallest carelessness or weakness to work its fatal poison. Special standards, special laws and regulations are called for in such a time." The editors argued that the AEC was "in a position to be supremely aware of the mortal Communist threat. In the Oppenheimer case the majority of its members have acted so as to avoid, as far as humanly possible, any flaw in the security regulations that might betray us now or later."[119]

The *New York Times* columnist Arthur Krock greeted the AEC's finding as a conclusive refutation of the arguments by Oppenheimer's supporters that the benefits to government of his scientific advice outweighed the risks. Referring to findings by the PSB, the general manager, and the AEC, the *Times* stated, "This viewpoint has now been impressively rejected after three fair and painstaking examinations of Oppenheimer as 'the man himself,' and as an official from whom the nation, having entrusted him with its deepest confidence, had a right to expect very different conduct."[120] Editorial pages across the country came down overwhelmingly against Oppenheimer and in support of the AEC's verdict. It was widely accepted that Oppenheimer was guilty of disreputable conduct. The *New York Post* said, "Dr. Oppenheimer is clearly guilty of arbitrariness and deceit," and the *Philadelphia Inquirer* stated, "The tragedy is not in the decision. It was in Oppenheimer's conduct which made that decision necessary." Common themes were that the verdict was necessary and unavoidable and that the fate of the individual was subordinate to the security of the nation. Many journalists wrote of the verdict and even of the proceedings as a vindication of the security system and an instantiation of the democratic principle that all are equal before the law. The *New York Journal-American* said, "No man or woman is, or ever can be, greater than the security of the nation." The *Detroit News* wrote, "Acting as a law unto himself, Dr. Oppenheimer flouted certain . . . rules. The transgression was no more tolerable in him than in any lesser man if general respect for the system is to be preserved." The *Los Angeles Times* asserted that "he willfully broke the rules—broke them with assurance, even with arrogance, as if they were not made for the special breed of which he is a member."[121]

Reaction regarding the issue of the GAC's opposition to the H-bomb was more complex. In revealing the struggles of 1949–50 for and against the H-bomb, the hearings pulled back the curtain on the political conflicts within the country's scientific elite. The *New York Times* reported, "The investigation merely brought into the open and intensified one of the most dramatic hidden conflicts of our times—the very wide schism that has split a part of the country's scientific community ever since World War II."[122] William L. Laurence wrote that the hearings had focused attention on the "momentous debates in the winter 1949–1950. Those behind-the-scenes arguments, of truly Homeric dimensions, raged over whether to proceed with a 'crash program' to develop and produce the hydrogen bomb with all possible speed."[123] The public response to these conflicts, and the complex intertwining of technical with moral and political issues revealed by the hearings in the case of the H-bomb, was worry about the relationship between science and politics and calls for policing the boundaries. If politics militated against disinterested inquiry, science was equally distorting of politics. A key lesson of the PSB's report, repeated widely in the press, was that the authority of scientists within the polity should be strictly delimited. Waldemar Kaempffert remarked, writing in the *New York Times Magazine* early on in the proceedings, that the case provided an "x-ray of the scientific mind." The scientific mind, he said, was characterized by "objectivity, curiosity, [and] skepticism." However, "the scientist does not necessarily apply these qualities with brilliance outside his own field." "The plain truth," Kaempffert asserted, "is that there is nothing unique about the 'scientific mind.' It has no monopoly on objectivity."[124] A *New York Times* headline after the release of the PSB's findings read, "Scientists' Views

Stir Panel Worry: U.S. Is Warned to Evaluate with Care Nontechnical Opinions of Experts."[125] Yet one scientist—Dr. John R. Schenken, president of the International Congress of Pathology—called for his colleagues not to be intimidated into being mere courtiers to the state. The *New York Times* reported that he warned scientists "not to become 'political eunuchs' and urged them to speak out forthrightly in the cause of freedom." He blamed "the 'modern tragedy' of Dr. J. Robert Oppenheimer, atomic scientist, on Oppenheimer's 'political celibacy.'"[126]

The hearings crystallized tensions between competing understandings of the legitimate place of scientists and scientific expertise in the American polity. After World War II, science was valued both as a manufactory of military and technological power and as a source and symbol of political legitimacy. It was frequently portrayed as embodying core values of liberal democratic civil society, and it was looked to as a model of rationality, efficiency, and objectivity for state administration and policy making. But these multiple material and ideological uses of science coexisted uneasily. The PSB's case rested on locating Oppenheimer as a bureaucratic official within the state, charged with efficient execution of the public will, with no authority to make decisions about what ends the government ought to pursue. Oppenheimer's supporters, notably Conant and Bush in the proceedings and the Alsops in the press, instead drew on norms of liberal democratic civil society or the public sphere, in particular freedom of speech and freedom of conscience. According to these supporters, Oppenheimer had every right to exercise his conscience as a citizen in opposing the development of the hydrogen bomb. These competing arguments revealed the tension between a conception of science as an instrumental resource of the state and the image of scientific knowledge and community as free and autonomous components of civil society. Both conceptions were important components of American political culture during this period, but they were never reconciled.

Put on the defensive, Strauss and the AEC were at pains after the hearings to say that Oppenheimer had not been simply purged because he uttered unpopular opinions. But that remained a widespread reading of the significance of the hearings. Scientists at Argonne National Laboratory released a statement warning that "if the consequences to the individual of an unpopular or unwise decision are the same as the consequences of a disloyal act, then the making of decisions . . . will be shunned, and two of the most important ingredients of national strength—faith in the individual's honesty of judgment and willingness to back one's opinions with action—will become increasingly rare." The Alsops said that the lesson of the hearings was "Don't argue!"[127] Dr. James R. Killian Jr., who served as presidential science advisor between 1957 and 1959, said, when looking back on the case more than ten years later, "One of the frightening aspects of the Oppenheimer case was the fear it created . . . that technical advice, when not in support of some current military or political policy, might be condemned."[128]

At the time of the hearings, it was widely remarked that the action against Oppenheimer could jeopardize the entire relationship between scientists and the federal government that had developed from the war and that was essential to the country's national security. Ultimately, however, there was no scientific boycott of weapons work or government service in the aftermath of the case. In fact, though many prominent physicists and other scientists came to Oppenheimer's defense, the scientific community was split in its response to the verdict. Journalist William Laurence reported from the annual meeting of the American Physical Society at the end of April 1954 that the physicists were divided over whether the withdrawal of Oppenheimer's clearance was justified. There were "diametrically opposing views among his colleagues," Laurence said. He quoted one "leading scientist" as saying, "We dismiss generals if they make mistakes, and sometimes even court-martial them. Why should scientists who give bad advice not be subject to the same treatment?" Samuel Goudsmit of Brookhaven National Laboratory commented at the meeting, "It may surprise many of you that there are colleagues, physicists, scientists, who sincerely believe in the possibility of Oppenheimer being a security risk. According to their line of reasoning, Oppenheimer's alleged obstruction of the hydrogen bomb is proof of his disloyalty."[129]

Teller's testimony in the hearings enraged much of the physics community, and antagonisms were brought to the boil in the fall of 1954 by the publication of a book by journalists James Shepley and Clay Blair Jr., *The Hydrogen Bomb: The Men, the Menace, the Mechanism.* Shepley and Blair's hero was Teller, resiliently pursuing his scientific-technological vision and safeguarding the nation's security against an Oppenheimer coterie distracted by "moral" concerns that were tantamount to disloyalty. Nevertheless, although the book provoked a great deal of anger within the scientific establishment, the issues at stake in this controversy were rather narrow. The anger directed at Shepley and Blair essentially concerned how they assigned credit for the achievement of the H-bomb. Los Alamos scientists were incensed by the book's presentation of Teller, and his and Ernest Lawrence's second weapons laboratory at Livermore, California, as almost solely responsible for the H-bomb.

They were angered further by the book's suggestion that Los Alamos, under Oppenheimer's influence, had blocked the weapon.[130] Los Alamos director Norris Bradbury released a rebuttal, asserting the fundamental contribution made by Los Alamos to the development of the weapon. Facing this angry reaction, Teller tried to distance himself from the book. In an effort to defuse the controversy, he published an article in *Science* attributing the successful development of the hydrogen bomb to "the work of many people."[131]

The scientists angered by the book did not publicly question whether the H-bomb's development was right or justified. The controversy concerned only who got the credit, the competing reputations of the two nuclear weapons laboratories, and criticism of the book's biases and vindictive tone. The most insightful criticism of the book was provided by journalist Nat Finney, who reviewed it for the *Herald Tribune*. He argued that the authors missed the point in their targeting of Oppenheimer and the GAC. The problem, Finney said, was that this body should never have been given responsibility for such an important decision in the first place: "The G.A.C. were invited to make a decision of state of transcendent magnitude. There did not seem to be a responsible political official in Washington who understood that the G.A.C. not only should not but could not make such a decision. The G.A.C. was, of course, unwise in the extreme to try to fill the policy vacuum in Washington. But what kind of statecraft permitted such a folly?" The scientific squabbling over the H-bomb highlighted the lack of political will and competence in "decisions where technical and political considerations must be weighed together." For Finney, it was obvious that what was required was a reassertion of political will, instead of passing the buck to scientists who had no business making political decisions.[132]

The transcript of the security hearings had provided the first real public glimpse into the politics and conflicts within the technocratic sanctum of the AEC, and many were disturbed by the revelation of the complex intermixing of politics with technical decision making in the H-bomb controversy. Political fractures within the AEC became increasingly hard to contain in the wake of the hearings. As early as June, while the AEC was still deliberating on the case, the *Herald Tribune* alleged that the Oppenheimer case highlighted a division in the AEC between a committee style of management and the more centralized, executive style that Strauss was introducing: "Chairman Strauss, by the force and vigor of his personality and by his special access to the President, has acquired a commanding position. The team operation of the past has been badly shaken."[133] On May 21, the AEC's director of classification, Dr. James G. Beckerley, a critic of Strauss's "conservative" security policy and an advocate of relaxing some security rules, announced that he would be resigning to join private industry.[134] Smyth, who had dissented in favor of Oppenheimer, resigned in mid-September to return to teaching at Princeton.[135] The following January, General Nichols stepped down from his position as AEC general manager, publicly saying that his resignation was not due to "any conflict of any kind" and that he simply wanted to go into the private sector in order to earn more money.[136]

For Oppenheimer, the significance of the hearings lay both in the withdrawal of his clearance and in the public humiliation to which it subjected him. As one AEC official put it in conversation with Teller, the point had been to "unfrock [Oppenheimer] in his own church."[137] The ceremonial quality of the hearings was key here. Oppenheimer's authority was embodied and performative, and the destruction of this authority had to be similarly dramatic. Because of the media attention the hearings received, the publication of Nichols's letter of charges and Oppenheimer's reply, and the release of the transcript, the hearings were a dramatic public event. In a sense, whatever the verdict, Oppenheimer could not have continued to occupy his former role. The release of the transcript made Oppenheimer's humiliation at the hands of Robb and the PSB publicly available in every detail, as well as opening for public view the "derogatory information" of the previously closed files and interrogations. Readers of the transcript and of the commentary in newspapers and journals saw an Oppenheimer they had not known before: passive, self-contradictory, reduced to admitting that he had been "an idiot."

Midway through the proceedings, Alistair Cooke of the Manchester *Guardian* stated that the case would "test, as no other has done, whether a very distinguished reputation, in the most secret counsels of the Government, can survive the publication of grave charges even if his loyalty and reliability are afterwards affirmed."[138] As Cooke suggested, the very process of these invasive hearings was enough to destroy Oppenheimer's reputation and authority. *The Reporter* magazine's Max Ascoli observed, "All that he did, the common and uncommon part of it, is now in the public domain." Oppenheimer, Ascoli said, was now "at once invulnerable and doomed"— invulnerable because there was no secret left to expose and he had no power left to be taken away from him, doomed because he would never again be trusted with public office. "For no man who has been the object of prolonged, widely publicized security investigation," Ascoli pointed out, "has ever succeeded in gaining a decisive, unalterable clearance. That is the first principle of the

Jurisprudence of Security."[139] In the wake of the PSB's verdict, it was widely reported that Oppenheimer had told an Australian journalist, "Maybe this is the end of the road for me"—though Oppenheimer denied the comment.[140]

The security hearings were certainly the end of the road on which Oppenheimer had been traveling since the war. The denial of security clearance, barring him from involvement with the agencies of the state, meant that although he might speak on contemporary issues, he would thereafter do so as an outsider. In the Cold War, being an expert meant knowing secrets. As he was cut off from state secrets, and from the inner chambers of technoscientific and military decision making, he could no longer speak as an expert with intimate knowledge of nuclear programs and the affairs of state. He was free to make moral or political pronouncements about nuclear weapons, but these would not be confused with, or carry the weight of, expert opinion.

Oppenheimer's fate was read as having broader significance for the position of scientists within the polity. Future scientific advisers would not seek to follow in Oppenheimer's footsteps or emulate his example. The particular combination that Oppenheimer had embodied after the war—of specialized expertise, a powerful position within the state, and broad moral and cultural authority—was now closed off and discredited. If the atomic scientists emerged from Los Alamos and the other wartime laboratories as Promethean figures, with the collective charisma of bearing the mysterious power of the atom, the hearings signified the end of this world-making role.[141] They marked the integration of science into the apparatus of the state, and the routinization and bureaucratization of the scientific role. Sociologist Daniel Bell observed, almost two decades after the hearings, that "what the Oppenheimer case signified was that the messianic role of the scientists . . . was finished."[142]

CHAPTER EIGHT

The Last Intellectual?

The loyalty-security hearings of 1954 altered, but did not end, Oppenheimer's public role. Since he was now excluded from the inner circles of nuclear policy, he could no longer lay claim to technocratic expert authority. But he was able to reconstitute his public and intellectual role and refashion an authoritative self-presentation by drawing on an alternative repertoire. Oppenheimer's uniqueness as a leader of Cold War science had consisted in his ability to hold together the roles of cultivated man and scientific-technical expert. Now, in reconstituting his public role outside the state, he fell back more heavily on repertoires of humanistic cultivation. In doing so, he was able to turn his outsider status into ascetic virtue, providing renewed moral authority. And in line with this new self-presentation, he wove the hearings and his own trajectory into a broader narrative of the tragic fate of the humanistic intellectual in modernity.

This tragic role was often seen as being symbolically manifested by Oppenheimer's body. His physical condition was widely ascribed moral meaning. Journalists Robert Coughlan and Alfred Friendly wrote that the pain of being denounced in the hearings had left him "a thin, gray, shrunken ghost" and that he "grew gray and withdrawn," losing his former "preternatural youthfulness."[1] Victor Weisskopf blamed Oppenheimer's early death on that trauma: "He was a broken man. It was really terrible to see how he sort of sagged after the trial and how he was melancholic and he had no longer the verve and all the qualities he had before. Then he got sick, of course, but in my mind it was a psychological disease, a psychosomatic disease. It was a complete breakdown due to the trial that made him die."[2] Yet the air of suffering that surrounded Oppenheimer in his later years added to his moral authority as ascetic outsider. Abraham Pais commented that Oppenheimer's "charisma" in these years was "enhanced by his now ascetically frail looks."[3]

Some were skeptical about the depth and significance of Oppenheimer's personal transformation. One of his former students said, "I'm afraid he has only assumed a new role in his big repertoire. Just now he appears to be, of necessity, saint and martyr, but if the wind ever changes, he'll be busy again in Washington with the rest of them."[4] But Oppenheimer was never again busy in Washington. When in 1955 the General Advisory Committee was debating who should chair an international conference on the peaceful uses of atomic energy, Rabi commented bitterly, "I guess we've killed cock robin."[5]

In November 1957, a member of the congressional Joint Committee on Atomic Energy, Senator Henry M. Jackson of Washington, stated that it would be "entirely proper" for the AEC to reconsider the Oppenheimer case. In December, the *Washington Post* and *Times-Herald* conducted a poll of members of Eisenhower's Science Advisory Committee, finding that a majority favored reinstating Oppenheimer's clearance and that none specifically objected to it. Later that month, AEC commissioner Thomas Murray, who had found against Oppenheimer in 1954, said that he now neither advocated nor opposed his reinstatement and that his earlier decision had been made "within the exigencies of the moment." In 1962, Oppenheimer was invited as a guest to a Kennedy White House dinner for Nobel laureates. Glenn Seaborg, now chairman of the AEC, asked Oppenheimer whether he would submit to another security hearing, this time to clear his name. Oppenheimer replied, "Not on your life." Instead, in late 1963, he was symbolically rehabilitated by being presented with the AEC's prestigious Fermi Award. But the significance of this return from exile was unclear. As Robert Coughlan observed, "Rehabilitated or not, even if his security clearance were restored (it was not) he could never be that man again. His case had caused too much bitterness, raised too many doubts, congealed too many protagonists in historic postures and attitudes ... Oppenheimer could never again have any leading role in government or government-affiliated science." In hindsight, Alfred Friendly described the award as an "anticlimax."[6]

Nevertheless, with the support of the trustees of the Institute for Advanced Study in Princeton, Oppenheimer was able to continue as director of the institute until June 1966, when, in worsening health, he stepped down.[7] He also traveled widely in the United States and internationally, giving lectures on the philosophy of science and on problems of culture. And he integrated himself into new social and intellectual networks, in particular becoming increasingly active in the Congress for Cultural Freedom. By examining the milieux in which he operated and the cultural repertoires on which he drew, this chapter traces how, in his later years, Oppenheimer was able to maintain cultural and intellectual authority separate from governmental and technocratic power.

MAKING SENSE OF THE OPPENHEIMER CASE: SCIENTISTS, INTELLECTUALS, AND ANTI-COMMUNISM

What was the meaning of the Oppenheimer case? The proceedings themselves and the findings against Oppenheimer were inherently confusing.[8] In his *New York Herald Tribune* column "Today and Tomorrow," Walter Lippmann wrote, "The one intolerable result is the result we have got, a divided, confused, contradictory verdict that raises enormous issues and settles none of them." From the outset, this outcome was "almost unavoidable," he said, "for the allegations were so vaguely defined, the issues were so carelessly posed, that they invited an indecisive result." The AEC's conclusion—that Oppenheimer was loyal to the United States but, due to "defects of character," was a security risk—was paradoxical. Oppenheimer was not a Hiss, a Fuchs, or a Rosenberg, but what was he? Could someone branded a "security risk" still legitimately play a role in public life? The terms of the findings allowed for multiple interpretations. Partly due to this ambiguity, the case became a symbolic peg onto which a variety of social and cultural conflicts were hung. And Oppenheimer himself became an emblem of the broader cultural condition. Even as the hearings were under way, *Time* magazine wrote, "However he came to his present ordeal, J. Robert Oppenheimer's life is a bitter parable of a bitter time."[9]

The journalists Joseph and Stewart Alsop saw Oppenheimer as an American Dreyfus, and they titled their defense of him after Emile Zola's famous 1898 letter "J'accuse." The Alsops' polemic was a passionate assertion of liberal democratic principles threatened by the expansion of the security apparatus. But their defense of Oppenheimer's liberal freedoms went only so far. Their willingness to champion him was dependent on his having repudiated Communism as a youthful error. Novelist and critic Waldo Frank pointed out in *The Nation* that the Alsops' defense divided Oppenheimer's biography into an acceptable and unacceptable part—the young leftist is firmly rejected so that the later pillar of the liberal elite can be rehabilitated.[10]

Most criticism of the hearings was framed in terms of worries over the decline of institutional decency and civility in the culture of accusation and investigation. Historian Arthur Schlesinger Jr. admitted that Oppenheimer "was doubtless at moments a cocky, irritating, even arrogant man." But he condemned the greater arrogance shown by the AEC in claiming the right to "search . . . the soul of an individual." "The government which claims to do this," Schlesinger argued, "would hardly seem a government for Americans." To others, the case dramatized the anti-intellectual undercurrent in American culture. One of Oppenheimer's former Berkeley colleagues, psychologist Edward C. Tolman, told the International Congress of Psychology that the Oppenheimer case was an illustration of America's "blind and stupid anti-intellectualism." Scientists should defy this, he said, by wearing the label "egg-head" with pride.[11]

Social scientists and literary intellectuals pointed to the hearings as indicative of the collective fate of intellectuals in modern America, thereby aligning their own position with that of Oppenheimer.[12] In an article for the journal *The Twentieth Century,* Philip Rieff argued that the hearings represented the breakdown of liberal norms that had hitherto protected intellectuals against political interference and discipline. They also dramatized the separation between humanists and scientists and between intellectuals and the public, indicating a vacuum in America's cultural life. For Rieff, Oppenheimer was a victim of this collapse of a common culture.[13]

In the pages of the *Partisan Review,* the case touched off a debate over how intellectuals should respond to McCarthyism. Diana Trilling wrote that "ever since Los Alamos, Dr. Oppenheimer had . . . been something of a culture hero for American intellectuals." Despite this, her attitude to Oppenheimer was ambivalent. Five years earlier, in response to the Hiss case, Trilling had advised liberals to maintain "a very delicate position which neither supports a McCarthy nor automatically defends anyone whom a McCarthy attacks." She appeared to condone HUAC's strategy of trying Communists for their beliefs, arguing that "the Communist idea must be judged as a Communist act." Trilling's first instinct had not been to jump to Oppenheimer's defense. She wrote that as a "conventional anti-Stalinist," her initial reaction to the H-bomb allegations had been to assume that Oppenheimer was indeed "wrongly motivated"—why else would he have worked for the A-bomb against Germany and Japan, but held back when later it came to developing the H-bomb against the Soviet Union? Her reading of the transcript changed her mind on that score: there were plenty of good reasons for opposing the H-bomb, and Oppenheimer's political attitudes were "wholly irrelevant to his H-bomb position." But she nevertheless thought that there were important lessons to be drawn from Oppenheimer's prewar political involvements. Oppenheimer was, she said, "*par excellence* the Popular Front fellow-traveler." His mistake during the hearings was in assuming that his own trajectory "is so very special." Rather, she suggested, Oppenheimer's defense should have been that his story "is the story of countless high-minded persons of liberal impulse who came to maturity with him." His slowness in recognizing the "totalitarian" nature of the Soviet Union, and his remaining loyalty to the "movement," were symptoms of a broader condition of the liberal-left milieu at that time. Trilling used the Oppenheimer case to launch a critique of what she regarded as the naive attitude of American liberals toward the Soviet Union before and during World War II and the history of American intellectuals' involvement with the Communist Party and the Popular Front. Trilling's conclusion was that Oppenheimer's political naïveté was only that of the liberal intelligentsia in general—and, she suggested, that is why "the intellectual does not belong in the active world of politics."[14]

The Oppenheimer case became a symbolic moment for embattled liberal intellectuals. The dredging up of Oppenheimer's prewar political commitments foregrounded the sensitive relationship between 1950s liberals and the 1930s Left. At the same time, the case forced liberals to confront their own complicity with postwar Red-baiting, which now appeared to be rebounding on them.[15] Identifying themselves as "anti-Stalinist," liberal intellectuals had often been unwilling to condemn, or had passively supported, HUAC's attacks on Communist Party activists, reserving their criticism for "irresponsible" excesses rather than the core ideology of Cold War anti-Communism. Oppenheimer's own position was ambivalent. He was himself associated with a liberal, anti-Communist group of intellectuals, the American Committee for Cultural Freedom (ACCF), affiliated with the international Congress for Cultural Freedom (CCF). The congress, which arose from a conference of intellectuals held in Berlin in June 1950, aimed to solidify the cultural relationship between American and European intellectuals and to promote liberal pluralist ideas as a way of countering the cultural and intellectual influence of Communism. In 1953, Oppenheimer was a sponsor of the CCF's Conference on Science and Freedom, held in Hamburg. The conference was organized by the Hungarian-born chemist and philosopher Michael Polanyi, based at Manchester University; he was a strong critic of the Soviet Union and a campaigner against British socialists' calls for the planning of science.[16]

In early March 1954, in the midst of preparing for the security hearings, Oppenheimer had taken time to try to persuade Einstein to dissociate himself from the Emergency Civil Liberties Committee, which was holding a gathering in honor of the physicist's seventy-fifth birthday. Oppenheimer had been tipped off by the ACCF of their view that the group was a Communist front organization. Sol Stein, the executive director of the ACCF, wrote to Oppenheimer that "leaders of the American Jewish community" were "very much concerned lest Dr. Einstein be sucked into another Communist-inspired occasion. Such an occasion will again tie up Judaism and Communism . . . [and] will help to spread the notion one hears so often nowadays about physical scientists being political babes-in-the-woods." Oppenheimer did the ACCF's bidding and persuaded Einstein not to participate in the meeting of what he called "this goddamn outfit."[17] The last thing that Oppenheimer wanted as he faced the Personnel Security Board was a political controversy involving the Institute for Advanced Study, of which Einstein had been a faculty member since the 1930s.

Einstein, as a matter of both principle and political strategy, took the position that anti-Communist attacks should be countered head-on with straightforward statements of principle defending civil liberties. In 1949, he had advised David Bohm to refuse to appear before HUAC, even if it meant a jail term. Einstein thought that the right course of action for Oppenheimer in facing the PSB was simple: he should just tell the officials that they were fools and then go home.[18]

Oppenheimer, however, could not contemplate this sort of direct opposition. Instead, since testifying before HUAC in 1949, he had pursued the more cautious strategy of trying to cooperate with and, up to a point, accommodate anti-Communism, in the hope that its sharper edges could be blunted. Oppenheimer worried that Einstein's views on civil liberties were, given the political climate of the time, "inflammatory and certainly most unpopular," and he urged Einstein to acknowledge "the harm that communists had done in this country."[19]

Pleased with Oppenheimer's intervention with Einstein, the ACCF duly sent a message to the AEC's Personnel Security Board vouching for Oppenheimer's anti-Communist credentials.[20] Oppenheimer agreed with the ACCF's stance of trying to foster a responsible, liberal anti-Communism as an inoculation against the excesses of McCarthyism. The security hearing, however, cast doubt on this strategy. The persecution of such a prominent scientist and cultural figure as Oppenheimer by the executive branch of the federal government was taken by ACCF intellectuals as a sign that anti-Communism had gone too far. In 1955, Schlesinger criticized the group as too fanatical, and he resigned within a year—as did, among others, David Riesman, John Kenneth Galbraith, and Diana Trilling. Historian Richard Pells wrote, "Each offered different explanations for resigning, but a major catalyst was clearly the persecution of Oppenheimer."[21]

Ironically, however, the hearings immediately propelled Oppenheimer into the ACCF. It was the sociologist Daniel Bell who put forward Oppenheimer's name and sponsored him for membership at the July 1954 meeting of the ACCF's Executive Committee, shortly after the AEC verdict. On November 9, Oppenheimer accepted the invitation to join the organization. The ACCF's press release announcing his membership proudly stated that "in March of 1954, he was . . . instrumental in helping the cultural freedom group discourage eminent scientists from cooperating with a Communist-line group in this country." Oppenheimer remained with the organization to the end of his life. The international CCF, in particular, provided him with a milieu in which he fashioned his intellectual identity of these later years.[22]

The Oppenheimer case symbolically tapped into and reignited a variety of simmering cultural conflicts. This symbolic function of the case was particularly evident in American universities, where academic freedom had already been greatly compromised by loyalty tests, oaths, and investigations. In 1955, the University of Washington, which had been the site of a number of very important battles over academic freedom since the late 1940s, became the focus of national attention for bowing to pressure from the state governor and the conservative local press and withdrawing its offer to Oppenheimer of a visiting professorship.[23] In protest, a number of prominent academics canceled visits and lectures there, and an informal boycott of the university began. Seven biochemists backed out of a medical school symposium, forcing the meeting to be canceled, and six physiologists refused to attend a symposium at the university's zoology department.[24] Cornelius Wiersma of Caltech explained his reasons for not attending this conference: "The whole principle of academic freedom is under severe attack these days. I feel strongly that if the academic world does not in every way protest whenever an obvious violation of academic freedom is proposed or executed, the idea of a university as a place in which freedom of speech is maintained at its highest level will soon belong to the past."[25]

Oppenheimer himself remained somewhat detached from the University of Washington controversy. He told a journalist that he had only heard of it via the *New York Times* and that he had never been formally invited to the university. (He and Edwin Uehling, a former postdoctoral student of Oppenheimer's at Berkeley and now acting chairman of the physics department at Washington, had informally worked out a date when he was available to come.) To the question of whether the university was violating academic freedom, Oppenheimer responded, "That's not my problem." When asked if the scientists' boycott might embarrass the university, he said, "It seems to me that the University has already embarrassed itself."[26] Instead of being an active participant in the controversy, Oppenheimer was, rather, a symbol around which others rallied and around which existing tensions coalesced into actual conflict. The divisions on campus festered for more than a year until a compromise was worked out in the summer of 1956: the physics department would host the National Science Foundation's International Congress on Theoretical Physics, which would include Oppenheimer but would involve no university funds.[27]

A large section of the wartime physics elite, particularly those who had spent the war at Los Alamos, rallied to Oppenheimer's defense during and after the hearings. Oppenheimer's defenders strove to present the image of a homogeneous scientific community unified behind him, despite the fact that he was vigorously opposed by those scientists who were prominent H-bomb advocates, including Ernest Lawrence, Luis Alvarez, and Edward Teller. An attack on Oppenheimer, it was frequently repeated, was an attack on the whole scientific community. These defenders contrived to make Oppenheimer a martyr for science. Weisskopf wrote to him during the hearings, "Somehow Fate has chosen you as the one who has to bear the heaviest load in this struggle . . . Who else in this country could represent better than you the spirit and philosophy of all that for which we are living."[28] The *Bulletin of the Atomic Scientists,* as well as publishing articles criticizing the proceedings, included letters of support for Oppenheimer written by colleagues at Princeton, Los Alamos, and Chicago.[29]

Caltech president Lee DuBridge, who had served with Oppenheimer on the GAC, testified in Oppenheimer's defense during the hearings and afterward expressed strong public support for the physicist. For example, he took the opportunity of a luncheon meeting of the American Institute of Electrical Engineers on June 23, 1954, to speak out against the hearings, and his comments were quoted in the Los Angeles press. For his support of Oppenheimer, DuBridge was bombarded with criticism from alumni and the general public in Pasadena and Los Angeles. One critic wrote,

By his own personal misconduct, [Oppenheimer] has ended his future usefullness [*sic*] to his country in any official capacity. Tragic as that may be, the political situation in the world to-day makes it imperative that our policies, both domestic and foreign, take on almost a "black or white" concept . . . As you must admit, Doctor, our educational and scientific circles have long been suspect. Often unfairly. Yet, opinions such as you voiced . . . do not make the nation any less willing to tolerate or approve or underwrite such sentiments . . . The country is in an ugly mood and will make short shrift of any leader that proves unfaithful to the confidence that we have the right to expect. This is no time to condone alleged "mistakes in judgment" that are, in fact, treason. No matter how innocently such "mistakes" may have been made.

DuBridge replied, "You are apparently right that the country is in an ugly mood and your letter is one of the ugliest parts of it."[30]

Anti-Communism had multiple meanings and was often connected in complex ways with other cultural struggles concerning the role of science. One woman wrote to DuBridge that

the American people are getting fed-up on the over-emphasis of "science." The most recent and most spectacular contribution to science is a creation that so far, has brought the world nothing but fear and destruction. Let us have more men like Lincoln and MacArthur and less of men like Fuchs and Rosenberg . . . the Atomic Energy Commission had disposed of Mr. Oppenheimer to the satisfaction of the American People.[31]

Anti-Communist discourse was often bound up with nativism, anti-intellectualism, and encoded anti-Semitism. Press portrayals of Oppenheimer as an archetypally amoral scientist tapped into deep-rooted conflicts in American society over cultural homogeneity versus pluralism and secular versus Christian bases for culture and morality.[32]

In November 1954, the president and Mrs. Eisenhower attended a service at St. John's Episcopal Church in Washington and heard a sermon by Dr. Charles Lowry, chairman of the Foundation for Religious Action in the Social and Civil Order. The Reverend Doctor told the congregation that "Robert Oppenheimer was symptomatic of a very large number of top flight scientists in the evident vacuum of his soul" and that "if by some evil chance our globe is destroyed, this will be the real reason—the tyranny of science and the poverty and defensiveness of the forces of salvation."[33] The right-wing *American Mercury* derided Oppenheimer as the "long-time glamor-boy of the atomic scientists." Attacking DuBridge for his public support of Oppenheimer, the *Mercury* asked, "If educators continue to coddle potential traitors . . . will not students be encouraged to feel that intellectual achievement carries with it an unrestrained license to pursue unmoral, if not immoral behavior? Have they not already before them the pattern whereby patriotism, moral and spiritual values are excluded from the realm of science?"[34] Hoping to head off this kind of attack, DuBridge replied to one letter from a member of the public by affirming his own Protestant upbringing:

> As the son of a YMCA secretary and a lifetime member of a Protestant church, it was rather a shock to have you imply that I was one of the people who did not "feel that high moral character is essential" . . . I know Oppenheimer personally and intimately and I can assure you that whatever terrible mistakes he may have made in his younger years, he has repented and atoned for them a thousand times over in recent years. My upbringing may have been old-fashioned, but I was taught that in cases like this the sins of youth might be forgiven . . . I would suggest that we not punish St. Paul for the sins of Saul.[35]

The letters DuBridge received from the public were overwhelmingly hostile and even included some anonymous hate mail. But a rumor that DuBridge had set up a defense fund for Oppenheimer also brought in some contributions from well-wishers, which DuBridge duly returned. DuBridge confided to his colleague Edward Condon that

> it is probably quite impossible for anything to be done about the Oppenheimer case itself. The term "security risk" is such a broad one that you can start out accusing a fellow of treason and end up by convicting him of fibbing, but still impose the same punishment. I guess there is no doubt that Robert did do some fibbing, and in the public mind now anybody who fibbed and also once was a "Communist" is clearly an unforgivable character.[36]

He also felt obliged to provide the Caltech board of trustees with a point-by-point statement of his views on the Oppenheimer case, as well as a defense of his decision to speak out publicly on the matter. He told the board, "I think I reflected in my opinions the views of the overwhelming majority of scientists."[37]

It was important to the scientists who came to Oppenheimer's defense to present themselves as speaking for a homogeneous and unified scientific community. Ostracizing Teller as a disruptive presence was part of the process of defining this community of opinion.[38] Teller remarked bitterly, "One of the things that happened to me is not only that I lost my friends, but I believe I lost my status as an intellectual. You know, an intellectual, as I found out to my grief, is not necessarily a man who is intelligent, but a man who agrees with other intellectuals. He is a man with whom it is acceptable for other intellectuals to associate. I lost my membership in that club."[39]

One "world-renowned physicist," when asked by Teller's biographers whether the AEC trial had destroyed Oppenheimer, answered "No. I think it made Oppenheimer. I think it destroyed Teller."[40] More specifically, if the hearings closed off to Teller the status of intellectual, they served to confirm Oppenheimer in that role. The "intellectual" was an available cultural image that was immediately invoked to make sense of Oppenheimer's identity and the events surrounding the hearings, and one that Oppenheimer could draw upon in reconstructing his public identity. In doing so, he was able to maintain a public role beyond the relatively calm enclave of academia.

ACADEMIA AND PUBLIC LIFE

Even while assuming a technocratic advisory role after World War II, Oppenheimer had maintained his position as an academic scientist. As director of the Institute for Advanced Study after 1947, Oppenheimer played a crucial role in nurturing young talent—a new generation of theoretical physicists including Freeman Dyson, Murray Gell-Mann, T. D. Lee, C. N. Yang, and Abraham Pais.[41] Oppenheimer occupied a central place in the postwar physics community. The biographer of Murray Gell-Mann wrote, "When a physicist came up with a new discovery, it was customary to make a pilgrimage to the Institute and try out the new idea on Oppenheimer and his young geniuses." This was a trial by fire; Oppenheimer was notorious for his ability to demolish faulty arguments and, on occasion, even some good ones.[42] He continued to play this symbolic presiding role in the physics community after 1954. Dyson thought that Oppenheimer became a better director of the institute in those later years, because without his governmental duties, he was free to devote more time to the place. Oppenheimer saw the purpose of the institute as being to take on young scholars for short periods of time so that they could work intensively, free from other obligations. He was in many ways the focal point of this community of scholars.

The Institute for Advanced Study was the most prestigious academic institution in the country and the archetypal ivory tower. Just as Oppenheimer's leadership of Los Alamos had fixed him indelibly in the public mind as the man behind the bomb, his directorship of the institute confirmed him as the chief embodiment, representative, and spokesman for "pure science." And he cultivated this image in his professional and public appearances. His lecturing style was, according to Pais, "priestly . . . It was, one might say, as if he were aiming at initiating his audience into Nature's divine mysteries." It was characteristic that at the end of the talk, the sense of mystery would remain. Pais recalled that after a lecture on mesons that Oppenheimer gave at the American Physical Society in January 1947, "I tried to play back what he had just said, and I recall my thought: What the hell do I remember about his talk? I had been intrigued, nay moved, by his words, but now found myself unable to reconstruct anything of substance. I would now say that this was not just a matter of stupidity on my part." Oppenheimer self-consciously cultivated an oracular style, whether talking about physics or public affairs. He had attempted to tutor David Lilienthal in the art, praising one of his speeches as "very sound and deep and with just the right lightness of touch in pointing to the great human and ethical substrata that determine our way of life without handling them in such an explicit way that the touch destroys." As Pais put it, Oppenheimer was "a rhetor rather than a speaker."[43]

Jeremy Bernstein heard Oppenheimer give a public lecture on quantum mechanics at Harvard in 1957: "His use of language was somewhat opaque, often poetic, and he had an odd, clipped diction that commanded attention." Bernstein noticed in the audience two frail-looking "classic blue-haired Boston dowagers," both listening in awe to Oppenheimer's speech. At one point, the physicist wrote an equation on the board, whereupon "the two old ladies clutched each other for reassurance. Perhaps they thought that the formula was going to explode." Oppenheimer's aura combined the raw power of the atomic bomb with the mystique of "pure science." That fusion, together with his eloquence, made him an "electrifying public figure" and, as one journalist put it, "the acknowledged spokesman for his profession."[44]

Yet Oppenheimer's authoritative public presence as a spokesman for science and his aggressive intellectual style in discussions with colleagues and students masked anxiety about his increasing distance from new creative work in physics. In the summer of 1952, when his term as GAC chairman was coming to an end, he wrote to his brother, "Physics is complicated and wondersome, and much too hard for me except as a spectator; it will have to get easy again one of these days, but perhaps not soon."[45] In an interview with Thomas Kuhn late in 1963, Oppenheimer strongly hinted at his sense of isolation. He said he missed the interaction with experimental physicists that he had had before the war at Berkeley (and, of course, during the war at Los Alamos). He did not miss teaching, since he thought he had lost his vocation for this: "I think that the charm went out of teaching after the great change of the war . . . I was always called away and distracted because I was thinking about other things, but actually I don't think I ever taught well after the war. I have a feeling that what my job was was to get a part of the next generation brought up and that job was done when I came here."[46]

In his later years, Oppenheimer was not content with a purely academic role and was eager to find spheres of intellectual discourse and influence beyond the university. He was increasingly concerned with the constitution of the public sphere but was pessimistic about the possibility of successfully connecting science with public discourse. Whereas Harvard president James Conant had tried to describe "science in the making" for the lay reader and introduced his popular book *Science and Common Sense* as a "citizen's guide to the methods of experimental science," Oppenheimer was suspicious of these populist sentiments.[47] When he read Conant's *On Understanding Science* in 1947, Oppenheimer wrote to Conant, "With its fundamental tenet— that one can understand science only intensively, not extensively, and that we in fact know too little of how scientific progress is made to theorize about it and know only barely enough in a few instances to describe it—with this I deeply agree."[48] But Oppenheimer drew from this the elitist lesson that "understanding science" was possible only for those initiated through practice and membership in the scientific community. He doubted that it could be achieved as part of a "general education." There was necessarily "something fake" about the controlled environment of the high school or undergraduate laboratory, where there is never real uncertainty about what is the correct experimental outcome. And to Oppenheimer's mind, Conant's project of teaching science through history ran the risk of becoming "corrupt with antiquarianism."[49] In contrast to Conant's optimism, Oppenheimer dwelled on what he saw as the inescapable gulf of misunderstanding between professional and layman, and between professionals in different fields.

This problem was at the heart of the Reith Lectures, which Oppenheimer delivered for the British Broadcasting Corporation (BBC) in November 1953. In them, he described science as a form of life with its own particular traditions, skills, and practices. The ideas of any branch of science were likely to be misinterpreted without immersion in the cumulative development of these traditions. New knowledge and techniques transcended but also incorporated past knowledge. Without being oneself a part of that process of change, one could not hope to fully understand the new knowledge. The cumulative character of science was archetypal for modern ideas of progress. But Oppenheimer presented this cumulativeness as having fundamentally premodern characteristics. He described science as a craft—performed by master artisans, organized in guilds, steeped in tradition.[50]

Yet if science was disconnected from public life and common culture, this held the danger that science would be valued for, and would come to rely entirely for its maintenance on, its instrumental utility as a source of material power. Public support for "pure science" depended on the ability of scientific communities to connect with a broader human community in the public sphere.

It was this connection with a wider community that the Reith Lectures were an attempt to foster. But it was clear to Oppenheimer that ideas would always be distorted by the process of translation from one community of experience to another. The problem was therefore in what sense one could speak of there being a common culture. A rapprochement of science with public culture was rendered particularly problematic by the ways in which the theories of relativity and quantum mechanics ran counter to commonsensical ideas about time, space, and causality.[51]

Oppenheimer presented this gap between science and common culture in stark terms. Nevertheless, he did hope to rescue some sense in which science could contribute to broader human culture. In this connection, he emphasized the notion of *analogy*. He saw the influence of Newtonianism within the Enlightenment as an ideal model for the central place that scientific ideas could occupy within the culture (however at odds this influence might be with the original meanings of the ideas). The grand mathematical synthesis accomplished by Newton provided for the philosophes an analogy through which to develop their aspirations for a unified science of man and for the power of reason to understand and control human affairs.

The situation in the mid-twentieth century was different: just as Newton's overarching synthesis had been overturned, so, for different reasons, had the vision of progress and unity that characterized the Enlightenment lost its luster. The question Oppenheimer was grappling with was whether quantum mechanics and atomic physics could play a cultural role in the present age comparable to that which Newtonianism had played in the eighteenth century. If quantum physics was to have such an impact, it would be not through direct application, but through analogy. It was in this indirect and imaginative sense that Oppenheimer sought to draw from physics moral lessons for human affairs. "The story of atomic discovery," Oppenheimer said, is "so full of instruction for all, for layman as well as specialist. For it has recalled to us traits of old wisdom that we can well take to heart in human affairs."[52]

The chief embodied source of such wisdom, for Oppenheimer, was Niels Bohr. Oppenheimer's old friend Jeffries Wyman recalled that when he was with Oppenheimer in Paris in late 1953, following the Reith Lectures,

Oppenheimer was talking excitedly about Bohr. It seemed to Wyman that "Bohr was his idol ... He spoke of Bohr almost as of a god."[53] Bohr's notion of complementarity was the central motif through which, Oppenheimer thought, the cultural role of physics could be developed. Oppenheimer employed complementarity as a metaphor for existential dilemmas, speaking, for example, of the complementarity of the eternal and the transient in human life. He also applied it as a principle of pluralism and liberal tolerance: there is not one overarching truth, but many truths, each of which is appropriate to a different dimension of experience. The notion of complementarity allowed Oppenheimer to move easily between physics and moral philosophy, while reminding his listeners of the multifaceted nature of the human spirit and of the diversity of human experience.[54]

Some years earlier, Oppenheimer had said, "Science is not all of the life of reason; it is a part of it."[55] In his final Reith lecture, "The Sciences and Man's Community," Oppenheimer again asserted the unbridgeable pluralism of modern culture. This condition, he said, ruled out the traditional ideal of the cultivated general intellect: "Even the best of us knows how to do only a very few things well; and of what is available in knowledge of fact, whether of science or of history, only the smallest part is in any one man's knowing." Applying the principle of complementarity to the life of the individual suggested that in a "man's life ... he may be any of a number of things; he will not be all of them."[56] The goal of cultivation, the formation of an individual self that would be equal to the scope of human culture, was no longer feasible.

Rather than through modes of self-formation, integration could now be achieved only through new forms of solidarity and community: "Each of us knows ... how much even a casual and limited association of man goes beyond him in knowledge, in understanding, in humanity, and in power ... Each of us knows how much he has been transcended by the group of which he has been or is a part."[57] In place of cultivation, Oppenheimer substituted an ideal of dialogue between diverse communities. The problem of the unity of knowledge and the problem of human solidarity were identical.

For the solution to this dual problem of knowledge and solidarity, Oppenheimer turned to Tocqueville's understanding of American democracy as being composed of pluralistic civic associations. This communitarian pluralism, Oppenheimer suggested, connected the trajectory of American democracy with that of modern science: these "fluid and yet intense communities ... form a common pattern for our civilization. It brought men together in the Royal Society and in the French Academy and in the Philosophical Society that Franklin founded."[58]

As he articulated a conception of science as an instantiation of liberal democratic values, Oppenheimer presented himself as a spokesman for these values
and constituted his *own* authority as interpreter of the cultural meaning of
science. In giving the Reith Lectures, Oppenheimer was defining his personal
role as an intellectual. The lectures were a performance, and it mattered that
it was Oppenheimer who was delivering them. Though he announced the
death of the cultivated man, these lectures were nothing if not a display of
cultivation; and while he pointed to the splintering of culture into specialized
segments, he demonstrated his own ability to transcend such divisions. The
Times wrote that Oppenheimer "can speak with authority, for he combines
the highest technical competence with administrative experience and wide
interests," and added that "a touch of the poet gives him the power to express
the scientist's situation in our time."[59] American journalist John Mason Brown
said, "It is the work of a scientist who is an artist and an artist who is a poet."[60]
Oppenheimer constituted himself as embodied solution to the very problems
of cultural fragmentation that he was pointing out.

Yet his lectures were in some ways a practical example of his point about
the gap between science and common understanding. Many among the British
audience who listened to the lectures on the radio complained that they could
not follow the speaker. Amid the controversy of the security hearings, the
Guardian recalled that the Reith Lectures had been "controversial only in
one sense: they could not be understood, so many complained, by the common
educated man."[61] Those who found the secular sermon uplifting were moved
not so much by the content as by the voice of the speaker. One listener wrote to
Oppenheimer, "Your voice, so full of the effect of wisdom and consciousness
of the Infinite, was a delight past defining in words." Another said, "I loved
the nobility of your utterance and the wisdom and beauty of the language. I
rose from my chair with a purity of mind and an elevation of emotion; I rose,
if I may with reverence say it, as from a Sacrament."[62]

One reviewer suggested that Oppenheimer's ability to hold together diverse intellectual elements, and to embody a kind of cultural unity, was a
trick of his voice, an effect that disappeared when the text was divorced from
the speaker. The reviewer of *Science and the Common Understanding* for
the British *Universities Quarterly* observed, "When I started to read these
lectures, they seemed inescapably associated with Dr. Oppenheimer's voice
and intonation which gave the broadcasts a flavour unique in modern scientific
exposition. But when I had got to the end and began to re-read in bits, I found
that the spoken word had gone and that the book divided itself into two

parts ... Two men wrote this book, Dr. Oppenheimer, Scientist [and] Dr. Oppenheimer, Romantic."[63]

EXILE AND HOPE

The Reith Lectures prepared the part that Oppenheimer would assume in the wake of the security hearings and supplied him with the vocabulary that he would use to remake himself in this new role. The hearings closed off to Oppenheimer the possibility of a technocratic role in the state but left available to him the cultural role that he had begun to map out in his BBC appearance. A *Herald Tribune* reviewer of *Science and the Common Understanding* told readers that "a few paragraphs" of this book "may tell more of the essential faith and nature of this man than columns of testimony."[64] It was in terms of the cultural problems addressed in the Reith Lectures that Oppenheimer made sense of his new position. Instead of underlining that he had been unfairly attacked by specific political enemies, he presented his new status as victim as manifesting the more general defeat of the intellectual in a fragmented and degraded modern culture.

On December 26, 1954, some six months after the AEC's finding against him, Oppenheimer gave a lecture titled "Prospects in the Arts and Sciences" for Columbia University's bicentennial. The talk was broadcast nationwide by the Columbia Broadcasting System (later CBS). He recapitulated many of the ideas that he had first articulated in the Reith Lectures. But this new version was without the optimistic notions of pluralism and community that he had outlined just a year earlier. Instead, Oppenheimer emphasized personal feelings of weakness and of being overwhelmed by a world beyond one's control. His communitarianism now took the form of a defensive retreat: "This is a world in which each of us, knowing his limitations, knowing the evils of superficiality, will have to cling to what is close to him, to what he knows, to what he can do, to his friends and his tradition and his love, lest he be dissolved in a universal confusion and know nothing and love nothing." The lecture was shot through with the fear of the dissolution of self, of being unable to hold one's own against others: "If a man tells us that he sees differently than we, or that he finds beautiful what we find ugly, we may have to leave the room, from fatigue or trouble; but that is our weakness and our default. If we must live with a perpetual sense that the world and the men in it are greater than we and too much for us, let it be the measure of our virtue that we know this and seek no comfort." Oppenheimer painted a picture of the individual confronting others as alien and hostile and experiencing the social world itself as a vast and opposing power. He presented not only the individual, but culture itself as on the defensive against this great anonymity. "Never before today," he said, "has the integrity of the intimate, the detailed, the true art, the integrity of craftsmanship and the preservation of the familiar, of the humorous and the beautiful stood in more massive contrast to the vastness of life, the greatness of the globe, the otherness of people, the otherness of ways, and the all-encompassing dark."[65]

Pervading the lecture was a profound sense of homelessness, of the "artist's loneliness" and of the scientist working at the boundaries of knowledge who finds himself "a very long way from home." Oppenheimer described this sense of homelessness above all in relation to mass culture. This meant the ersatz products channeled through what he called the "superhighways" of the "mass media." Echoing the views of liberal social scientists, Oppenheimer linked popular culture with totalitarianism. The "superhighways" ranged from "the loudspeakers in the deserts of Asia Minor and the cities of Communist China to the organized professional theater of Broadway. They are purveyors of art and science and culture for the millions upon millions." While making us aware of events across the globe, these "superhighways" were ultimately destructive of genuine solidarity: "They are also the means by which the true human community . . . [is] being blown dry and issueless, the means by which the passivity of the disengaged spectator presents to the man of art and science the bleak face of unhumanity."[66] Degraded mass culture left no room for an authentic intellectual role.

In his critique of popular culture, Oppenheimer was echoing a dominant theme in mid-twentieth-century intellectual discourse. His language was particularly evocative of the defensive pessimism of the 1930s cultural critics José Ortega y Gasset and Clement Greenberg. However, there was a key difference. Most critics of mass culture dwelled on the opposition between popular culture and avant-garde art and literature and were inclined to see science, particularly technological "big science," as itself symptomatic of bureaucratic and mass society. Oppenheimer's account stood out by the prominence he gave to science, rather than literature, as a locus of high cultural values. Oppenheimer placed the figure of the scientist at the center of the preexisting narrative of the lonely and embattled modern intellectual.[67]

Alistair Cooke, who had followed the security hearings for the Manchester *Guardian*, wrote that in his Columbia address, Oppenheimer both spoke about and himself instantiated and embodied "the isolation of the specialist" in modern society. From "the loneliness of his own exile," Oppenheimer articulated the condition of "the lonely man in society." "So in the end Oppenheimer came to identify the world's plight with his own: that of a natural recluse too much bruised by the public world which repulsed the great gifts he felt he owed it."[68] The Edinburgh *Scotsman* elected Oppenheimer as its man of the year, explaining, "It seems to us that Dr Oppenheimer has emerged as a kind of new human prototype—the brilliant intellectual shorn of his roots . . . A respected leper, lingering in the outer purlieus of a strange citadel he knew so well he seems to us to be the sad symbol of an age that can take nothing on trust. He is the man who knew too much and therefore, to us, the man of the year."[69]

On December 16, 1954, television cameras invaded Oppenheimer's own anchoritic retreat—the Institute for Advanced Study—and a half-hour conversation between Oppenheimer and journalist Edward R. Murrow was broadcast on the popular television program *See It Now.* Avoiding discussion of the hearings, the program focused instead on Oppenheimer's role as director of the institute. Oppenheimer appeared as the linchpin holding the scholars of the institute together. The man who at Los Alamos had been the only one able to understand in detail each part of the laboratory's work and to synthesize it into a whole now oversaw and appeared able to fuse together the intellectual endeavors of the institute's solipsistic inhabitants. Describing the life of institute, he gave brief summaries of the work of each of the major scholars who were sequestered there, from young physicists Abraham Pais and Freeman Dyson to the psychologist Jean Piaget, the art historian Erwin Panofsky, and the medievalist Ernst Kantorowicz. Only when it came to Einstein did Oppenheimer have little to say. Asked by Murrow, "And Professor Einstein is still here too, isn't he?" Oppenheimer replied, "Oh, indeed he is. Indeed he is. He's—he's one of the most lovable of men."[70]

Tensions in Oppenheimer's relationship with Einstein were in large part due to their competing conceptions of the proper role of scientists as intellectuals. Though he has become almost archetypal of the solitary and disengaged life of the mind, in his later years Einstein came to see direct political engagement (for example, in defense of civil liberties) as the intellectual's duty. This was anathema to Oppenheimer's conception both of the scientific vocation and the function of the institute. Oppenheimer described the institute to Murrow as shutting out the world: "We are here as an institution . . . to take away from men the cares, the pleasures that are their normal excuse for not following the rugged road of their own—own life and need and destiny." At the institute, he said, "they can't run away . . . from the job that it is their destiny to do."[71]

Oppenheimer presented the life of the scientist as ascetic and disciplined: it was not only a retreat from the world, but also a means of steeling oneself with the mental toughness required to cope with a harsh and chaotic world. Paradoxically, however, this intensely inward concern with shaping the self was at the same time being enacted and displayed for public consumption. Oppenheimer provided the television audience with a glimpse into this ascetic community and way of life that he at the same time presented as utterly anathema to mass culture.

A condition of the Murrow interview was that there be no questions about the hearings.[72] It was three years before Oppenheimer was willing to talk publicly about his "case." In 1957, he granted an interview to Victor Cohn, a reporter from the *Minneapolis Tribune.* The article opened with a quotation from the physicist: "I have tried to prove that a security risk can survive." He said, "I had to establish . . . that what was put out as a final judgment about me wasn't the final judgment. And the only way to do this was by surviving." This survival owed a great deal to his continuing role as director of the Institute for Advanced Study. But the article also dwelled heavily on Oppenheimer's scientific and social philosophy. Oppenheimer survived as an interpreter of science and of the condition of the culture.[73]

A few months later, on October 4, 1957, the Soviet Union's successful launch of *Sputnik I,* the world's first artificial satellite, sent shock waves through the United States. The Eisenhower administration demanded a new heightened state of scientific mobilization, this time to catch up with the Soviets in the space race. Just as the explosion of the Soviet atomic bomb in August 1949 had demolished the security of America's atomic monopoly, *Sputnik* evaporated the complacent view, commonly expressed by American scientists and liberal intellectuals, that science and technology could not prosper outside Western democracies. And if relations between scientists and the federal government had been soured by the Oppenheimer affair, it was clear that these had to be mended. A side effect of the crisis was that it led to calls for Oppenheimer's reinstatement. The director of the United States' own satellite program, for example, when asked by journalists "whether 'a nation in first place' could afford to waste the services of Dr. J. Robert Oppenheimer," replied that "'a nation in first or last place' could not afford it."[74]

Blame for America's lagging behind the Soviets in space quickly focused on the education system. A new drive for improved science and technology education, and broader access to that education, led to the passing in 1958 of the National Defense Education Act. The increased federal funding for education was broadly welcomed. But at a talk at Pingry School, a private preparatory school in New Jersey, Vannevar Bush expressed worry about the narrow focus on technical education. Instead, he called for a revival of the ideal of the "gentleman of culture," so that "youth will . . . seek to emulate the full man."[75]

For Bush, the model of the "cultured gentleman" was necessary for "the modern select group upon whom the continuance and further development of our free way of life ultimately depends." In a time of crisis, the nation needed to be able to look to an elite who would stand above the diverse interests and pressure groups of the pluralist society, an elite with "a certain aloofness from the crowd" who could act "as trustees of the common weal." This elite was also necessary for waging the Cold War: "We cannot compete effectively in a complex world of air transport, guided missiles, and satellites by being merely tough and practical." Even if *Sputnik* had weakened the certainty in the scientific and cultural superiority of democracy and the free market, Bush argued, the Soviets had another weakness, and that was the "narrowness" of their education. A well-rounded gentlemanly elite, Bush hoped, would assert American cultural authority and superiority by integrating science, intellectual life, and culture for the benefit of the nation.[76]

It was that very integration in Oppenheimer, his ability to embody culture, that was the source of his continuing fascination for Cold War liberal intellectuals and was essential to his survival after the hearings. Yet even as he took on the role of "cultured gentleman," Oppenheimer also bore witness to the decline of this ideal in society at large. Specialization in science was inevitable, he said, and had to be accepted. In his 1958 lecture "Knowledge and the Structure of Culture," delivered at Vassar College in upstate New York, Oppenheimer argued that scientists' responsibilities were purely vocational. A commitment to "know something [rather] than not know it," he asserted, was "the only clear simple answer to the question . . . 'What is the responsibility of the scientist?'"[77] It was an ethic of awareness, rather than of action. Even this injunction to be aware was limited. Oppenheimer maintained that it did not extend to a responsibility for Soviet scientists to criticize the political system of their country, nor for American scientists to criticize theirs. It was not a generalized responsibility to speak truth to power. Oppenheimer saw awareness as necessarily restricted within the professional domain of scientists' expertise. Scientists were specialists, not general intellectuals.

Oppenheimer presented a defense of scientific specialization as a bulwark against the dilution of knowledge by mass culture: "We who live in universities . . . have a kind of high duty to insist on being difficult . . . and insist on being recondite and honest and intimate."[78] But that still was an inadequate remedy for the essential lack of commonality that Oppenheimer saw in modern life. He told an audience at Chapel Hill, North Carolina, the following year that the very intimacy of small, specialized communities, while bringing these few people together, by the same token rendered them "isolated" from the rest of culture and humankind. Oppenheimer contrasted the public sector "not only . . . with what in the privacy of a man's life or his family may be dear to him, but what is held as a guild or trade secret by small communities of men."[79]

Science was in some important senses anathema to the public sector. The image of objectivity that accompanied science, and the fetishism of objectivity arising from the prestige of science, corroded the kind of intersubjective discourse that constitutes the public sector. Oppenheimer argued that "we have to some extent lost the confidence in the value of talking with one another in a common discussion where verifiable truth in the sense of the sciences, and objectivity in the very special sense of the sciences, is not attainable."[80]

Oppenheimer was now centrally concerned with the limits of scientific rationality and objectivity. Science, he recognized, was incapable of providing answers about what goals society ought to pursue, about meanings, and about the purposes to which scientific developments should be applied. The crowding out of moral discourse by scientific instrumentality and objectivity was most brutally apparent in the case of atomic weapons. "Public discourse and common discourse," Oppenheimer argued, were "cryingly needed" if the threat of atomic warfare was to be addressed in moral, and not just instrumental, terms.[81] The AEC's Personnel Security Board had ruled that Oppenheimer's opposition to the hydrogen bomb was illegitimate insofar as he was an expert advising the state. Now, having been disbarred from that role, Oppenheimer found that this kind of technocratic thinking and the fetishism of a narrowly constructed scientific objectivity were eliminating any potential sphere in which such issues could be meaningfully addressed.

Oppenheimer's worries about the culturally fragmenting effects of specialization connected with wider concern and debate on both sides of the Atlantic about the relationship between science and the humanities. In 1959, C. P. Snow delivered his massively influential Rede Lecture at Cambridge University, lamenting the division between the "two cultures" of science and the humanities and attacking the dominance of a traditionalist literary culture in British academic life.[82] Also in 1959, historian of ideas Jacques Barzun wrote on the divided condition of the "House of Intellect," criticizing academic specialization that had made "abundance of information . . . into a barrier between one man and the next."[83]

Some years later, University of California president Clark Kerr argued for a new academic "multiversity": "The faculty world seems to sense a loss of unity—intellectual and communal unity and . . . what Robert Oppenheimer calls 'a thinning of common knowledge.' Knowledge is now in so many bits and pieces and administration so distant that faculty members are increasingly figures in a 'lonely crowd,' intellectually and institutionally."[84] Kerr presented the multiversity as a pluralistic solution, allowing loose cooperation between different knowledge-communities. It was an image strikingly close to Oppenheimer's portrayal in the Reith Lectures of the many rooms in the ramshackle "House of Science." But there were also worries that the ramshackle structure of the multiversity would be, as historian David Kaiser aptly puts it, just a sprawling academic "suburb." Many among the older generation of physicists worried that the rapid postwar expansion of their discipline meant that physics was no longer the close-knit community they went into, nor the intellectual retreat imagined by Oppenheimer, but an increasingly impersonal, routinized mass profession—a lonely crowd.[85]

Some months after Snow's lecture, Oppenheimer gave a talk in Rheinfelden, Switzerland, for the Congress for Cultural Freedom in which he returned to the theme of the decline of the "public sector" and "public discourse." The diplomat George Kennan was there, as was sociologist Edward Shils, who described Kennan and Oppenheimer as two "icily lofty American saints."[86] The French political philosopher and writer Raymond Aron had initially asked Oppenheimer to give a talk introducing "Western policy in the atomic age." But Oppenheimer said, "I would prefer the somewhat wider and more interesting theme of the effects of contemporary science on Western culture and politics, of which the problem of nuclear weapons is a principal but a rather special example."[87]

In contrast to Snow's call for more science against Oxbridge humanism, Oppenheimer argued that public culture had been stunted by "an overemphasis . . . of the role of certitude," based on the prestige of science. This was particularly the case regarding atomic weapons, which were now usually addressed in terms of rational-choice models rather than ethics. "What are we to think of such a civilization," he asked, "which has not been able to talk about the prospect of killing almost everybody, except in prudential and game-theoretic terms?"[88]

Yet it was unclear what, if any, solution Oppenheimer was offering. He spoke wistfully of the Hindu dedication to ahimsa, or doing no harm, and the nonviolent ethic "which you find in Jesus—as well as . . . in Socrates." But he apparently did not see such an ethic as translatable into action in the world. He dismissed antinuclear campaigners such as Bertrand Russell by saying, "These people want heaven and earth too. They are not in any way talking about deep ethical dilemmas, because they deny that there are such dilemmas. They say that if we behave in a nice way, we will never get into trouble. But that, surely, is not ethics."[89]

Russell was too superficial because he would not accept the nuclear dilemma as intractable, whereas Oppenheimer saw the world as basically corrupt and resistant to reform. Oppenheimer offered to weep for the world but not to help change it. The philosopher Karl Jaspers, commenting later on Oppenheimer's quasi-religious appeal at the end of "Prospects in the Arts and Sciences" to "love one another," wrote, "In such sentences I can see only an escape into sophisticated aestheticism, into phrases that are existentially confusing, seductive, and soporific in relation to reality."[90]

Novelist Mary McCarthy took Oppenheimer's frequent but vague talk about love as a sign that he had finally lost his marbles. Writing to Hannah Arendt about the CCF conference "Progress in Freedom," held in Berlin in June 1960, she said, "Another feature of the Congress was Oppenheimer, who took me out to dinner and is, I discovered, completely and perhaps even dangerously mad. Paranoid megalomania and sense of divine mission." At one point, according to McCarthy, Oppenheimer turned to CCF secretary-general Nicolas Nabokov "and said the Congress was being run 'without love'. After he had repeated this several times, I remarked that I thought the word 'love' should be reserved for the relation between the sexes."[91]

Aron recalled his impressions of Oppenheimer from the 1960 conference: "Devoured by an internal flame or by the battles he was fighting with himself, he tended to take any episode of his existence not seriously but tragically. I can see him in his room, with his wife, discussing the latest conference presentations with me, as though he were disturbed by their possible banality."[92] To others, however, it appeared that Oppenheimer and his compatriots had avoided banality by succumbing to melodrama.

If the CCF's goal was to engage in cultural cold-warfare, fighting Communism on the cultural front and asserting Western hegemony, it seemed to *Time* magazine that it was doing a poor job, particularly when it came to advertising the benefits of Western capitalism to African and Asian leaders. *Time* reported from Berlin that "the Afro-Asians came expecting leadership, and found only hand-wringing." They came "want[ing] only some cars and some irrigation ditches and some good technical ideas from the gloomy Westerners." Instead, "the spokesmen for the sophisticated societies spent most of their time reproaching themselves or apologizing." The critic Friedrich Luft announced that amid West German affluence, "culture is dead." Raymond Aron lamented the weakness of parliamentary rule in France. Mary McCarthy "moaned" that "Western literature is the mirror on the ceiling of the whorehouse," and Oppenheimer "apologized for all the wrongs he said science has done."[93]

DEFENDER OF THE FAITH

Oppenheimer aimed his criticisms at the condition of the culture, rather than at politics or policy. He was leery of any involvement in protest or political opposition. David Lilienthal recorded in his diary that in late March 1955, at a meeting of the Twentieth Century Fund, chairman Adolf Berle raised the issue of the crisis over the islands of Quemoy and Matsu in the Taiwan (Formosa) Strait and of the Eisenhower administration's threats to resort to the atomic bomb if China attacked the islands. Berle proposed the circulation of a petition warning against nuclear brinkmanship. According to Lilienthal,

> Oppenheimer explained that he didn't think he should sign the statement. Though agreeing with it, because of the to-do this would cause. But his voice was insistent on the point that we should not take a position that war over Formosa was necessarily a worse alternative than peace under all circumstances; nor did he believe that using A-bombs for tactical purposes was not possible, i.e. that they might be used only for a limited military purpose rather than spread, necessarily, into mass bombing of cities; nor did he think the statement ought to imply that thoughtful and careful and intelligent attention to the relevant issues was not already being given, in Washington.[94]

Despite his exclusion from government, it seems that Oppenheimer easily fell back into the habits of his earlier role as scientific-military strategist of the winnable nuclear war and apologist for the powers that be. Oppenheimer's intervention deflated the political hopes of the assembled group and prevented any action. Lilienthal noted, "All these [arguments] were ameliorative of a critical tone in the statement, so much so that it was pointed out that if we said that we assumed that the Executive Department was proceeding wisely, was there any point in making a statement of caution and concern." As a result, no statement was issued.[95]

Oppenheimer also declined an invitation in 1957 from Bertrand Russell to attend the first Pugwash conference. This meeting, billed as an "international exchange of scientists" aimed at reducing Cold War geopolitical tensions by building trust between East and West, surely came close to a realization of Oppenheimer's expressed ideal of scientific internationalism as a path to peace. But he told Russell that he found himself "somewhat troubled when I look at the proposed agenda." He thought it included too many problems of the sort that, "if they can be answered at all, call for the wisdom of historians and philosophers rather than the technical knowledge of scientists." He also apparently took umbrage at what he saw as the antinuclear tenor of the meeting's agenda: "Above all, I think that the terms of reference 'the hazards arising from the continuous development of nuclear weapons' prejudges where the greatest hazards lie, and what course left open to us by the recent past still has the best hope of assuring man's survival and freedom." Russell replied that he could not understand this latter point. "I can't think," he said, "that you would deny that there are hazards associated with the continued development of nuclear weapons."[96] Although he had been excommunicated from the inner circle of the nuclear state, Oppenheimer remained, it seems, a supporter of the fundamental direction of its policies.

That Oppenheimer's stance in his later years was essentially conciliatory and supportive of the American status quo can be seen also in the way in which he acted as a kind of cultural envoy for the United States. Despite the fact that the U.S. government viewed him as a dangerous Communist, and FBI and CIA agents worried that he might defect to the Soviet Union, Oppenheimer presented to the rest of the world the civilized face of American intellect and culture. Between April and June 1958, Oppenheimer gave six lectures on physics at the University of Paris.[97] For French liberals, Oppenheimer represented a model of the non-Marxist scientific intellectual. Particularly in regard to France, this fit with the CCF's goal, which political scientist Giles Scott-Smith described as being "an effort to bolster the damaged European tradition of the free-thinking 'universal' intellectual and re-launch it in new circumstances," in opposition to Marxism.[98]

The most significant of Oppenheimer's foreign trips was to Japan for two weeks in September 1960. The trip was sponsored by the Committee for Intellectual Interchange (CII) of the International House of Japan. On arrival in Tokyo on September 5, Oppenheimer gave a crowded and chaotic impromptu press conference. When asked by a journalist whether he felt remorse for his role in building the bomb, he replied, "I do not regret that I had something to do with the technical success of the atomic bomb. It isn't that I don't feel bad. It is that I don't feel worse tonight than I did last night."[99] In reply to questions about whether he would be visiting Hiroshima, he said that, though he would like to, he would not have time and that it was not on his itinerary: "I have no quasi-official duties in Hiroshima." The atomic bomb was not what he wanted to talk about. Instead, he asked the reporters to "stress that I am glad to be here to talk about the things I know we can work on together: science and common cultural problems which underlie political problems."[100]

Oppenheimer's trip took place against the recent background of large left-wing protests and riots in Tokyo in May and June against the signing of a new security treaty with the United States. The treaty brought to the fore nationalist sentiments and undercurrents of resentment, largely suppressed since the end of the war, over the position of Japan as a U.S. protectorate. This boiled up in anti-American street protests. Eisenhower had been planning to make a state visit to Japan in June. However, when his press secretary, James Hagerty, arrived with an advance party, a crowd of eight to ten thousand demonstrators blocked his car on the way out of Tokyo airport. Hagerty was trapped in the car while demonstrators hammered on it and shouted for him to go home. In mid-June, amid American newspaper reports that Japan was on the brink of a Communist revolution, and because of the Japanese government's admission that it could not guarantee his safety, Eisenhower canceled his trip. The unpopular treaty was ratified in June; immediately afterward, Prime Minister Nobusuke Kishi announced his resignation, leading to a restoration of outward calm.[101] But tensions in U.S.-Japanese relations would not have been far from the surface during Oppenheimer's visit less than three months later. It is hardly surprising, therefore, that he preferred to avoid the topic of the atomic bombings and focused on scientific and "common cultural" concerns.

In addition to this highly symbolic timing, Oppenheimer's visit was potentially a direct and painful reminder both of wartime defeat and of postwar subordination to the United States. This was especially the case because of the central place in postwar Japan's national consciousness of its being the only nation to have been the victim of atomic weapons. The Peace Memorial Ceremony, held at Hiroshima on August 6 annually from 1947, expressed this postwar consciousness and national identity. The sense of atomic victimization was reinforced by an incident in 1954 when a Japanese fishing boat, the *Lucky Dragon*, was covered with fallout from an American nuclear test at Bikini. Less than three months after the incident, more than a million signatures were collected in Japan for a petition calling for a nuclear test ban.[102]

During Oppenheimer's trip, a twenty-one-year-old American antinuclear protestor, Tim Reynolds, traveled from Hiroshima to Osaka to meet the physicist and express disappointment that he would not be visiting the city on which the first atomic bomb had been dropped. Oppenheimer then explained that his sponsoring body feared that he would receive a hostile reception in the city.[103] The young American was the son of Earle Reynolds, who, with his family, had made a landmark protest against nuclear testing by sailing their yacht, the *Phoenix,* into the area around the Marshall Islands restricted for American nuclear tests. Tim Reynolds's appeal to Oppenheimer to visit Hiroshima made explicit the connection between Oppenheimer's visit to Japan, the current controversy around nuclear testing, and the atomic bombings at the end of World War II. And the issue of nuclear weapons was, for the Japanese, inseparable from questions about nationhood and their relationship to the United States.

But Oppenheimer hoped to distance his visit from the issue of the atomic bomb. He was welcomed to Japan not as the father of the bomb, but as a representative of science, of the "pure science" ideal, and of liberal intellectual culture. Science was a key motif in the postwar reconstruction of Japan as a Western-oriented democracy, allied with America.[104] Oppenheimer's visit was symbolic of that connection. At a time when Japanese intellectuals had been at the forefront of opposition to the security treaty, Oppenheimer put forward and embodied a model of the intellectual role fusing scientific modernity with liberal cultural values.[105] His lectures in Japan repeated the lessons of the dangers of cultural fragmentation and mass culture and of the necessity of elite intellectual community, which he extrapolated from the American experience. While Oppenheimer's vision of the role of the intellectual was ostensibly apolitical, its meaning was in fact highly political in the context of Cold War culture and relations between the United States and Japan.

Oppenheimer gave a two-hour lecture on the physics of elementary particles at the Kyoto Institute for Fundamental Physics, but most of his speeches during the trip were on the cultural meaning of science and the relationship between science and the broader culture. For example, to a gathering of more than a thousand in Osaka, he gave a two-and-a-half-hour talk titled "Tradition and Discovery"; in Tokyo, he spoke on "the future of civilization in the scientific age." Oppenheimer was recognized as a spokesman for science and an interpreter of scientific culture. He was described by one Japanese newspaper as a "slender and somewhat aristocratic scholar." A student who attended Oppenheimer's Kyoto talk on elementary particles remarked, "He spoke in a deep tone as if he were reading a poem."[106]

On September 9, Oppenheimer spoke in Tokyo's Bunkyo Public Hall for the International House, on the topic of science and culture. The talk repeated Oppenheimer's standard themes—the growth and specialization of the sciences, the increasing divorce between science and common understanding, and the decline of the "public sector"—and concluded with a discussion of "intellectual community" versus "mass culture." Condemning the cultural gluttony of mass society, he said that cultural values needed to be nurtured within an intellectual community. This meant, first, protecting communities of specialists (scientific or artistic): "We have, all of us, to preserve our competence in our own professions . . . This is, in fact, our only anchor in honesty." Second, it meant bridging the divisions between specialists to "reknit" an overarching intellectual community. As he began his speech, Oppenheimer apologized that he would be talking about what he knew: American and Western European, rather than Japanese, culture. But he suggested that as science and advanced industrialism became global, this Western experience would be increasingly generalizable. America's cultural problems were common ones, as were the solutions. Oppenheimer appropriated the discourse of the New York intellectuals and American Cold War liberalism and made it a universal discourse. As a gift to his Japanese host, CII chairman Yasaka Takagi, Oppenheimer sent a copy of the *Partisan Review*.[107]

The paradox of Oppenheimer as security risk at home and American cultural ambassador abroad was partially resolved when he was officially rehabilitated during President Kennedy's term. Oppenheimer was a figure of admiration among the young technocrats who flocked to work for the new administration. He was granted the AEC's Fermi Award on December 2, 1963. The assassination of Kennedy the previous month meant that the prize was conferred by Lyndon Johnson, who called the decision to award it to Oppenheimer "one of President Kennedy's most important acts."[108] However, though officially rehabilitated, Oppenheimer was still barred from secret government work. As the *New York Herald Tribune* put it, the award was "an honor, not an indemnity."[109] Nevertheless, it signaled that he was no longer a pariah. Oppenheimer described his reaction to being chosen for the award by saying, "Most of us look to the good opinion of our colleagues and to the good will and confidence of our Government. I am no exception."[110] Oppenheimer never saw himself in an outsider role of critical intellectual. Rather, he saw himself as a servant of science, and a servant of power. Looking down from Olympian heights on mass society, Oppenheimer aligned himself with the elite and courted the powers and the powerful.

In these final years of his life, Oppenheimer constituted himself as a public moralist, albeit a highly abstract one when compared with figures such as Einstein or Russell. He told a *Look* magazine journalist in 1958 of his concern about "the erosion of the Puritan ethic, which has been replaced by a deeply complacent view of human nature and of ourselves."[111] And in 1963, he commented on the immorality of war for the same magazine, in an article entitled "Morality USA: Have Bigness, the Bomb, and the Buck Destroyed Our Old Morality?"[112]

In shaping his persona as a moralist, Oppenheimer looked to the figure of Niels Bohr as a source of inspiration. Bohr presented for Oppenheimer an example of how the scientist could have an authority beyond the merely technical. Bohr was a visionary, but he was not overtly political. He acted not as part of a campaign, but on his own, his force deriving from his own personal moral authority and intellectual insight. Oppenheimer, socially and politically isolated, must have been attracted by this image of a lone figure, a philosopher in exile (as Bohr was during the war) who nevertheless wove science into a vision of moral and social renewal.[113]

In an article for the *New York Review of Books,* Oppenheimer emphasized Bohr's prescience during the war in foreseeing an arms race with the Soviet Union and his failed attempt to forestall it through the advocacy of an "open world."[114] Bohr had offered a way to avoid the trap of the Cold War and arms race. Oppenheimer wrote, "If we had acted wisely, clearly, and discreetly in accordance with his views, at the least we might have been freed of our rather blasphemous sense of omnipotence and secrecy. We might have turned our society and our life toward a healthier vision of a future worth living for, an increased dedication to knowledge and truth."[115]

Bohr was misunderstood by the politicians, and therefore this opportunity was lost. Margaret Gowing's official history of the British atomic energy program, published shortly before Oppenheimer's article, had revealed that Churchill and Roosevelt regarded Bohr as a potential traitor.[116] The parallel between these suspicions and the fate of Oppenheimer was not lost on the press at the time of these revelations and as Oppenheimer emerged as Bohr's public champion.[117] Oppenheimer's views on Bohr were widely reported. But Oppenheimer was embittered by what he perceived to be his lack of influence. He thought that his views were being suppressed. Despite the fact that the *New York Review* had featured the title of his piece in bold red characters on its front page, Oppenheimer wrote to the editor accusing him of having done "your best to conceal the Bohr story." The editor wrote back assuring Oppenheimer that it was "the most important thing that we've ever published" and pointing out that stories about the article had been printed "in dozens of papers all over the country."[118] But although Oppenheimer could still make headlines, he lacked a clear a clear sense of what sort of interventions he should make in public life, what role he could play. He was unwilling to engage in political controversy and instead concerned himself with issues of culture and morality. But he seemed unclear about how this role as moralist could be related to his professional identity as a scientist.

In his January 1962 Whidden lecture titled "War and the Nations," at McMaster University in Ontario, Canada, Oppenheimer still regarded the responsibility of the scientist as a narrowly restricted ethic of awareness, the duty "to give an honest account of what we all know together, know in the way in which I know about the Lorentz contraction and wave-particle duality, know from deep scientific conviction and experience. We think that we should give that information openly whenever that is possible, that we should give to our governments in secret when the governments ask for it, or, even if the governments do not ask for it, that they should be made aware of it."[119] As an example of this responsibility, he reminded his audience of Einstein's letter to Roosevelt in 1939 apprising the president of the possibility of atomic weapons. It is ironic that Oppenheimer should have taken this as a model of responsibility, for Einstein himself deeply regretted even this small involvement in the making of the atomic bomb.[120] Einstein's understanding of responsibility led him, after the war, to take on the role of an outsider intellectual critic, in open dissent against Cold War politics. In contrast, Oppenheimer was skeptical about whether a public sphere in which one could participate politically as an intellectual even existed. Portraying public culture as anti-intellectual and stultifying, he was instead attracted by the notion of retreating into a small and elite circle that could protect and carry high culture through a dark time.

The CCF provided for Oppenheimer some realization of this ideal. Its meetings were a key forum for his musings on the relationship between science and culture.[121] The group actualized his ideal of intimate discussion among elites as a way of repairing cultural fragmentation. The most complete expression of this was the Seven Springs Farm conferences at the Mount Kisco, New York, estate of Agnes Meyer, widow of Eugene Meyer, the former owner of the *Washington Post*. Oppenheimer was taken with the "open, hilly country" and with the house, which he described to Nicolas Nabokov as "extremely fine, in perfect taste, embellished with what can only be regarded as a great deal of loot in the way of art from China, and a good deal of modern European art as well."[122] Organized by Oppenheimer and partly supported by the CCF, the first "intimate and informal 'Rencontre'" took place in the summer of 1963, with the Princeton Jefferson scholar Julian Boyd; philosophers Stuart Hampshire from Oxford and Jeanne Hersch from Geneva; University of Edinburgh psychiatrist Morris Carstairs; diplomat George Kennan; architect

Wallace K. Harrison; Los Alamos veteran George Kistiakowsky, the first head of the presidential Scientific Advisory Committee; the poet Robert Lowell; Nicolas Nabokov, who helped plan the event; and hostess Agnes Meyer.[123] Numbers were kept below fifteen in order to, as Meyer put it, "maintain intimacy of discussion."[124]

The theme of the conference was allusively suggested to invitees with an extract of Oppenheimer prose:

> For myself, there are at least two things that I should look for: an image of the world that responds to the changing needs and hopes of men without resort to the institution of war, indeed without its possibility; and a world in which, despite the vastness, the complexity, the rapid change in the circumstances of our life, and the knowledge which underlies it, the scope and depth of human responsibility and human nobility grow instead of shrinking.[125]

Oppenheimer chaired the discussion, but, as Hersch put it, he "was hardly a 'chairman' in any conventional sense. He simply, from time to time, suggested a neglected aspect of a question, softened a particular dogmatism, cast doubt on such and such an alleged bit of evidence or dropped a word indicating a complexity which might lie hidden beneath an over-simplification resulting from facile hope or impatient pessimism." Meyer had no doubt that the event relied on Oppenheimer's personal and intellectual "leadership." And Oppenheimer's talk, in which he dwelled on Bohr and the meanings of complementarity, was punctuated by her comments of "That's lovely Robert" and "That's wonderful Robert." She wrote to him that when she later read and reread the script of his talk, "I re-lived the deep emotions with which I heard you speak. And yet, there was something lost because the warmth of your voice and personality gave the experience of listening to you an ambience which cannot be recaptured." She later wrote to Kitty about the annual Seven Springs Farm event that "I cannot imagine our meeting without Robert."[126]

The meetings at Seven Springs Farm instantiated Oppenheimer's ideal of a small and intimate intellectual community in close contact with men of power, able to exert behind-the-scenes influence on the cultural-political agenda of the times. It was this ideal of intellectual life that was behind the critical tone of Oppenheimer's UNESCO lecture of December 1965, given in honor of Einstein on the fiftieth anniversary of the general theory of relativity. Oppenheimer began by saying that he wanted to "dispel the clouds of myth" that surrounded Einstein. Most of his talk was an attempt to break through the mystique of genius by showing how Einstein's originality was located in and grew out of his appreciation for the tradition of physics as a science. Oppenheimer sought to relocate Einstein's genius within an understanding of science as a community with history and traditions. He also discussed Einstein's arguments with Bohr and his skepticism about quantum mechanics; in Oppenheimer's view, this skepticism led him to "failure" in the intellectual project of his later years, in which, Oppenheimer argued, Einstein "lost most contact with the profession of physics." His portrait of Einstein was informed by his view that science was necessarily the work of community, rather than of the individual genius or "lone worker."[127]

Even as Oppenheimer paid his respects, it was clear that Einstein was a discomforting figure for him. Oppenheimer could not help but come across as snobbish and condescending when he said that Einstein "was wholly without sophistication and wholly without worldliness." Einstein has often been described as having a childlike innocence; coming from someone else, this kind of description might have been a compliment. Coming from Oppenheimer, so obviously proud of his own sophistication, the observation about Einstein's lack of worldliness was far from being praise. "I think," Oppenheimer said, "that in England people would have said that he did not have much 'background', and in America that he lacked 'education.'" Pais wrote to Oppenheimer that the remarks had made him "slightly uncomfortable." Oppenheimer regretted that "a number of colleagues," since reading of his comments in the *New York Times,* "have suggested that I had been out of my mind." The executor of Einstein's estate, Otto Nathan, said that he was "seriously disturbed" by the talk. In light of Oppenheimer's assertion that Einstein's early papers were "full of errata," Nathan no longer trusted Oppenheimer and the Institute for Advanced Study to oversee publication of a new edition of Einstein's papers. The *New York Times* reported that Oppenheimer said that correcting these errors had delayed publication of the volume for ten years. Nathan was outraged by the idea that the papers might need "correcting." Oppenheimer wrote to Einstein's secretary, Helen Dukas, that when he saw the *New York Times* article, "I shuddered for you."[128] He enclosed the original text of his speech, but while it did not include the "ten years" remark, it is rather unlikely that reading it would have made her feel a great deal better.

Oppenheimer was wrestling with the example of Einstein, who, for a great many people, embodied the moral qualities—particularly the quality of ahimsa—that Oppenheimer claimed had vanished from the culture. But Einstein was, as Oppenheimer put it, "without power": "He had a deep distrust of power; he did not have that convenient and natural converse with states-men and men of power that was quite appropriate to [Ernest] Rutherford and to Bohr."[129] That, from Oppenheimer, was also a criticism. Einstein was in many ways a living contradiction of Oppenheimer's conception of the role of the intellectual. Yet in another sense, the problem was that Einstein was the realization of the uncompromised ideal, an ideal that Oppenheimer could not himself achieve. Einstein made Oppenheimer's more ethically equivocal example look like a poor compromise. In particular, Oppenheimer was made uncomfortable by the fact that Einstein did not apparently value the thing for which Oppenheimer himself had given up so much in order to court—power.

Bohr had failed to convert the wartime British and American leadership to his vision for world order; rather, he had incurred their suspicion. Op-penheimer had been humiliated and cast aside by the state that he served. Yet Oppenheimer continued to aspire to meaningful and influential "con-verse" with the powerful, and this aspiration shaped the intellectual, social, and political allegiances of his later years. The portrait of the communitarian characteristics of scientific and intellectual life that Oppenheimer painted in his writings and speeches was ostensibly apolitical. But its meaning was in fact political through and through. Oppenheimer's view of science fit perfectly with pluralism, the form of liberal political thought dominant in American academic and intellectual life after the war and at least until the late 1960s. To list the key thinkers associated with pluralism is to give a roll call of the major American political and social scientists of the period—including figures such as Daniel Bell, Nathan Glazer, Richard Hofstadter, William Kornhauser, Seymour Martin Lipset, Talcott Parsons, David Riesman, and Edward Shils, among others.[130] Pluralism did not mean simply tolerance for diversity. Rather, as political theorist Michael Rogin argued, it was a theory of social order that expressed the anxieties of a generation of American liberal intellectuals, who saw liberal democracy as a fragile accomplishment, threat-ened by the totalitarian impulses of mass society. The pluralists were obsessed with how to promote a stable capitalist economy and a vibrant civil society composed of associations, rather than a more volatile mass.[131]

In particular, pluralism was a response to the rise of McCarthyism—which the pluralists interpreted as a radical movement, deeply rooted historically in American populism or agrarian radicalism, and an attack by disgruntled "nouveaux" on established elites. The intellectuals saw it as a movement of mass society against the establishment and as hostile to the existing social order. As in Hofstadter's famous study of anti-intellectualism, the pluralists saw the problem of McCarthyism as one of defending the cultural authority of the elite against the pressures of popular Know-Nothingism. A disinterested elite was needed to preserve liberal democracy against the tyranny of the majority. The pluralists looked to people like themselves as the vanguards of social order and civilized values.[132]

Oppenheimer's defensive view of science as a fragile accomplishment of a cultural elite, threatened by the homogenizing trends of mass society, meshed perfectly with the pluralist worldview. The intimate converse of a cultural and political elite, which Oppenheimer found in the CCF and at Seven Springs Farm, was the realization of his ideal of an elite preserving cultural values against a hostile mass society. Rogin argued that the pluralist view of McCarthyism as a populist revolt of the masses was a distortion. In fact, McCarthyism drew on traditional American conservatism and relied on the acquiescence, and very often the support, of the elite. The Oppenheimer case clearly instantiates this elite acquiescence and support. McCarthy lurked in the background. But it was Eisenhower who ordered Oppenheimer's dismissal, Admiral Strauss's Atomic Energy Commission that carried out the purge, and scientific colleagues, such as Teller and Alvarez, who twisted the knife. Yet almost immediately after the hearings, Oppenheimer wove the event into what was essentially the liberal pluralist narrative of the condition of the intellectual, besieged by mass society. In so doing, he proceeded both to muddy the understanding of his own "case" and to choose a largely conservative rather than a critical intellectual role.

In the pluralist diagnosis, the intellectual, threatened from below, needed to be nurtured within the elite. The CCF was the home in which many pluralist intellectuals, such as Shils, Riesman, and Bell, among others, found this elite community. As Oppenheimer became a spokesman for the cultural condition of the intellectual, he also cemented his ties with that organization. However, the cultural, political, and military conflicts of the 1960s threatened to fragment this liberal elite milieu in which Oppenheimer felt comfortable in his later years.[133]

The Vietnam War gave rise not only to increasing revulsion against American military might, but also to difficult questions regarding the compact between the liberal intelligentsia and American global power, and between ideological anti-Communism and aggressive militarism. On April 27, 1966, as part of a series of exposé articles on CIA covert operations, the *New York Times* revealed the extent to which the agency had used the distribution of funds to influence academic and intellectual activity. Among the revelations was that the CCF and its magazine, *Encounter,* had been bankrolled by the CIA. In the ensuing controversy, Oppenheimer came to the defense of the

CCF. In a letter published in the *New York Times* on May 9, Oppenheimer, Kennan, Galbraith, and Schlesinger stated that in their experience with the CCF, "there has been no question regarding the independence of its policy, the integrity of its officials, or the value of its contribution. In our experience the congress . . . has been an entirely free body, responsive only to the wishes of its members and collaborators and the decisions of its Executive Committee."[134] The statement gave the impression of denying CIA involvement without literally doing so. Journalist Dwight Macdonald called it "an evasion, not a lie, but not meeting the issue either." Stuart Hampshire, in Princeton at the time, said that "Oppenheimer was amazed that I was amazed, and amazed that I was upset at the *New York Times* revelations . . . Oppenheimer wasn't amazed because he was half in it himself. He knew full well. He was part of the apparat [Hampshire's term for the CCF's controlling group]. I don't think it bothered him morally. If you're imperially-minded, which the Americans were at the time, you don't think much about whether it's wrong or not. It's like the imperial British in the Nineteenth Century. You just do it."[135]

Oppenheimer's complacency regarding CIA involvement in intellectual life is revealing, for the CCF was the closest instantiation for Oppenheimer of his ideal of intimate, cohesive, and influential intellectual community. This intimacy and cohesiveness was disrupted by the Vietnam War. Agnes Meyer worried that the Seven Springs Farm rencontre of 1966, aimed at understanding the "growing lack of sympathy" of European intellectuals toward the United States, might be marred by a "knockdown fight about Viet Nam." Since a number of the attendees, most prominently Senator William Fulbright, were against the war, Meyer was worried about the meeting appearing "like an anti-administration meeting." She herself was against the war, but she felt unable to speak out because of ties to the Johnson administration. Oppenheimer promised her "not to say a word on the subject," although he, too, was uneasy about U.S. involvement in the war.[136]

In December 1966, in an article for a special edition titled "Europe and the United States Today" of the *Herald Tribune*'s European edition, Oppenheimer gave voice to his concerns. Characteristically, his wording was oblique, and the argument was framed in terms of culture and morality rather than as direct political criticism. His premise was that the cultural problems of the age were shared by Europe and America. At a time when the world was divided along the Cold War axis of East and West, Oppenheimer traced the historical roots of America and American freedom to the cultural and religious tradition of Western Europe. But this tradition was now being tested both in Europe and in America. Oppenheimer said, "I doubt whether, at this writing, there is among us any wide conviction that France, or the United States, Germany or England, now has a government even remotely competent to the problems of the time, or in fact has available for those problems the human resources, the insight, the wisdom and skill, and the underlying stoic confidence for which they call." The article was suffused with a sense of entropy and decline. So it was with "haunting nostalgia" that Oppenheimer quoted Lincoln's belief that "the past actions and influences of the United States were generally regarded as having been beneficial toward mankind." Oppenheimer observed that few were likely to think that about current U.S. foreign policy. Vietnam represented both the sacrifice of America's moral innocence and a historical connection between new American and old European colonialism. No longer an immaculate "City upon a Hill," America now shared in the sins of the Old World. Oppenheimer said, "The sense of what it is like to live with a government held wrong in a moral matter is a new and desperate bond between European peoples and those of the United States."[137]

It was in the cultural and intellectual bond between America and Western Europe that Oppenheimer placed his faith. In a reference to the traditional role of the intellectual in speaking truth to power, Oppenheimer concluded, "This is a time in need, not of delusion but of hope, when we, not least in Europe and America, who have taught hope so willingly and widely, must bear true witness." Oppenheimer portrayed the intellectual as a carrier of moral values forgotten in a secular world. His article was a reaffirmation of his view of the defensive role of an enlightened elite, which had been central to his thought since the hearings.[138]

Oppenheimer accepted a position on the CCF's board of directors in autumn 1966 and was due to attend the board meeting in March of the following year. Indeed, he told CCF executive director Michael Josselson, "There is nothing to which I more look forward than playing a part in the Congress' new life with you and other friends." Josselson was eager that Oppenheimer's *Herald Tribune* article provide a basis for discussion at the meeting. In a memorandum, he informed board members that "recently, Robert Oppenheimer (in a statement criticizing U.S. involvement in Vietnam) castigated most of the Western governments for their incompetence in the face of major problems of our time." Josselson suggested to Oppenheimer that they distribute in advance of the meeting a copy of Oppenheimer's article and at the meeting "have you comment as briefly as you wish on how you see the Congress's role in the framework of your statement."[139] By February, however, Oppenheimer, undergoing treatment for throat cancer, was in too much pain, and his speech and hearing were impaired. On February 13, only five days before his death, he wrote to Aron, "I think that it would be appropriate to regard my seat on the Board as empty, though I hate to put these words in writing."[140]

KNOWING SIN

Oppenheimer's lasting public image has been shaped above all by the notion that he suffered from profound guilt and self-reproach for his work on the bomb. However, Oppenheimer was always deeply uncomfortable with this perception. As guest of honor at a reception hosted by Norris and Lois Bradbury at Los Alamos in 1955, Oppenheimer was told to his face by the teenage daughter of one of the laboratory's explosives experts, "I think you're a saint." When he asked why she would say that, the young woman replied, "Well, because you had second thoughts." As she remembered the incident, "He was just stricken. I could see his face change . . . The idea that I had touched him really scared me."[141] Oppenheimer was to frequently confront the notion that, of all the atomic scientists, it was he who experienced real moral pain and who struggled with inner demons.

Robert Jungk's account of the building of the atomic bomb, *Brighter Than a Thousand Suns,* which appeared in English in 1958, most widely popularized the image of Oppenheimer as regretting his work on the bomb. Oppenheimer was the central figure of the book, and the idea that "the physicists have known sin" provided the core theme in Jungk's narrative. The book is also most responsible for bringing to public consciousness Oppenheimer's quotation "I am become Death," as an iconic statement of the nuclear age. For Jungk, Oppenheimer represented the tragic corruption of science by power. Jungk frequently turned to Oppenheimer for a diagnosis of the disease; he quoted Oppenheimer as saying, "We did the devil's work."[142] But he saw Oppenheimer—asserting that the scientist must follow what is "technically sweet"—also as the chief symbol of that sin. Oppenheimer, for Jungk, symbolized the duality of modern scientific culture, in the conflict between its liberal humanist ideal and its increasingly technological and authoritarian direction.

In Jungk's narrative, there was a golden time and place—"the beautiful years" of Göttingen in the 1920s. But this was destroyed by the rise of Nazism in Germany and the flight of physicists into the hands of the U.S. Army. For Jungk, it was the German scientists who remained true to the humanistic spirit. In a chapter titled "The Strategy of Prevention," Jungk uncritically presented Werner Heisenberg's statement that he and his colleagues had secretly sabotaged the German effort to produce an atomic bomb. He also published Heisenberg's claim that on his visit to Copenhagen in the autumn of 1941, he had attempted to persuade Bohr to cooperate in what he hoped would be an international scientific conspiracy to prevent either side from acquiring atomic weapons. The brotherhood of physics would guard its secret knowledge from those who would use it for ill. According to Jungk, Heisenberg and his colleagues "obeyed the voice of conscience and attempted to prevent the construction of atomic bombs, while their professional colleagues in the democracies, who had no coercion to fear, with very few exceptions concentrated their whole energies on production of the new weapon."[143]

As part of the research for his book, Jungk met with Haakon Chevalier, then living in Paris. On reading the transcript of the security hearings, Chevalier had become convinced that the root of his having been blacklisted since 1943 was the "cock-and-bull story" that Oppenheimer had told to Manhattan Project security officers. What the transcript suggested, and what Chevalier was now sure of, was that when Oppenheimer named Chevalier as the intermediary, he had not retracted his fabrication that three Manhattan Project scientists had been approached for information by this intermediary. In other words, Oppenheimer had named Chevalier only to implicate him in a complex espionage conspiracy, which Oppenheimer had himself invented.

Chevalier had all but worshipped Oppenheimer before the war and continued to admire him afterward, often seeking the physicist's advice and help. For example, Oppenheimer put him in touch with Jeffries Wyman, who was scientific adviser to the U.S. embassy in Paris, for advice on his passport situation. In December 1953, Chevalier introduced Oppenheimer to André Malraux, whom Chevalier had translated into English. Wyman had the impression that "Chevalier's two idols at that time were Malraux and Oppenheimer, and he brought them together, as he thought, as an almost astronomical conjunction." The result was disappointing; Oppenheimer said, "Malraux has some understanding as to what science *isn't*. But he has no conception of what science *is*."[144] After the hearings, however, Chevalier was left with a far greater disappointment. During all the years when Chevalier had sought Oppenheimer's counsel and understanding about his security troubles, joblessness, and marital breakdown, Oppenheimer had never told him about the "tissue of lies" that had implicated Chevalier in a serious espionage conspiracy. Chevalier was left with a smoldering sense of betrayal. He opened to Jungk his correspondence with Oppenheimer and gave him his side of the "Chevalier incident."[145]

Chevalier also channeled his bitterness into two books of his own. The first was a novel, *The Man Who Would Be God*, published in 1959. The protagonist, physicist Sebastian Bloch—clearly modeled on Oppenheimer—was a

Dr. Frankenstein, dehumanized by the atomic "monster" that he had created. Chevalier followed this in 1965 with *Oppenheimer: The Story of a Friendship,* an account of their relationship that gave his own side of the notorious "Chevalier incident" and his own interpretation of Oppenheimer's "defects of character." Chevalier surmised that buried deep somewhere in Oppenheimer's youth was the source of his combined insecurity and arrogance, a deep-rooted psychological flaw that set him on a tragic path. It was Jungk, Chevalier said, "who helped me to achieve the final and decisive insight into Oppenheimer's character . . . He forced upon me the recognition that the defect of character . . . must have existed in him from the beginning."[146] For both Jungk and Chevalier, Oppenheimer was a classic tragic figure. Both aimed to find and describe the psychological flaw in Oppenheimer's personality at the same time that they treated Oppenheimer as a symbol of a Faustian tragedy of the atomic scientists in general. To understand Oppenheimer, they were saying, was to understand the situation of the modern scientist.

Jungk was also an important influence for Heinar Kipphardt, whose play *In der Sache J. Robert Oppenheimer* (In the matter of J. Robert Oppenheimer) first appeared in German in 1964. The play was a dramatization of the hearings, composed almost entirely of verbatim extracts from the official transcript. However, Kipphardt was able to skillfully transform the meaning of the event by careful clipping of the transcript, as well as by the interjection of fictionalized soliloquies by the major figures and the addition into the proceedings of relevant statements made in other contexts by these figures. Kipphardt transformed the hearings from a matter of loyalty and disloyalty to a matter of morality. He constructed a trial in which the central issue was the immorality of the scientists' role in producing the bomb and in the destruction of Hiroshima and Nagasaki. Whereas in the actual proceedings Oppenheimer was driven to admit that he had been "an idiot" insofar as he had deceived military security officers, more important for Kipphardt was Oppenheimer's statement, now injected into the proceedings themselves, that "we [the physicists] have known sin."[147]

The play ends with a soliloquy by Oppenheimer in which he wonders aloud "whether we were not perhaps traitors to the spirit of science when we handed over the results of our research to the military, without considering the consequences." To Kipphardt's Oppenheimer, it seemed that "the actions the Board hold against me were closer to the idea of science than were the services which I have been praised for." And the playwright had his character express deep regret for his work on the bomb: "We have spent years of our lives in developing ever sweeter means of destruction, we have been doing the work of the military and I feel in my very bones that this was wrong."[148]

Kipphardt therefore transformed Oppenheimer from a relatively passive figure, which he appeared to be in the hearings themselves, into a far more forceful and powerful character, a humanist critic of the bomb that he himself had created. Kipphardt achieved this transformation by using statements that Oppenheimer had made in various other contexts. In Kipphardt's rendition, Oppenheimer was placing himself on trial. Oppenheimer the humanist was wrestling with, and ultimately exorcising, Oppenheimer the technocrat.

In a tense correspondence with Kipphardt, Oppenheimer accused the playwright of misrepresenting him and of being generally cavalier with the historical record. Conveying a deferential and respectful tone in formal German, the playwright was at pains to win Oppenheimer's goodwill. Kipphardt insisted on the difference between a play and a historical document. And he tried to impress on Oppenheimer that the playwright's concern had to be with drama and with capturing the essence, rather than all the factual details, of the historical events.[149] Above all, Oppenheimer was incensed at the play's soliloquy in which Kipphardt had Oppenheimer express regret for working on the bomb. Oppenheimer wrote to Kipphardt,

> You make me say things which I did not and do not believe. Even this September in Geneva, during a conference of the Recontres de Genève, I was asked by the Canon van Kamp whether now, knowing the results, I would again do what I did during the war: participate in a responsible way in the making of atomic weapons. To this I answered *yes*. When a voice in the audience angrily asked "Even after Hiroshima?" I repeated my *yes*.

And he told the playwright, "It seems to me you may well have forgotten Guernica, Dachau, Coventry, Belsen, Warsaw, Dresden, Tokyo. I have not." He also threatened to sue.[150]

Oppenheimer described Kipphardt's text as "really dreadful," "miserable," "sordid," and even "anti-American." He was marginally happier with the 1965 French-language adaptation of the play, *Le Dossier Oppenheimer*, by Jean Vilar; he thought it took less dramatic license. But he nevertheless felt that "the playwrights have failed to study the [original security hearings] transcript, and that they have been guided more by Robert Jungk's mendacious book, and perhaps by some things Chevalier has written."[151] He told a *Washington Post* interviewer that the play "turned the whole damned farce into a tragedy."[152] Oppenheimer's scientific colleagues and associates from the CCF also publicly came to his defense. The representation of Oppenheimer was a matter in which these groups had a collective stake. Victor Weisskopf, now director of the European nuclear physics laboratory CERN, emphasized to Oppenheimer that "your name . . . after all is considered a symbol of the scientific community."[153]

By the end of 1965, Oppenheimer's reaction had mellowed somewhat. He told Kipphardt that "the passage of time" had made him regret being "unduly harsh and unkind" in his initial reaction, and he acknowledged that "you meant me no harm."[154] Nevertheless, Oppenheimer continued to be privately bitter about the play.

Oppenheimer's former student David Bohm wrote to him from England late in 1966, expressing concern about Kipphardt's play (which had recently appeared in a London theater) and about the fact that Oppenheimer often appeared in public to be suffering from bad conscience: "I have also seen some television programmes in which you appeared, in an account of the Los Alamos project and its consequences. I was rather disturbed especially by a statement you made, indicating a feeling of guilt on your part. I feel it to be a waste of the life that is left to you to be caught up in such guilt feelings." Bohm tried to impress on Oppenheimer the futility of this sort of angst, "since whatever happens, you are what you are."[155]

Oppenheimer replied to Bohm, "The play and such things have also been rattling around for a long time. What I have never done is to express regret for doing what I did and could at Los Alamos; in fact, on varied and recurrent occasions, I have reaffirmed my sense that, with all the black and white, that was something that I did not regret." Oppenheimer drafted but did not send a more intimate letter, in which he added, "My principle [*sic*] remaining disgust with Kipphardt's text is the long and totally improvised final speech I am supposed to have made, which indeed affirms such regret. My own feelings about responsibility and guilt have always had to do with the present, and so far in this life that has been more than enough to occupy me."[156]

Oppenheimer projected a generalized guilt without a firm object. In his talk at Seven Springs Farm in 1963, he veered from a discussion of Bohr's notion of complementarity to an assessment of his own personal failings: "I hardly took any action, hardly did anything, or failed to do anything, whether it was a paper on physics, or a lecture, or how I read a book, how I talked to a friend, how I loved, that did not arouse in me a very great sense of revulsion and of wrong."[157] Oppenheimer's sense of guilt was intensely personal and self-absorbed. He made apparent that he was a man wracked by guilt and self-doubt, yet he pulled back from a repudiation of the bomb. It was a guilt that he would not allow to be attached to any particular source or site. Hence his comment in Japan that he felt no worse on arriving there than previously. In 1961, he even went so far as to say, "I carry no weight on my conscience," in reference to the atomic bombings. For, he said, the use of science was a problem for government, not for the scientists.[158]

Oppenheimer seemed to be in a pendulum swing of guilt for the scientist's original sin, periodically assuaged by a sense of the redeeming qualities of the scientific vocation. Wyman thought that Oppenheimer in his later years "was almost reveling in his feeling of guilt about the bomb." But at the same time, "it was obviously a great achievement and a very important thing in his life . . . and there was a kind of play back and forth between these two attitudes."[159] This oscillation is understandable when one considers that it was the public identification of Oppenheimer with the bomb that was largely the source of his postwar status and power. Even after the hearings, his past role at Los Alamos was essential to his status as spokesman for modern science and, more broadly, for the condition of humanity in the modern "atomic age." For that reason, he could not, without maiming himself, separate himself from the weapon that had so strongly defined who he was. Paradoxically, his professions of guilt further cemented his public association with the bomb. As John von Neumann observed, "Some people profess guilt to claim credit for the sin."[160] Yet from the end of the war onward, Oppenheimer had also sought to be a spokesman for the values of "pure science," and after the hearings he had laid claim to a role as general intellectual and cultural critic. These roles, as spokesman for science and for intellectual culture more generally, required that he not be identified wholly with the bomb—hence his frequent defenses of a notion of vocation that made the "uses" of science a matter for the politician, not the scientist. Thus, Oppenheimer both embraced and repudiated the bomb, often simultaneously.

In a CBS television interview to mark the twentieth anniversary of Hiroshima, Oppenheimer reiterated the official justification for the bombing— that it was done to prevent "a slaughter of Americans and Japanese on a massive scale"—and that the wartime American leaders, such as General Marshall and Secretary Stimson, had arrived at the decision "in good faith with the best evidence that they then had." When the interviewer suggested that "you and many like you who brought the bomb into being still seem to suffer . . . from [a] bad conscience about it," Oppenheimer answered by returning to his statement that "the physicists have known sin." "I didn't mean by that," he said, "that the deaths were caused as a result of our work. I meant that we had known the sin of pride. We had turned to effect . . . the course of man's history. We had the pride of thinking we knew what was good for man . . . This is not the natural business of a scientist." Their business was, rather, "studying nature, learning the truth about it."[161] So even while in his later years Oppenheimer's public authority was based on his self-presentation as cultivated humanist and general philosopher, when it came to the atomic bomb, he retreated into a narrow conception of responsibility as the scientist's responsibility to his vocation.

A professor of literature who watched the program wrote to Oppenheimer, "The depth of your compassion as it came through your voice was such that I found myself weeping before a television set as I had not since the assassination of President Kennedy." But he was also surprised at Oppenheimer's definition of sin as pride: "I do not agree; it is inevitable in this scientific age that the scientist become a part of government and thus forced to assume responsibility for decisions that affect nations. That this has happened is . . . potentially a very hopeful thing. It is the time indeed for all men to cease to be just specialists, just irresponsible cogs in the wheel, and to become fully human, that is responsible for the goals and values of his society."[162]

In late 1966, journalist Thomas B. Morgan spent an afternoon with Oppenheimer at Princeton. He found the physicist in a contemplative mood: "Oppenheimer would gaze with pleasure across the deep-green lawn rolling down from the Georgian cluster of Institute buildings to the still pond and the autumn woods beyond. When I arrived that afternoon, this was the first thing he said: 'We're having a beautiful season.'" Oppenheimer was dying, ravaged by his cancer—an illness that he bore, as Morgan put it, "with stoic grace . . . He was very frail . . . There were deep lines in his face. His hair was hardly more than a white mist. And yet, he prevailed with that grace." He had given up his pipe, on his doctor's orders. Morgan's visit found Oppenheimer ruminating on the question of "responsibility." When Morgan suggested that Oppenheimer's "devotion" to that word "seemed almost religious," the physicist replied, "The use of the word 'responsibility' . . . is almost a secular device for using a religious notion without attaching it to a transcendent being . . . I don't know how to describe my life without using some word like 'responsibility' to characterize it, a word that has to do with choice and action and the tension in which choices can be resolved."[163]

This way of speaking about responsibility contrasted with Oppenheimer's earlier formulations in which he treated responsibility as limited and relative to the demands of vocation, in particular the scientist's responsibility to his science. Now, in the last months of his life, Oppenheimer was groping for a way of transcending the specialized ethos of scientific vocation and the limitations of office and coming to terms in a more personal and human way with his responsibility for the atomic bomb. But despite his contemplative mood, Oppenheimer was wary of separating responsibility from the context of action in the world; he said that responsibility is always "limited by what one can do." This suggested a paradox. When Oppenheimer was in positions of institutional power, he found himself unable to sustain a sense of responsibility beyond a narrow bureaucratic or soldierly conception of duty. Now, in his exile from state power, he began to formulate a broader and more "fully human" conception of moral responsibility. Yet as he defined his responsibility in these broader terms, and as in these later years such thoughts were separate from political action and consequence, this responsibility seemed indistinct and lacking in substance. Oppenheimer's conclusion was revealing: "There is no meaningful responsibility without power."[164]

SINNER AND SAINT

When Oppenheimer died at age sixty-two, on February 18, 1967, he was immediately, and by now predictably, labeled the tragic embodiment of the cultural crises of the era. In Japan, he was called a "symbol of the tragedy of the modern nuclear scientists." Alfred Friendly wrote in the *Washington Post,* "His tragedy was also the epitome of the tragedy of the age." His victimization during the hearings was central in this narrative. Physicist Hideki Yukawa was quoted in one obituary as saying, "He was so sensitive and differences of opinion with the Government put him under severe psychological pressure . . . I may say that this might have shortened his life." George Kennan said at the memorial service held at Princeton, "I know of nothing more tragic than the series of mistakes (in part, no doubt his own, but what small part!) . . . that obliged him to spend the last decade and a half of his life eating out his heart in frustration over the consciousness that the talents he knew himself to possess, once welcomed and used by the official establishment of his country to develop the destructive possibilities of nuclear science, were rejected when it came to the development of the great positive ones he believed that science to possess." At the same service, physicist Henry D. Smyth (the only member of the AEC to dissent in favor of Oppenheimer) referred to the AEC's action: "Such a wrong can never be righted, such a blot on our history never erased," adding that "we share his deep regret that a brilliant discovery of science has been perverted to an appalling weapon." Friendly said, "If the bomb was sin,

Oppenheimer, as its chief creator, was presumably chief sinner. But in the end he was more sinned against than sinning."[165]

At the time of Oppenheimer's death, embroilment in Vietnam was fast eroding the optimism and triumphalism of postwar American culture and was spurring new questions to be asked by the country's youth. Obituaries of Oppenheimer ran alongside articles on covert CIA funding of cultural and educational organizations—including the CCF—and beside photographs vividly portraying the violence and brutality of the Vietnam War.[166] On campuses across the country, faculty and students were beginning to challenge the very compact between science and the state that Oppenheimer had helped to construct.[167] Their far-reaching criticisms of American foreign policy and of Cold War culture challenged those institutions to which Oppenheimer and the postwar liberal scientific elite had been so careful to accommodate themselves.[168]

Oppenheimer's anxieties about cultural fragmentation took on new meaning as the scientific and academic elite of the Truman and Eisenhower eras found their authority and worldview challenged by rising tides of political protest, for which the university was a locus. Publishing in 1967 a revised version of his talk "The Gentleman of Culture," Vannevar Bush added a jibe against "our beatniks, who would substitute protest for hard work."[169] *Sputnik*-era worries about the "two cultures" continued to resonate for Cold War liberals as they responded to the new context of the social and cultural struggles of the late 1960s. The concerns of liberal academics and intellectuals about cultural fragmentation expressed a profoundly conservative obsession with preserving a cultural elite and maintaining its social and political influence.[170]

For many liberals, Oppenheimer's appeal was as a living model of the "gentleman of culture"—combining scientific with humanistic elements of Western culture and successfully uniting liberal culture with power in the Cold War state. The 1954 security hearings helped to shatter this unity. Cultural fragmentation and the breach of faith between the state and the intellectual classes led, on this view, from the anti-intellectual populism of the 1950s to the conflicts of the 1960s. In his 1969 collection *On Intellectuals,* sociologist Philip Rieff lamented that "America is without a cultural elite."[171]

Oppenheimer's self-presentation as the last intellectual has been the dominant mode in which he has been remembered. He came to embody, and to stand as a proxy for, a particular vision of liberal cultural and political order. Atomic energy and the atomic bomb had represented a great technological hope for the resurrection of liberal democratic order after the global crises of the Great Depression and World War II. Atomic fission, as a source of might and energy, promised liberal democracy a new era of stability and modernity. The atomic bomb even promised, briefly, to be a lever toward a new international order, one modeled on the social order of science: in Bohr and Oppenheimer's utopian phrase, an "open world." In Snow's famous dictum, scientists carried "the future in their bones"—an appealing image for a world escaping a recent past of fascism and war.[172]

The postwar dawn was quickly darkened by the Cold War, and the security of mutually assured destruction was fragile indeed compared with the kind of peace that had been hoped for in the immediate aftermath of the war. In the paranoid obsession with protecting the (arguably mythical) "atomic secret," scientific openness itself was conceived as a new threat to security. The integration of scientists into the apparatus of the state raised the specter of undemocratic influence by these unelected advisers and of technocracy narrowing the scope of democratic decisions. Above all, the barbarism of Hiroshima and Nagasaki hung in the background, haunting liberal visions of rational, technological, and scientific progress.

Oppenheimer was a focal point for these senses of "peril and hope." Despite his protestations that he never regretted his role in the atomic bombings, Oppenheimer remains imprinted on the collective memory as a tragic figure tormented by his sense of moral failure. He could never separate himself from the image—seemingly a necessary fiction, but one that he himself helped to construct—of the guilt-ridden atomic scientist. The legend of Oppenheimer's guilt and atonement gave dramatic expression to the central tension in Western liberal culture between science and humanistic values. His personal struggles indicated how science and technology simultaneously promised to realize and threatened to undermine postwar liberal dreams.

APPENDIX:

INTERVIEWS BY THE AUTHOR

Argo, Harold, December 16, 1997
Bagley, Charles H., June 22, 1998
Balagna, John, January 13, 1998
Bethe, Hans A., May 12, 1998
Bradner, Hugh and Marge, October 15, 1997, and August 9, 2001
Brixner, Berlyn, January 14, 1998
Dabney, Winston, January 15, 1998
Davis, Neil, January 13, 1998
Diven, Benjamin, December 12, 1997
Francis, William C., January 21, 1998
Hammel, Edward, December 8, 1997
Hawkins, David, November 16, 1998
Hull, McAllister, Jr., January 16, 1998
Mark, Kathleen, January 14, 1998
Morrison, Philip, December 19, 2001
Rasmussen, Roger, December 16, 1997
Roensch, Arno, January 16, 1998
Rotblat, Joseph, April 14, 1999
Rynd, Ed, January 22, 1998
Schreiber, Raemer, January 15, 1998
Teller, Edward, June 20, 1998
Wahl, Arthur, January 20, 1998
Wechsler, Jacob J., December 18, 1997
York, Herbert F., March 14, 1998, and October 27, 1998

NOTES

AIP   American Institute of Physics
LAHM   Los Alamos Historical Museum
LAHS   Los Alamos Historical Society
LANL   Los Alamos National Laboratory
MED   Manhattan Engineer District
MIT   Massachusetts Institute of Technology

PREFACE

1. J. Robert Oppenheimer, public lecture (1948) to the Rochester Association for the United Nations and the Rochester Foreign Policy Association, published as "The Open Mind," in *The Open Mind* (New York: Simon and Schuster, 1955), 45–57, on 49–50; George Sarton, "The History of Science," in *The Life of Science: Essays in the History of Civilization* (New York: Henry Schuman, 1948), 55; Robert K. Merton, "The Normative Structure of Science" (1942), in Merton, *The Sociology of Science: Theoretical and Empirical Investigations,* ed. Norman W. Storer, 267–78 (Chicago: University of Chicago Press, 1973). See also David A. Hollinger, *Science, Jews, and Secular Culture: Studies in Mid-Twentieth-Century American Intellectual History* (Princeton, NJ: Princeton University Press, 1996); and Charles Thorpe, "Violence and the Scientific Vocation," *Theory, Culture and Society* 21, no. 3 (June 2004): 59–84.

2. Vannevar Bush, "Report to the President: OSRD in War" [ca. October 1945], Papers of Vannevar Bush, box 139, Manuscript Division, Library of Congress, Washington, DC.

3. Ian Welsh, *Mobilising Modernity: The Nuclear Moment* (London: Routledge, 2000), 17–23, 34–67.

4. Max Weber, "Science as a Vocation," in *From Max Weber: Essays in Sociology,* ed. H. H. Gerth and C. Wright Mills (New York: Oxford University Press, 1958), 129–56, on 152; Sanford A. Lakoff, "Ethical Responsibility and the Scientific Vocation," in *Science and Ethical Responsibility: Proceedings of the U.S. Student Pugwash Conference, University of California, San Diego, June 19–26, 1979,* ed. Sanford A. Lakoff (Reading, MA: Addison-Wesley, 1980), 20.

5. Herbert Marcuse, "Industrialization and Capitalism," *New Left Review* 30 (March–April 1965): 11.

6. See also Sidney M. Willhelm, "Scientific Unaccountability and Moral Accountability," in *The New Sociology: Essays in Social Science and Social Theory in Honor of C. Wright Mills,* ed. Irving Louis Horowitz, 181–87 (New York: Oxford University Press, 1964).

7. Zygmunt Bauman, *Modernity and the Holocaust* (Ithaca, NY: Cornell University Press, 1989).

8. Daniel S. Greenberg, *Science, Money, and Politics: Political Triumph and Ethical Erosion* (Chicago: University of Chicago Press, 2001).

9. E-mail to the author from the National Archives, May 20, 2005.

10. Harold Garfinkel, "Conditions of Successful Degradation Ceremonies," *American Journal of Sociology* 61, no. 5 (March 1956): 420–24.

11. Chandra Mukerji, *A Fragile Power: Scientists and the State* (Princeton, NJ: Princeton University Press, 1989), 190–203, esp. 203.

12. Kai Bird and Martin J. Sherwin, *American Prometheus: The Triumph and Tragedy of J. Robert Oppenheimer* (New York: Alfred A. Knopf, 2005); Priscilla McMillan, *The Ruin of J. Robert Oppenheimer and the Birth of the Modern Arms Race* (New York: Viking Press, 2005).

13. Silvan S. Schweber, *In the Shadow of the Bomb: Bethe, Oppenheimer, and the Moral Responsibility of the Scientist* (Princeton, NJ: Princeton University Press, 2000), xv, 23, 93.

14. David C. Cassidy, *J. Robert Oppenheimer and the American Century* (New York: Pi Press, 2005).

15. Norbert Elias, *The Society of Individuals,* ed. Michael Schröter, trans. Edmund Jephcott (Oxford: Basil Blackwell, 1991), 75.

16. Welsh, *Mobilising Modernity.*

17. Anthony Giddens, *The Giddens Reader,* ed. Philip Cassell (London: Macmillan, 1993), 149–53.

18. Charles Horton Cooley, "Sociability and Personal Ideas," in *Human Nature and the Social Order,* rev. ed. (New York: Charles Scribner's Sons, 1922), 95.

19. Bryan C. Taylor, "The Politics of the Nuclear Text: Reading Robert Oppenheimer's *Letters and Recollections,*" *Quarterly Journal of Speech* 78 (1992): 431.

CHAPTER ONE

1. Vincent C. Jones, *Manhattan: The Army and the Atomic Bomb* (Washington, DC: United States Army Center of Military History, 1985), esp. 344; Kevin O'Neill, "Building the Bomb," in *Atomic Audit: The Costs and Consequences of U.S. Nuclear Weapons since 1940*, ed. Stephen I. Schwartz (Washington, DC: Brookings Institution Press, 1998), esp. 53–59; Thomas P. Hughes, *American Genesis: A Century of Innovation and Technological Enthusiasm, 1870–1970* (New York: Viking, 1989), 381–442; Peter Bacon Hales, *Atomic Spaces: Living on the Manhattan Project* (Urbana: University of Illinois Press, 1997); Richard G. Hewlett and Oscar E. Anderson Jr., *The New World, 1939/1946*, vol. 1 of *A History of the United States Atomic Energy Commission* (University Park: Pennsylvania State University Press, 1962); Richard Rhodes, *The Making of the Atomic Bomb* (New York: Simon and Schuster, 1986); Welsh, *Mobilising Modernity*, chap. 1.

2. Journalistic accounts, historical studies, and reminiscences include Hans A. Bethe, "Oppenheimer: 'Where He Was There Was Always Life and Excitement,'" *Science* 155 (1967): 1080–84; John Mason Brown, *Through These Men: Some Aspects of Our Passing History* (New York: Harper, 1956); Haakon Chevalier, *Oppenheimer: The Story of a Friendship* (New York: George Braziller, 1965); Charles Pelham Curtis, *The Oppenheimer Case: The Trial of a Security System* (New York: Simon and Schuster, 1955); Nuel Pharr Davis, *Lawrence and Oppenheimer* (New York: Simon and Schuster, 1969); Peter Goodchild, *J. Robert Oppenheimer: Shatterer of Worlds* (Boston: Houghton Mifflin, 1981); Gregg Herken, *Brotherhood of the Bomb: The Tangled Lives and Loyalties of Robert Oppenheimer, Ernest Lawrence, and Edward Teller* (New York: Henry Holt and Co., 2002); Robert Jungk, *Brighter Than a Thousand Suns: A Personal History of the Atomic Scientists*, trans. James Cleugh (1958; Harmondsworth, UK: Penguin, 1965); J. Alvin Kugelmass, *J. Robert Oppenheimer and the Atomic Story* (New York: Julian Messner, 1953); James W. Kunetka, *Oppenheimer: The Years of Risk* (Englewood Cliffs, NJ: Prentice-Hall, 1982); Rebecca Larsen, *Oppenheimer and the Atomic Bomb* (New York: F. Watts, 1988); John Major, *The Oppenheimer Hearing* (New York: Stein and Day, 1971); Peter Michelmore, *The Swift Years: The Robert Oppenheimer Story* (New York: Dodd, Mead, 1969); Isidor I. Rabi et al., *Oppenheimer* (New York: Charles Scribner, 1969); Mary Palevsky, *Atomic Fragments: A Daughter's Questions* (Berkeley and Los Angeles: University of California Press, 2000); Rhodes, *The Making of the Atomic Bomb*; Denise Royal, *The Story of J. Robert Oppenheimer* (New York: St. Martin's Press, 1969); Schweber, *In the Shadow of the Bomb*; Alice Kimball Smith and Charles Weiner, eds., *Robert Oppenheimer: Letters and Recollections* (Stanford, CA: Stanford University Press, 1980); Philip M. Stern, *The Oppenheimer Case: Security on Trial*, with Harold P. Green (New York: Harper and Row, 1969); Thomas Williams Wilson, *The Great Weapons Heresy* (Boston: Houghton Mifflin, 1970); and Herbert F. York, *The*

Advisors: Oppenheimer, Teller, and the Superbomb (San Francisco: W. H. Freeman, 1976). Novels include Haakon Chevalier, *The Man Who Would Be God* (1959; Berlin: Seven Seas, 1963); Martin Cruz Smith, *Stallion Gate* (New York: Ballantine Books, 1986); and Joseph Kanon, *Los Alamos* (New York: Broadway Books, 1997). *The Day after Trinity: J. Robert Oppenheimer and the Atomic Bomb* is a documentary film by Jon Else (Santa Monica, CA: Pyramid Films, 1980).

3. See Roslynn D. Haynes, *From Faust to Strangelove: Representations of the Scientist in Western Literature* (Baltimore: Johns Hopkins University Press, 1994), esp. 246–63, 285–88; and John Canaday, *The Nuclear Muse: Literature, Physics and the First Atomic Bombs* (Madison: University of Wisconsin Press, 2000). See also Taylor, "Politics of the Nuclear Text."

4. Jungk, *Brighter Than a Thousand Suns*, 295, 266. See also Herbert Knust, "From Faust to Oppenheimer: The Scientist's Pact with the Devil," *Journal of European Studies* 13, nos. 1–2 (1983), esp. 129; and Shiv Visvanathan, "Atomic Physics: The Career of an Imagination," in *Science, Hegemony and Violence: A Requiem for Modernity,* ed. Ashis Nandy, 113–66 (1988; Oxford: Oxford University Press, 1996).

5. Chevalier, *Oppenheimer,* 117. See also Freeman Dyson, quoted in Knust, "From Faust to Oppenheimer," 129.

6. Lewis S. Feuer, *The Scientific Intellectual: The Psychological and Sociological Origins of Modern Science* (New York: Basic Books, 1963), 394.

7. Lewis A. Coser, *Men of Ideas: A Sociologist's View* (New York: Free Press, 1965), 309, 311, 312.

8. Philip Rieff, "The Case of Dr. Oppenheimer," in *On Intellectuals: Theoretical Studies, Case Studies,* ed. Philip Rieff (New York: Doubleday/Anchor, 1970), 341–69, quoting 346, 369.

9. Giorgio de Santillana, *The Crime of Galileo* (London: Heinemann, 1958), viii; de Santillana, "Galileo and Oppenheimer," in *Reflections on Men and Ideas* (Cambridge, MA: MIT Press, 1968; first published in *The Reporter,* December 26, 1957); de Santillana, "Notes sur la science en Amérique," *La Table Ronde* 105 (September 1956): 62–66; Jerome Ravetz, "Tragedy in the History of Science," in *Changing Perspectives in the History of Science: Essays in Honour of Joseph Needham,* ed. Mikuláš Teich and Robert Young (London: Heinemann, 1973), esp. 214.

10. Sanford A. Lakoff, "The Trial of Dr. Oppenheimer," in *Knowledge and Power: Essays on Science and Government,* ed. Sanford A. Lakoff (New York: Free Press, 1966), 79, 84.

11. Max Weber, "Bureaucracy," in *From Max Weber,* 196–244, on 216, 243.

12. Weber, "Science as a Vocation," in *From Max Weber,* 143, 152. See also Harvey Goldman, *Politics, Death, and the Devil: Self and Power in Max Weber and Thomas Mann* (Berkeley and Los Angeles: University of California Press, 1992), esp. 9–18, 51–86; Schweber, *In the Shadow of the Bomb,* 93; and Kathryn Mary Olesko, *Physics as a Calling: Discipline and Practice in the Königsberg Seminar for Physics* (Ithaca, NY: Cornell University Press, 1991).

13. Michel Foucault, "Truth and Power," in *Power/Knowledge: Selected Interviews and Other Writings, 1972–1977,* ed. Colin Gordon (New York: Pantheon, 1980), 131, 129, 127–28. See also William Ray Arney, *Experts in the Age of Systems* (Albuquerque: University of New Mexico Press, 1991), esp. 26–27, 151–75; Silvan S. Schweber, "Reflections on the Sokal Affair: What Is at Stake?" *Physics Today* 50, no. 3 (March 1997): 73–74; Andrew Feenberg, *Questioning Technology* (London: Routledge, 1999), 123; and Paul Rabinow, *French Modern: Norms and Forms of the Social Environment* (Cambridge, MA: MIT Press, 1989).

14. See also Thorpe, "Violence and the Scientific Vocation."

15. Gerald Holton and Yehuda Elkana, eds., *Albert Einstein: Historical and Cultural Perspectives* (Princeton, NJ: Princeton University Press, 1982); Fred Jerome, *The Einstein File: J. Edgar Hoover's Secret War against the World's Most Famous Scientist* (New York: St. Martin's Press, 2002).

16. Bohr, quoted in Rhodes, *The Making of the Atomic Bomb,* 530. See also Martin Sherwin, *A World Destroyed: The Atomic Bomb and the Grand Alliance* (New York: Vintage Books, 1977), 91–98; Sherwin, "Niels Bohr and the First Principles of Arms Control," in *Niels Bohr: Physics and the World; Proceedings of the Niels Bohr Centennial Symposium, Boston, MA, USA, November 12–14, 1985,* ed. Herman Feshbach, Tetsuo Matsui, and Alexandra Oleson, 319–29 (London: Harwood, 1988); and Margaret Gowing, "Niels Bohr and Nuclear Weapons," in *Niels Bohr: A Centenary Volume,* ed. A. P. French and P. J. Kennedy, 266–77 (Cambridge, MA: Harvard University Press, 1985).

17. Szilard, quoted in Gar Alperovitz, *The Decision to Use the Atomic Bomb* (New York: Vintage, 1995), 190–91. See also William Lanouette, *Genius in the Shadows: A Biography of Leo Szilard; The Man behind the Bomb,* with Bela Silard (Chicago: University of Chicago Press, 1992); and Alice Kimball Smith, *A Peril and a Hope: The Scientists' Movement in America, 1945–1947* (1965; Cambridge, MA: MIT Press, 1970). On the limited character of such opposition within the bomb project, see Peter N. Kirstein, "False Dissenters: Manhattan Project Scientists and the Use of the Atomic Bomb," *American Diplomacy,* March 2001, http://www.unc.edu/depts/diplomat/archives_roll/2001_03-06/kirstein_manhattan/kirstein_manhattan.html.

18. Edward Shils, "Science and Scientists in the Public Arena," *American Scholar* 56 (Spring 1987): 197.

19. Susan Landau, "Joseph Rotblat: The Road Less Traveled," *Bulletin of the Atomic Scientists* 52, no. 1 (January–February 1996): 47–54; Joseph Rotblat, "Leaving the Bomb Project," *Bulletin of the Atomic Scientists* 41 (August 1985): 15–19; Rotblat, interviewed by the author, April 14, 1999.

20. Schweber, *In the Shadow of the Bomb,* 149–77, esp. 170.

21. Edward Teller, "The Role of the Scientist," in *Better a Shield than a Sword: Perspectives on Defense and Technology* (New York: Free Press, 1987), 234.

22. Edward Teller, interviewed by the author, June 20, 1998.

23. Cf. Hugh Gusterson, *Nuclear Rites: A Weapons Laboratory at the End of the Cold War* (Berkeley and Los Angeles: University of California Press, 1996), esp. 38–67.

24. Albert Einstein, "A Message to Intellectuals," in *Ideas and Opinions,* 147–51 (1954; New York: Crown, 1982), on 148.

25. Visvanathan, "Atomic Physics," 113. Cf. Sheldon Ungar, "The Great Collapse, Democratic Paralysis and the Reception of the Bomb," *Journal of Historical Sociology* 5, no. 1 (March 1992): 84–103.

26. Karl Popper, *The Open Society and Its Enemies* (London: Routledge, 1945); Michael Polanyi, *Science, Faith and Society* (Chicago: University of Chicago Press, 1946). See also David A. Hollinger, "The Defense of Democracy and Robert K. Merton's Formulation of the Scientific Ethos," in *Science, Jews, and Secular Culture,* 80–96.

27. Oppenheimer, "The Open Mind," in *The Open Mind,* 49–50.

28. Oppenheimer, quoted in Gerald Holton, "Heisenberg, Oppenheimer, and the Transition to Modern Physics," in *Transformation and Tradition in the Sciences,* ed. Everett Mendelsohn (Cambridge: Cambridge University Press, 1984), 166; Glenn T. Seaborg, "Public Service and Human Contributions," in Rabi et al., *Oppenheimer,* 45–59, quoting 55; William L. Laurence, *Men and Atoms: The Discovery, the Uses and the Future of Atomic Energy* (New York: Simon and Schuster, 1959), 118; Goodchild, *J. Robert Oppenheimer,* 279.

29. Chevalier, *Oppenheimer,* 21, 20; Diracs to Oppenheimers, October 25, 1964 (written by Dirac's wife), Papers of J. Robert Oppenheimer, box 30, Manuscript Division, Library of Congress; Brown, *Through These Men,* 286; Thomas B. Morgan, "A Visit with J. Robert Oppenheimer," *Look,* April 1, 1958, 35–38, on 38.

30. Isidor I. Rabi, "Introduction," in Rabi et al., *Oppenheimer,* 3–8, on 5; Charles Critchfield, "The Oppenheimer I Knew," in *Behind Tall Fences: Stories and Experiences about Los Alamos at Its Beginning,* 169–77 (Los Alamos, NM: Los Alamos Historical Society, 1996), quoting 171.

31. Chevalier, *The Man Who Would Be God,* 14–15.

32. Chevalier, *Oppenheimer,* 21; Victor F. Weisskopf, "The Los Alamos Years," in Rabi et al., *Oppenheimer,* 23–28, quoting 28; Tuck, quoted in Davis, *Lawrence and Oppenheimer,* 187.

33. R. R. Wilson to Oppenheimer [ca. 1963], Oppenheimer Papers, box 78; Rabi, "Introduction," 6–7; Jungk, *Brighter Than a Thousand Suns,* 125.

34. Canaday, *The Nuclear Muse,* 183–203; Rhodes, *Making of the Atomic Bomb,* 571–72, 676; Richard Rhodes, *Dark Sun: The Making of the Hydrogen Bomb* (New York: Simon and Schuster, 1995), 205–6; James A. Hijiya, "The *Gita* of J. Robert Oppenheimer," *Proceedings of the American Philosophical Society* 144, no. 2 (June 2000): 123–67; Herken, *Brotherhood of the Bomb,* 150; J. Robert Oppenheimer, "Physics in the Contemporary World" (1947), in *The Open Mind,* 81–102, on 88.

35. Max Weber, "The Sociology of Charismatic Authority," in *From Max Weber*, 245–52, quoting 245–46; Magali Sarfatti Larson, "The Production of Expertise and the Constitution of Expert Power," in *The Authority of Experts: Studies in History and Theory*, ed. Thomas Haskell (Bloomington: Indiana University Press, 1984), esp. 58. See also Donald MacKenzie and Boelie Elzen, "The Charismatic Engineer" (1991), in *Knowing Machines: Essays on Technical Change*, ed. Donald A. MacKenzie (Cambridge, MA: MIT Press, 1996), esp. 133–34; and Charles Thorpe and Steven Shapin, "Who Was J. Robert Oppenheimer? Charisma and Complex Organization," *Social Studies of Science* 30 (2000), esp. 548–51.

36. Daniel Bell, *The Coming of Post-industrial Society: A Venture in Social Forecasting* (New York: Basic Books, 1976), 398, 408.

37. Rabi, "Introduction," 8; John S. Rigden, *Rabi: Scientist and Citizen* (New York: Basic Books, 1987), 243; Robert R. Wilson, "A Recruit for Los Alamos," *Bulletin of the Atomic Scientists* 31 (March 1975): 45; Seaborg, "Public Service and Human Contributions," 56; Edward Teller (1982 interview), quoted in Rhodes, *The Making of the Atomic Bomb*, 539. See also historians' descriptions of Oppenheimer as "charismatic": Smith and Weiner, *Robert Oppenheimer*, 221; Rhodes, *Dark Sun*, 308, 206; James G. Hershberg, *James B. Conant: Harvard to Hiroshima and the Making of the Nuclear Age* (Stanford, CA: Stanford University Press, 1993), 165; Rigden, *Rabi*, 159; and Schweber, *In the Shadow of the Bomb*, esp. 14–16.

38. Goodchild, *J. Robert Oppenheimer*, 29; Bernice Brode, *Tales of Los Alamos: Life on the Mesa, 1943–1945* (Los Alamos, NM: Los Alamos Historical Society, 1997), 5; Bell, *Post-industrial Society*, 398; Raymond Aron, *Memoirs: Fifty Years of Political Reflection* (New York: Holmes and Meier, 1990), 175; Robb, quoted in Goodchild, *J. Robert Oppenheimer*, 248. See also Katrina R. Mason, *Children of Los Alamos: An Oral History of the Town Where the Atomic Age Began* (New York: Twayne, 1995), 146; Eleanor Jette, *Inside Box 1663* (Los Alamos, NM: Los Alamos Historical Society, 1977), 123; and Rabi, "Introduction," 5.

39. Chevalier, *Oppenheimer*, 11; Eleanor (Jerry) Stone Roensch, *Life within Limits* (Los Alamos, NM: Los Alamos Historical Society, 1993), 32.

40. Jane S. Wilson, "Not Quite Eden," in *Standing By and Making Do: Women of Wartime Los Alamos*, ed. Jane S. Wilson and Charlotte Serber (Los Alamos, NM: Los Alamos Historical Society, 1988), 43–55, quoting 50; Rabi, "Introduction," 8; Leona Marshall Libby, *The Uranium People* (New York: Crane Russak, 1979), 109; Brown, *Through These Men*, 286. See also Kugelmass, *J. Robert Oppenheimer*, 4–5; William Lawren, *The General and the Bomb: A Biography of General Leslie Groves, Director of the Manhattan Project* (New York: Dodd, Mead, 1988), 96; and Rhodes, *The Making of the Atomic Bomb*, 444.

41. Steven Shapin, "The Philosopher and the Chicken: On the Dietetics of Disembodied Knowledge," in *Science Incarnate: Historical Embodiments of Natural Knowledge*, ed. Christopher Lawrence and Steven Shapin, 21–50 (Chicago: University of Chicago Press, 1998).

42. See, for example, Lawren, *The General and the Bomb,* 287: "But with his figurative death on the sacrificial altar of McCarthyism, the Oppenheimer legend began to take on almost mythical proportions."

43. In describing Oppenheimer's role at Los Alamos, including his charismatic leadership of the laboratory, it is frequently necessary to rely on reminiscences, written later by participants, and other post hoc accounts. It is in the nature of such a secret and highly mission-directed project as Los Alamos was that available archival sources tend to be official memoranda, rather than personal reflections. To capture the intimate details of life on the project, and for detailed reflections on Oppenheimer's personal characteristics and their consequences, it is therefore necessary to look to post hoc sources. These are a rich resource for understanding aspects of everyday life and culture not captured in official wartime reports. But the use of reminiscence and oral history also raises the question of whether charisma is being ascribed to Oppenheimer merely in hindsight, as a result of the success of the wartime project and his later persecution. It seems to me, however (from putting together a variety of historical sources), that an account of wartime Los Alamos in which Oppenheimer's charismatic authority was real and made a difference is far more plausible than any alternative construction in which such charisma is treated merely as post hoc attribution. If one is to reject a coherent portrait built up by many oral histories and reminiscences, one must assume the burden of giving good reasons for doing so, and of providing an at least equally plausible alternative account. For additional comments, see Thorpe and Shapin, "Who Was J. Robert Oppenheimer?" 580–81.

44. Robert Coughlan, "The Equivocal Hero of Science: Robert Oppenheimer," *Life,* February 1967, 34–34A, on 34.

45. David Bohm to Miriam Yevick [ca. 1953], quoted in F. David Pleat, *Infinite Potential: The Life and Times of David Bohm* (Reading, MA: Helix Books, 1997), 160; Chevalier, *Oppenheimer,* 20.

46. Coughlan, "The Equivocal Hero of Science," 34; Elting E. Morison, "Bomb-Builders," review of *Lawrence and Oppenheimer,* by Nuel Pharr Davis, *New York Times Book Review,* October 6, 1968.

47. Historians Alice Kimball Smith and Charles Weiner point out that Oppenheimer's postwar correspondence became increasingly formal: "By the end of the war Oppenheimer's defenses were up" (*Robert Oppenheimer,* ix–x).

48. Report by W. Rulon Paxman, "Julius Robert Oppenheimer aka Jerome Robert Oppenheimer," April 3, 1947, in *J. Robert Oppenheimer, FBI Security File* (Wilmington, DE: Scholarly Resources, 1978), microfilm, 116-2717-313; Coughlan, "The Equivocal Hero of Science," 34A.

49. Bloch and Rabi, quoted in Rigden, *Rabi,* 228, 229, 231; Abraham Pais, *A Tale of Two Continents: A Physicist's Life in a Turbulent World* (Princeton, NJ: Princeton University Press, 1997), 239. See also Bruno Rossi, *Moments in the Life of a Scientist* (Cambridge: Cambridge University Press, 1990), 70.

50. Jungk, *Brighter Than a Thousand Suns*, 324; Richard Rhodes, "'I Am Become Death': The Agony of J. Robert Oppenheimer," *American Heritage* 20, no. 6 (October 1977): 72; Schweber, *In the Shadow of the Bomb*, esp. 103, 186.

51. "Transcript of Hearing before the Personnel Security Board" (hereafter "Transcript"), in United States Atomic Energy Commission, *In the Matter of J. Robert Oppenheimer: Transcript of Hearing before Personnel Security Board and Texts of Principal Documents and Letters* (hereafter *Transcript and Texts*) (Cambridge, MA: MIT Press, 1971), 710. The transcript of the hearing was originally published in 1954.

52. Journal entry for July 24, 1946, in David E. Lilienthal, *The Atomic Energy Years, 1945–1950,* vol. 2 of *The Journals of David E. Lilienthal* (New York: Harper and Row, 1964), 69; Alfred Friendly, "Drama Packs Amazing Oppenheimer Transcript," *Washington Post and Times-Herald,* July 25, 1954, reproduced in *Oppenheimer, FBI Security File,* 100-17828, sec. 47; Heinar Kipphardt, *In the Matter of J. Robert Oppenheimer: A Play Freely Adapted on the Basis of Documents* (1964; New York: Hill and Wang, 1968).

53. Libby, *The Uranium People,* 105; Rabi, quoted in Rigden, *Rabi,* 221; Jeremy Bernstein, *The Life It Brings: One Physicist's Beginnings* (New York: Ticknor and Fields, 1987), 112. See also Taylor, "Politics of the Nuclear Text," 443.

54. On sociological biography, see Steven Shapin, "Personal Development and Intellectual Biography: The Case of Robert Boyle," *British Journal of the History of Science* 26 (1993), esp. 338; Shapin, "'A Scholar and a Gentleman': The Problematic Identity of the Scientific Practitioner in Early Modern England," *History of Science* 29 (1991): 279–327; and Shapin, "Who Was Robert Boyle? The Creation and Presentation of an Experimental Identity," in *A Social History of Truth: Civility and Science in Seventeenth-Century England,* 126–92 (Chicago: University of Chicago Press, 1994). See also Norbert Elias, *Mozart: Portrait of a Genius* (Cambridge: Polity Press, 1993); and C. Wright Mills, *The Sociological Imagination* (1959; Oxford: Oxford University Press, 2000), esp. 6–7. My understanding of the relationship between biography and social order, following Shapin, draws on interactionist sociology—in particular Erving Goffman, *The Presentation of Self in Everyday Life* (Garden City, NY: Doubleday Anchor, 1959); Howard S. Becker, "Notes on the Concept of Commitment," *American Journal of Sociology* 66 (1960): 32–40; and Becker, "Personal Change in Adult Life," *Sociometry* 27 (1964): 40–53. On special issues concerning scientific biography, see Michael Shortland and Richard Yeo, eds., *Telling Lives in Science: Essays on Scientific Biography* (Cambridge: Cambridge University Press, 1996). I also draw on the work of the 1998–99 Scientific Personae research group, directed by Lorraine Daston at the Max Planck Institute for the History of Science; this group has developed the notion of persona as a key concept for studying individual lives in relation to cultures of science.

55. See also the discussion of Oppenheimer as a "nodal signifier" in Taylor, "Politics of the Nuclear Text," 431.

56. For example, allegations of his disloyalty, however dubious, reemerged in the wake of the Cold War. Pavel Sudoplatov and Anatoli Sudoplatov, *Special Tasks: The Memoirs of an Unwanted Witness—a Soviet Spymaster,* with Jerrold L. and Leona P. Schecter (New York: Little, Brown, 1994).

CHAPTER TWO

1. Goodchild, *J. Robert Oppenheimer,* 10; Schweber, *In the Shadow of the Bomb,* 42; Bird and Sherwin, *American Prometheus,* 10; Cassidy, *J. Robert Oppenheimer,* 1–2.

2. Goodchild, *J. Robert Oppenheimer,* 10; Frank Oppenheimer, interviewed by Charles Weiner, February 9, 1973, 2, 1, Oral History Collection, Niels Bohr Library, American Institute of Physics, College Park, MD (hereafter AIP); Schweber, *In the Shadow of the Bomb,* 42. See also Raymond T. Birge, "History of the [University of California, Berkeley] Physics Department," vol. 3, "1928–1932" (n.d.), chap. 9, p. 26, Niels Bohr Library, AIP; and Cassidy, *J. Robert Oppenheimer,* 4–11.

3. "Physicist Oppenheimer," *Time,* November 8, 1948, 70–81, on 70; J. Robert Oppenheimer, interviewed by Thomas S. Kuhn, November 18, 1963, 10, 3, Archives for the History of Quantum Physics, AIP; Frank Oppenheimer, interviewed by Weiner, February 9, 1973, 2; Smith and Weiner, *Robert Oppenheimer,* 7, 66.

4. Oppenheimer, interviewed by Kuhn, November 18, 1963, 3.

5. Goodchild, *J. Robert Oppenheimer,* 10–11; Paul Horgan, interviewed by Alice Kimball Smith, March 3, 1976, 15–16, J. Robert Oppenheimer Oral History Collection, box MC 85, Archives and Special Collections of the Massachusetts Institute of Technology, Cambridge, MA (hereafter MIT Archives); "Physicist Oppenheimer," 70; Rhodes, "Agony of J. Robert Oppenheimer," 72.

6. Smith and Weiner, *Robert Oppenheimer,* 107.

7. "Physicist Oppenheimer," 70; Goodchild, *J. Robert Oppenheimer,* 11; Horgan, interviewed by Smith, March 3, 1976, 15; Horgan, quoted in Smith and Weiner, *Robert Oppenheimer,* 2; Frank Oppenheimer, interviewed by Weiner, February 9, 1973, 3–4; William C. Boyd, interviewed by Alice Kimball Smith, December 21, 1975, 11, Oppenheimer Oral History Collection, MIT Archives.

8. Frank Oppenheimer, interviewed by Weiner, February 9, 1973, 3–4; Oppenheimer, quoted in "Physicist Oppenheimer," 70; Rhodes, "Agony of J. Robert Oppenheimer," 72; Bird and Sherwin, *American Prometheus,* 13; Cassidy, *J. Robert Oppenheimer,* 15–18.

9. Oppenheimer, interviewed by Kuhn, November 18, 1963, 3; Smith and Weiner, *Robert Oppenheimer,* 6; Rhodes, "Agony of J. Robert Oppenheimer," 73; Jane Kayser, quoted in Smith and Weiner, *Robert Oppenheimer,* 6–7; Bird and Sherwin, *American Prometheus,* 21.

10. Oppenheimer, interviewed by Kuhn, November 18, 1963, 3; Schweber, *In the Shadow of the Bomb,* 42.

11. Frank Oppenheimer, interviewed by Weiner, February 9, 1973, 3; Smith and Weiner, *Robert Oppenheimer*, 3.

12. Robert S. Guttchen, *Felix Adler* (New York: Twayne Publishers, 1974), 15; Howard B. Radest, *Toward Common Ground: The Story of the Ethical Culture Societies in the United States* (New York: Frederick Ungar Publishing Co., 1969), esp. 97; Benny Kraut, *From Reform Judaism to Ethical Culture: The Religious Evolution of Felix Adler* (Cincinnati, OH: Hebrew Union College Press, 1979); Schweber, *In the Shadow of the Bomb*, 42–53. On the social-reformist agenda of Ethical Culture, in the context of the American Progressive era, see Cassidy, *J. Robert Oppenheimer*, 23–31.

13. George L. Mosse, *German Jews beyond Judaism* (Bloomington: Indiana University Press, 1985), 3; W. H. Bruford, *The German Tradition of Self-Cultivation: 'Bildung' from Humboldt to Thomas Mann* (Cambridge: Cambridge University Press, 1975); Norbert Elias, *The Germans* (New York: Columbia University Press, 1996), esp. 113–14, 128; Peter Gay, "Encounter with Modernism: German Jews in German Culture, 1888–1914," *Midstream* 21 (1975): 23–65; Harvey Goldman, *Max Weber and Thomas Mann: Calling and the Shaping of Self* (Berkeley and Los Angeles: University of California Press, 1988); Goldman, *Politics, Death, and the Devil*; Roy Pascal, *Culture and the Division of Labour*, Occasional Papers in German Studies, no. 5 (Warwick: University of Warwick, 1974); Fritz K. Ringer, *The Decline of the German Mandarins: The German Academic Community, 1890–1933* (1969; Hanover, NH: University Press of New England, 1990); David Sorkin, "Wilhelm von Humboldt: The Theory and Practice of Self-Formation (*Bildung*), 1791–1810," *Journal of the History of Ideas* 44 (1983): 55–73; Schweber, *In the Shadow of the Bomb*, 80–85.

14. Mosse, *German Jews beyond Judaism*, 1–54; Zygmunt Bauman, *Modernity and Ambivalence* (Oxford: Polity Press, 1991), esp. 102–59, 132–37; Sander L. Gilman, *Jewish Self-Hate: Anti-Semitism and the Hidden Language of Race* (Baltimore: Johns Hopkins University Press, 1986), 270–86. See also the relevant personal reflections by Victor Weisskopf in *The Joy of Insight: Passions of a Physicist* (New York: Basic Books, 1991), 1–29, esp. 7, 13, 26–27.

15. Abraham Barkai, *Branching Out: German-Jewish Immigration to the United States, 1820–1914* (New York: Holmes and Meier, 1994), 191–222; Reinhard Bendix, *From Berlin to Berkeley: German-Jewish Identities* (New Brunswick, NJ: Transaction Books, 1986); Stephen Birmingham, *"Our Crowd": The Great Jewish Families of New York* (New York: Harper and Row, 1967); Naomi W. Cohen, *Encounter with Emancipation: The German Jews in the United States, 1830–1914* (Philadelphia: Jewish Publication Society, 1984); Moses Rischin, *The Promised City: New York's Jews, 1870–1914* (Cambridge, MA: Harvard University Press, 1962); Irving Howe, *World of Our Fathers: The Journey of the East European Jews to America and the Life They Found and Made There* (1976; London: Phoenix Press, 2000); Arthur Liebman, "Anti-Semitism on the Left?" in *Anti-Semitism in American History*, ed. David A. Gerber, 329–59 (Urbana: University of Illinois Press, 1986).

16. Howard M. Sachar, *A History of the Jews in America* (New York: Alfred A. Knopf, 1992), 85–102, quoting 102; Gilman, *Jewish Self-Hate,* 354; Nathan C. Belth, *A Promise to Keep: A Narrative of the American Encounter with Anti-Semitism* (New York: Times Books, 1979); Leonard Dinnerstein, ed., *Antisemitism in the United States* (New York: Holt, Rinehart and Winston, 1971); Leonard Dinnerstein, *Antisemitism in America* (Oxford: Oxford University Press, 1994); Robert Singerman, "The Jew as Racial Alien: The Genetic Component of American Anti-Semitism," in Gerber, *Anti-Semitism in American History,* 103–28; Schweber, *In the Shadow of the Bomb,* 44–46; Cassidy, *J. Robert Oppenheimer,* 11–13.

17. Quoting Radest, *Toward Common Ground,* 185. See also Kraut, *From Reform Judaism to Ethical Culture,* 197; Cohen, *Encounter with Emancipation,* 275.

18. Oppenheimer, interviewed by Kuhn, November 18, 1963, 4; quoted also in Smith and Weiner, *Robert Oppenheimer,* 4–5. A note of appreciation that Oppenheimer wrote for Klock is attached to Oppenheimer to Dr. Black, November 7, 1963, Oppenheimer Papers, box 43. See also Schweber, *In the Shadow of the Bomb,* 53.

19. Smith and Weiner, *Robert Oppenheimer,* 17, 66.

20. Ibid., 5; Oppenheimer to Smith, January 12, 1923, ibid., 18–19, quoting 18.

21. Smith and Weiner, *Robert Oppenheimer,* 7–8; Cassidy, *J. Robert Oppenheimer,* 61–63.

22. Smith and Weiner, *Robert Oppenheimer,* 10; Horgan, interviewed by Smith, March 3, 1976, 7.

23. Horgan, interviewed by Smith, March 3, 1976, 7–8, 11–12; Oppenheimer to Smith, November 2, 1923, in Smith and Weiner, *Robert Oppenheimer,* 50–51, on 51.

24. Fergusson, quoted in Smith and Weiner, *Robert Oppenheimer,* 9; Schweber, *In the Shadow of the Bomb,* 54.

25. Oppenheimer to Smith, February 11, 1923, and February 18, 1923, in Smith and Weiner, *Robert Oppenheimer,* 22–23; Oppenheimer to Fergusson, September 16, 1923, ibid., 37–38.

26. Rhodes, "Agony of J. Robert Oppenheimer," 72; Smith and Weiner, *Robert Oppenheimer,* 1; Cassidy, *J. Robert Oppenheimer,* 1–2. That Oppenheimer was given his father's name goes against the usual European Jewish practice, which is to name children after a deceased, rather than a living, relative (see also Bird and Sherwin, *American Prometheus,* 11).

27. Serrell Hillman to Rob Hagy, "Oppenheimer Cover," October 22, 1948, Oppenheimer Papers, box 294; Smith and Weiner, *Robert Oppenheimer,* 9. See also Cassidy, *J. Robert Oppenheimer,* 13–14.

28. See, for example, Oppenheimer to Smith, October 2, 1922, in Smith and Weiner, *Robert Oppenheimer,* 13; and Oppenheimer to Fergusson, June 14, 1923, ibid., 31.

29. Oppenheimer, quoted in "Physicist Oppenheimer," 73.

30. Rabi and Bloch, quoted in Rigden, *Rabi,* 228–29. Similar observations about Oppenheimer were made by the Berkeley philosopher William Dennes in "Philosophy

and the University since 1915," Oral History Interview [ca. 1970] by Regional Oral History Office, University of California Oral Histories, Bancroft Library, University of California, Berkeley, 136–37. See also Schweber, *In the Shadow of the Bomb,* 54.

31. Rigden, *Rabi,* 17–29.

32. See the related discussions in Sander L. Gilman, *Smart Jews: The Construction of the Image of Jewish Superior Intelligence* (Lincoln: University of Nebraska Press, 1996); and Hollinger, *Science, Jews, and Secular Culture.*

33. Sachar, *History of the Jews in America,* 324–34, quoting 329. See also Morton Rosenstock, "Are There Too Many Jews at Harvard?" in Dinnerstein, *Antisemitism in the United States,* 102–8, esp. 107; Marcia Graham Synott, "Anti-Semitism and American Universities: Did Quotas Follow the Jews?" in Gerber, *Anti-Semitism in American History,* 233–71; Daniel J. Kevles, *The Physicists: The History of a Scientific Community in Modern America* (New York: Vintage Books, 1979), 210–13; and Cassidy, *J. Robert Oppenheimer,* 66–71.

34. Oppenheimer, quoted in "Physicist Oppenheimer," 71; Rabi, quoted in Rigden, *Rabi,* 32–33; Hawkins, quoted in Palevsky, *Atomic Fragments,* 103.

35. Wyman, quoted in Goodchild, *J. Robert Oppenheimer,* 15; Oppenheimer to Smith, October 2, 1922.

36. Smith and Weiner, *Robert Oppenheimer,* 12, 39–40; Fergusson to Smith, Saturday [May 15, 1922] and [ca. fall 1922], Oppenheimer Papers, box 294; Oppenheimer, interviewed by Kuhn, November 18, 1963, 4; Jeffries Wyman, interviewed by Charles Weiner, May 28, 1975, 2, Oppenheimer Oral History Collection, MIT Archives.

37. Smith and Weiner, *Robert Oppenheimer,* 12, 33; Frederick Bernheim, interviewed by Charles Weiner, October 27, 1975, 2, 6, Oppenheimer Oral History Collection, MIT Archives.

38. Betty Thomas to Smith, November 4, 1921, Oppenheimer Papers, box 294. On the intellectual seriousness of the Ethical Culture School, see also Cassidy, *J. Robert Oppenheimer,* 34.

39. Fergusson to Smith, January 22, 1922, Oppenheimer Papers, box 294. See also Cassidy, *J. Robert Oppenheimer,* 74.

40. Wyman, quoted in Goodchild, *J. Robert Oppenheimer,* 15; Oppenheimer, interviewed by Kuhn, November 18, 1963, 9.

41. Boyd, interviewed by Smith, December 21, 1975, 4. Boyd added, "I don't think he was really worried about his background"; and Smith and Weiner interpret his comments to mean that Oppenheimer had, as they put it, "a relaxed attitude toward his heritage" (*Robert Oppenheimer,* 61).

42. Horgan, interviewed by Smith, March 3, 1976, 16, 5, 20. Eagels's critically acclaimed performance was in 1922–23.

43. Boyd, interviewed by Smith, December 21, 1975, 19.

44. Cassidy, *J. Robert Oppenheimer,* 71–72; John Edsall, interviewed by Charles Weiner, July 16, 1975, 2, 4–7, Oppenheimer Oral History Collection, MIT Archives; Oppenheimer, interviewed by Kuhn, November 18, 1963, 9; Bird and Sherwin,

American Prometheus, 33; Oppenheimer to Smith, November 14, 1922, in Smith and Weiner, *Robert Oppenheimer,* 15–16, quoting 15.

45. Bernheim, interviewed by Weiner, October 27, 1975, 14; Fergusson to Smith, February 19 (?), 1922, Oppenheimer Papers, box 294.

46. Oppenheimer to Smith, December 1, 1923, in Smith and Weiner, *Robert Oppenheimer,* 52–53, on 53; Oppenheimer to Smith [ca. January 22, 1924], ibid., 58–60, on 59.

47. Oppenheimer to Smith [March 1924?], Oppenheimer Papers, box 294. This letter is printed, with this sentence omitted, in Smith and Weiner, *Robert Oppenheimer,* 63–64.

48. Oppenheimer to Smith [winter 1923–24], ibid., 54–55, on 54. See the references to *Crome Yellow* in Oppenheimer to Fergusson, Christmas 1923, ibid., 55–58, on 55; and Oppenheimer to Smith, November 14, 1922, 15.

49. Wyman, interviewed by Weiner, May 28, 1975, 6, 10, 15; Smith and Weiner, *Robert Oppenheimer,* 60.

50. Boyd, interviewed by Smith, December 21, 1975, 1. See also Smith and Weiner, *Robert Oppenheimer,* 33.

51. Bernheim, interviewed by Weiner, October 27, 1975, 15–16.

52. Oppenheimer to Smith [winter 1923–24], 54; Oppenheimer to Smith [ca. January 22, 1924], 59.

53. See, for example, his worries about the lack of correspondence from Fergusson in Oppenheimer to Smith [ca. March 1924?], in Smith and Weiner, *Robert Oppenheimer,* 64.

54. Oppenheimer to Smith, May 2, 1923, in Smith and Weiner, *Robert Oppenheimer,* 25–26, on 25.

55. See, for example, Oppenheimer to Smith [ca. January 1925], in Smith and Weiner, *Robert Oppenheimer,* 70–71, on 70. Oppenheimer wrote to Smith asking for a letter of recommendation to Cambridge: "The certificate should say that you have known me for -?- years and that you believe me to be of good moral character; that two assertions so contradictory in sense should be demanded in one certificate is additional proof, I think, of the proverbial irrationality of the British."

56. Oppenheimer to Smith [winter 1923–24], 54.

57. Bernheim, interviewed by Weiner, October 27, 1975, 7; Smith and Weiner, *Robert Oppenheimer,* 26; Oppenheimer to Smith, May 15, 1923, ibid., 27–28, on 27; Oppenheimer to Smith, Sunday [ca. 1923?], Oppenheimer Papers, box 294; Oppenheimer to Smith, February 18, 1923, 23.

58. Fergusson to Smith, Tuesday [ca. February 1923], Oppenheimer Papers, box 294. This letter has had the date January 20, 1923, written on it post hoc. But since Oppenheimer was not in the infirmary at that time, I believe there is good reason for thinking the letter coincides with the infirmary spell in February.

59. Oppenheimer to Smith, February 18, 1923, 22–23.

60. Oppenheimer to Smith, November 14, 1922, 15; Oppenheimer to Smith [late winter 1924], ibid., 62–63, quoting 62.

61. Oppenheimer to Fergusson, June 14, 1923, 31; Fergusson letter, quoted in Smith and Weiner, *Robert Oppenheimer,* 16; Horgan, quoted in Smith and Weiner, *Robert Oppenheimer,* 40. See also Oppenheimer to Horgan, October 6, 1923, and Horgan's commentary, ibid., 43.

62. Bernheim, interviewed by Weiner, October 27, 1975, 7; Boyd, interviewed by Smith, December 21, 1975, 12. Bernheim did not remember ever meeting Fergusson (Smith and Weiner, *Robert Oppenheimer,* 44)—though that is understandable, given that Fergusson had gone to England by the time Oppenheimer and Bernheim began to room together.

63. Dinnerstein, *Antisemitism in America,* 87; David Hollinger, "Academic Culture at the University of Michigan, 1938–1988," in *Science, Jews, and Secular Culture,* 121–54, on 122.

64. Schweber, *In the Shadow of the Bomb,* 62; Smith and Weiner, *Robert Oppenheimer,* 68–69; Wyman, interviewed by Weiner, May 28, 1975, 2.

65. Oppenheimer, interviewed by Kuhn, November 18, 1963, 9, 11. See also Maila L. Walter, *Science and Cultural Crisis: An Intellectual Biography of Percy Williams Bridgman (1882–1961)* (Stanford, CA: Stanford University Press, 1990), 28, 68, 72; and Smith and Weiner, *Robert Oppenheimer,* 68–69.

66. Percy Bridgman, *Reflections of a Physicist* (New York: Philosophical Library, 1955), 81; quoted also in Steven Shapin, "How to Be Antiscientific," in *The One Culture? A Conversation about Science,* ed. Harry Collins and Jay A. Labinger (Chicago: University of Chicago Press, 2001), 105; Oppenheimer, interviewed by Kuhn, November 19, 1963, 9.

67. Bridgman to Rutherford, June 24, 1925, quoted in Smith and Weiner, *Robert Oppenheimer,* 77. For similar examples, see Andrew S. Winston, "'As His Name Indicates': R. S. Woodworth's Letters of Reference and Employment for Jewish Psychologists in the 1930s," *Journal of the History of the Behavioral Sciences* 32, no. 1 (January 1996): 30–43.

68. George D. Birkhoff to Professor Hall, April 12, 1928, folder labeled "Oppenheimer official documents and 1954 clippings," Raymond Thayer Birge Papers [ca. 1928–54], BANC MSS 87/147 c, Bancroft Library. On Oppenheimer's studies with Birkhoff, see Smith and Weiner, *Robert Oppenheimer,* 68–69.

69. Horgan, quoted in Smith and Weiner, *Robert Oppenheimer,* 9; Goodchild, *J. Robert Oppenheimer,* 16; Birkhoff to Hall, April 12, 1928, Birge Papers.

70. Oppenheimer to Smith, October 6, 1923, in Smith and Weiner, *Robert Oppenheimer,* 46–47, on 47; Oppenheimer to Smith [ca. January 22, 1924], 60; Boyd, quoted in Smith and Weiner, *Robert Oppenheimer,* 45; Boyd, interviewed by Smith, December 21, 1975, 3. See also Wyman, interviewed by Weiner, May 28, 1975, 13.

71. See Winston, "'As His Name Indicates,'" esp. 35.

72. Oppenheimer, interviewed by Kuhn, November 18, 1963, 9, 15; quoted also in Smith and Weiner, *Robert Oppenheimer*, 45–46, 69.

73. Oppenheimer, interviewed by Kuhn, November 18, 1963, 4–5. Oppenheimer was referring to Alfred North Whitehead and Bertrand Russell, *Principia Mathematica* (Cambridge: Cambridge University Press, 1910).

74. Oppenheimer to Smith [ca. January 1925], 70. Whitehead gave Oppenheimer one of the few B grades he received. Smith and Weiner, *Robert Oppenheimer*, 74.

75. Smith and Weiner, *Robert Oppenheimer,* 45.

76. Oppenheimer to Fergusson, Christmas 1923, 57; Oppenheimer, interviewed by Kuhn, November 18, 1963, 13.

77. Oppenheimer to Fergusson, April 25 [1925], in Smith and Weiner, *Robert Oppenheimer*, 72–73, quoting 73. See also Cassidy, *J. Robert Oppenheimer*, 90.

78. Bridgman to Rutherford, June 24, 1925, quoted in Smith and Weiner, *Robert Oppenheimer*, 77.

79. Oppenheimer, interviewed by Kuhn, November 18, 1963, 13–14.

80. Ibid., 12, 14. See also Silvan S. Schweber, "The Empiricist Temper Regnant: Theoretical Physics in the United States, 1920–1950," *Historical Studies in the Physical and Biological Sciences* 17 (1988): 17–98; and Kevles, *The Physicists*, 168–69.

81. Oppenheimer to Fergusson, November 1, 1925, in Smith and Weiner, *Robert Oppenheimer*, 86–87, quoting 87; Oppenheimer to Fergusson, January 23, 1926, ibid., 91–92, quoting 92. On Oppenheimer's interest in metallic conduction, which began during his work under Bridgman, see Oppenheimer to R. E. Priestley [June 1925], in Smith and Weiner, *Robert Oppenheimer*, 76–77; Oppenheimer, interviewed by Kuhn, November 18, 1963, 14; and Walter, *Science and Cultural Crisis*, 69.

82. Robert Oppenheimer to Frank Oppenheimer [ca. late spring 1926], in Smith and Weiner, *Robert Oppenheimer*, 95–96, quoting 96.

83. Oppenheimer to Fergusson, November 1 [1925], in Smith and Weiner, *Robert Oppenheimer*, 86–87; Robert Oppenheimer to Frank Oppenheimer, October 7 [1933], ibid., 162–65, on 163.

84. Bernheim, interviewed by Weiner, October 27, 1975, 20. See also Oppenheimer to Fergusson, November 15, 1925, in Smith and Weiner, *Robert Oppenheimer*, 87–88.

85. Wyman, interviewed by Weiner, May 28, 1975, 19. The incident in Paris occurred during Christmas vacation 1925 (Smith and Weiner, *Robert Oppenheimer*, 91).

86. Wyman's recollection of Oppenheimer's words, in Wyman, interviewed by Weiner, May 28, 1975, 20–22, quoting 20; Edsall, interviewed by Weiner, July 16, 1975, 27. See also Smith and Weiner, *Robert Oppenheimer*, 93–94; Goodchild, *J. Robert Oppenheimer*, 17–18; and Mary Jo Nye, *Blackett: Physics, War, and Politics in the Twentieth Century* (Cambridge, MA: Harvard University Press, 2004), 44. Bird and Sherwin suggest that the poisoned-apple incident may not have been entirely a fantasy and may have occurred the previous year, making the panic on Corsica some

sort of flashback (*American Prometheus,* 46–47, 50). The story, however, remains mysterious.

87. Smith and Weiner, *Robert Oppenheimer,* 94; Horgan, interviewed by Smith, March 3, 1976, 18; Wyman, interviewed by Weiner, May 28, 1975, 21. Horgan remembered that in his late teens, Oppenheimer was prone to depressive episodes during which "he would seem to be incommunicado emotionally for a day or two at a time" (Horgan, interviewed by Smith, 17).

88. Gerald Holton, "Young Man Oppenheimer," *Partisan Review* 48 (July 1981): 380–88.

89. Oppenheimer, interviewed by Kuhn, November 18, 1963, 17, 21; Wyman, interviewed by Weiner, May 28, 1975, 17; Oppenheimer, quoted in Smith and Weiner, *Robert Oppenheimer,* 96. See also Holton, "Heisenberg, Oppenheimer, and the Transition to Modern Physics."

90. Oppenheimer to Fergusson, November 14 [1926], in Smith and Weiner, *Robert Oppenheimer,* 100–101, on 100; Holton, "Heisenberg, Oppenheimer, and the Transition to Modern Physics," 168; Oppenheimer, quoted in Smith and Weiner, *Robert Oppenheimer,* 98. See also Goodchild, *J. Robert Oppenheimer,* 19. For more on the development of Oppenheimer's physics at Göttingen and elsewhere, before his final return to the United States, see Cassidy, *J. Robert Oppenheimer,* 109–12.

91. Smith and Weiner, *Robert Oppenheimer,* 99; Oppenheimer, interviewed by Kuhn, November 20, 1963, 4–5.

92. Kennard, quoted in Kevles, *The Physicists,* 217.

93. Smith and Weiner, *Robert Oppenheimer,* 103–4; Oppenheimer to Fergusson, November 14 [1926], 100; Oppenheimer, interviewed by Kuhn, November 20, 1963, 4, 1, 6.

94. J. Robert Oppenheimer, "A Science in Change," in *Science and the Common Understanding* (London: Oxford University Press, 1954), 37–54, on 37; Kevles, *The Physicists,* 162–63; Oppenheimer, interviewed by Kuhn, November 20, 1963, 4; Smith and Weiner, *Robert Oppenheimer,* 98, quoting Kemble on 107; Jungk, *Brighter Than a Thousand Suns,* 33.

95. Bird and Sherwin, *American Prometheus,* 67.

96. Max Born, *My Life: Recollections of a Nobel Laureate* (London: Taylor and Francis, 1978), 229, 233–34. See also Kevles, *The Physicists,* 217.

97. Goodchild, *J. Robert Oppenheimer,* 20–21.

98. Barbara Marshall, "Politics in Academe: Göttingen University and the Growing Impact of Political Issues, 1918–33," *European History Quarterly* 18, no. 3 (July 1988), esp. 297–99, 305–7, 309–11; Oppenheimer, interviewed by Kuhn, November 20, 1963, 5; Oppenheimer to Fergusson, November 14 [1926], 100; transcript of conversation between J. Robert Oppenheimer and Edward R. Murrow, recorded December 16, 1954, for CBS television program *See It Now,* broadcast on January 4, 1955, in *Oppenheimer, FBI Security File,* 100-17828, sec. 55, quoting 29; Jungk, *Brighter Than a Thousand Suns,* 32; Cassidy, *J. Robert Oppenheimer,* 105.

99. Goudsmit, quoted in Goodchild, *J. Robert Oppenheimer,* 22.

100. Oppenheimer, interviewed by Kuhn, November 20, 1963, 18; quoted also in Smith and Weiner, *Robert Oppenheimer,* 114.

101. Oppenheimer, interviewed by Kuhn, November 20, 1963, 21–22; Ehrenfest, quoted in Smith and Weiner, *Robert Oppenheimer,* 122. Weisskopf's rendition of Ehrenfest's letter, and Weisskopf's comments on Pauli, are in Weisskopf, *The Joy of Insight,* 85. See also Cassidy, *J. Robert Oppenheimer,* 117–18, 126–29.

102. Rabi, "Introduction," in Rabi et al., *Oppenheimer,* 5.

103. Oppenheimer to Fergusson, November 14 [1926], 100; Dirac, quoted in Jungk, *Brighter Than a Thousand Suns,* 31; Goodchild, *J. Robert Oppenheimer,* 21. See also transcript of conversation between Oppenheimer and Murrow, December 16, 1954, 28.

104. Oppenheimer to Fergusson, November 14 [1926], 100. See also, for example, Max Weber's analysis in 1918 of the "Americanization" of German academic life ("Science as a Vocation," esp. 131).

105. Oppenheimer to W. J. Robbins, May 14 [1929], in Smith and Weiner, *Robert Oppenheimer,* 128–29, on 128.

106. J. S. Rigden, "J. Robert Oppenheimer—Before the War," *Scientific American* 273 (July 1995), esp. 70–71; Leo Ezaki, "Long Journey into Tunneling," Nobel Lecture, December 12, 1973, http://www.nobel.se/physics/laureates/1973/esaki-lecture.pdf, 1.

CHAPTER THREE

1. Raymond T. Birge, "Professor J. R. Oppenheimer" [ca. 1930], folder labeled "Oppenheimer official documents and 1954 clippings," Birge Papers; Smith and Weiner, *Robert Oppenheimer,* 130; Kevles, *The Physicists,* 215–18.

2. Felix Bloch, interviewed by Charles Weiner, August 15, 1968, 14, Niels Bohr Library, AIP; Edward U. Condon, interviewed by Charles Weiner, October 17 and 18, 1967, 23, Niels Bohr Library, AIP; Frederick Seitz, *On the Frontier: My Life in Science* (New York: American Institute of Physics, 1994), 119–20. See also Schweber, "The Empiricist Temper Regnant"; Charles Weiner, "A New Site for the Seminar: The Refugees and American Physics in the Thirties," in *The Intellectual Migration,* ed. Donald Fleming and Bernard Bailyn, 152–89 (Cambridge, MA: Harvard University Press, 1969); and Cassidy, *J. Robert Oppenheimer,* 117–31.

3. Raymond T. Birge, "Supplementary Remarks on Professor Oppenheimer," January 28, 1936, folder labeled "Oppenheimer official documents and 1954 clippings," Birge Papers. See also Rigden, "J. Robert Oppenheimer," 73; and John L. Heilbron and Robert W. Seidel, *Lawrence and His Laboratory: A History of the Lawrence Berkeley Laboratory,* vol. 1 (Berkeley and Los Angeles: University of California Press, 1989). For a comprehensive discussion of the work of Oppenheimer and his school of theoretical physics at Berkeley, see Schweber, *In the Shadow of the Bomb,* 61–75; and Cassidy, *J. Robert Oppenheimer,* 133–80.

4. Robert Wilson, interviewed by Spencer R. Weart, May 19, 1977, 51–52, Niels Bohr Library, AIP.

5. Schweber, *In the Shadow of the Bomb*, 14; Peter J. Kuznick, *Beyond the Labora tory: Scientists as Political Activists in 1930s America* (Chicago: University of Chicago Press, 1987), 316 n. 131; Birge, "Professor J. R. Oppenheimer"; Rabi, "Introduction," in Rabi et al., *Oppenheimer*, 8; Robert Serber, *Peace and War: Reminiscences of a Life on the Frontiers of Science*, with Robert P. Crease (New York: Columbia University Press, 1998), 42–43.

6. Schweber, *In the Shadow of the Bomb*, 69; David Hawkins, interviewed by the author, November 16, 1998; Pauli, quoted in Emilio Segrè, *A Mind Always in Motion: The Autobiography of Emilio Segrè* (Berkeley and Los Angeles: University of California Press, 1993), 138; Jungk, *Brighter Than a Thousand Suns*, 119; Goodchild, *J. Robert Oppenheimer*, 28; Wilson, interviewed by Weart, May 19, 1977, 46. See also Chevalier, *Oppenheimer*, 22.

7. Wilson, interviewed by Weart, May 19, 1977, 46–47; Serber, *Peace and War*, 36–40. In 1928–29, Robert and Frank Oppenheimer began leasing the New Mexico ranch, located near Pecos and Los Alamos, that they named Perro Caliente. They later bought it. Smith and Weiner, *Robert Oppenheimer*, 165; Bird and Sherwin, *American Prometheus*, 72–73.

8. Goodchild, *J. Robert Oppenheimer*, 28; Rudolf Peierls, *Bird of Passage: Recollections of a Physicist* (Princeton, NJ: Princeton University Press, 1985), 189; Wilson, "A Recruit for Los Alamos," 45; Smith and Weiner, *Robert Oppenheimer*, 133; Rhodes, *The Making of the Atomic Bomb*, 444; Stanley A. Blumberg and Gwinn Owens, *Energy and Conflict: The Life and Times of Edward Teller* (New York: G. P. Putnam's Sons, 1976), 75; Stanley A. Blumberg and Lewis G. Panos, *Edward Teller: Giant of the Golden Age of Physics* (New York: Charles Scribner's Sons, 1990), 45; Herken, *Brotherhood of the Bomb*, 65.

9. During the 1954 hearings, Oppenheimer estimated that his salary in 1942 was $5,000 and that he made $8,000 in dividends and interest for the year. The AEC's attorney stated, "I have looked at your income-tax return for, I think, 1942, and it seemed to me to be about $15,000" ("Transcript," 184). Schweber puts Oppenheimer's income even higher than this: upon his father's death on September 20, 1937, Oppenheimer inherited $392,602. He received from this an annual dividend income of between $10,000 and $15,000, in addition to an annual academic salary of $6,000. Schweber, *In the Shadow of the Bomb*, 209 n. 153.

10. Serber, *Peace and War*, 31.

11. Goodchild, *J. Robert Oppenheimer*, 27–28; Segrè, *A Mind Always in Motion*, 138–39.

12. Thomas Hager, *Force of Nature: The Life of Linus Pauling* (New York: Simon and Schuster, 1995), 152.

13. Memorandum, April 20, 1954, in *Oppenheimer, FBI Security File*, 100-17828, sec. 29. Bird and Sherwin outline reasons to be skeptical of these reports (*American*

Prometheus, 367). Abraham Pais later thought that "a strong, latent homosexuality was an important ingredient in Robert's emotional makeup" (*A Tale of Two Continents,* 241).

14. Robert Oppenheimer to Frank Oppenheimer, October 14, 1929, in Smith and Weiner, *Robert Oppenheimer,* 134–36, quoting 135.

15. Hijiya, "The *Gita* of J. Robert Oppenheimer," 129–30, 148; Rabi, "Introduction," 5; Robert Oppenheimer to Frank Oppenheimer, August 10 [1931], in Smith and Weiner, *Robert Oppenheimer,* 142–43, on 143.

16. On such models, see Shapin, "The Philosopher and the Chicken."

17. Robert Oppenheimer to Frank Oppenheimer, Sunday [ca. January 1932], in Smith and Weiner, *Robert Oppenheimer,* 151–52, quoting 151.

18. Robert Oppenheimer to Frank Oppenheimer, March 12 [1932], in Smith and Weiner, *Robert Oppenheimer,* 154–56, on 155; Robert Oppenheimer to Frank Oppenheimer, Sunday [ca. fall 1932], ibid., 157–60, on 157.

19. Robert Oppenheimer to Frank Oppenheimer, March 12 [1932], 156.

20. Robert Oppenheimer to Frank Oppenheimer, October 7 [1933], 165; Kugelmass, *J. Robert Oppenheimer,* 59; Oppenheimer, quoted in "Physicist Oppenheimer," *Time,* November 8, 1948, 75; Hijiya, "The *Gita* of J. Robert Oppenheimer," esp. 136–37; Bird and Sherwin, *American Prometheus,* 99–102.

21. Rabi, quoted in Rigden, *Rabi,* 229.

22. Rabi, "Introduction," 4.

23. Hawkins, interviewed by the author, November 16, 1998.

24. Airtel memorandum on interview with Dr. Harvey Hall (Technical Director, Naval Science Division, ONR), April 23, 1954, in *Oppenheimer, FBI Security File,* 100-17828, sec. 31; "Transcript," 8; Schweber, *In the Shadow of the Bomb,* 56–57; Laura Hein, "Learning about Patriotism, Decency, and the Bomb," in *Living with the Bomb: American and Japanese Cultural Conflicts in the Nuclear Age,* ed. Laura Hein and Mark Shelden (Armonk, NY: M. E. Sharpe, 1997), esp. 282–84; Kuznick, *Beyond the Laboratory,* 157; Goodchild, *J. Robert Oppenheimer,* 32; Oppenheimer to Professor Hall, June 5, 1932, and Hall to Oppenheimer, June 14, 1932, folder labeled "Oppenheimer. General," Birge Papers.

25. Rhodes, "Agony of J. Robert Oppenheimer," 75–76; "Transcript," 8, 154; Goodchild, *J. Robert Oppenheimer,* 31; Schweber, *In the Shadow of the Bomb,* 56, 197 n. 35; Bird and Sherwin, *American Prometheus,* 249–54.

26. *Daily People's World,* April 1938, in *Oppenheimer, FBI Security File,* 100-17828, sec. 42; Special Agent D. L. Johnson, memorandum, "Subject: Frank Friedman Oppenheimer," section titled "Police Check," July 22, 1943, Investigation Files, box 100, General Correspondence, 1943–47, Records of the Office of the Commanding General, Manhattan Project, Records of the Office of the Chief of Engineers, Record Group 77, National Archives and Records Administration, College Park, MD; *American Civil Liberties Union News* (San Francisco) 4, no. 1 (January 1939), in *Oppenheimer, FBI Security File,* 100-17828, sec. 42; "Transcript," 8; Chevalier,

Notes to Pages 55–59 313

Oppenheimer, 22–25; Kuznick, *Beyond the Laboratory,* 231; Condon, interviewed by Weiner, September 11 and 12, 1973, 215, Niels Bohr Library, AIP; Stern, *The Oppenheimer Case,* 22.

27. Chevalier, *Oppenheimer,* 19; "Transcript," 159, 10. See also Chevalier to Oppenheimer, July 23, 1964, and Oppenheimer to Chevalier, August 7, 1964, Oppenheimer Papers, box 26; "Transcript," 3, 8–9; Stern, *The Oppenheimer Case,* 15; Goodchild, *J. Robert Oppenheimer,* 206–7; and Kuznick, *Beyond the Laboratory,* 140, 316 n. 132. The strongest case that Oppenheimer was a Communist Party member is presented in Herken, *Brotherhood of the Bomb.* See also Daniel J. Kevles, "The Strange Case of Robert Oppenheimer," review of Herken, *Brotherhood of the Bomb, New York Review of Books,* December 4, 2003, 37–40; and Gregg Herken, "The Oppenheimer Case: An Exchange," with a reply by Kevles, in *New York Review of Books,* March 25, 2004, 46–47.

28. Philip Morrison, interviewed by the author, December 19, 2001.

29. Chevalier, *Oppenheimer,* 22; Condon, interviewed by Weiner, September 11 and 12, 1973, 215.

30. Morrison, interviewed by the author, December 19, 2001. See also Bird and Sherwin, *American Prometheus,* 104.

31. Chevalier, *Oppenheimer,* 24; Kamen, quoted in Ellen Schrecker, *No Ivory Tower: McCarthyism and the Universities* (New York: Oxford University Press, 1986), 134. See also Chevalier's fictionalized description of one of these parties in his novel *The Man Who Would Be God,* 21–42.

32. "Transcript," 8; Herken, *Brotherhood of the Bomb,* 29.

33. Schweber, *In the Shadow of the Bomb,* 56; Nathan Glazer, *The Social Basis of American Communism* (New York: Harcourt, Brace and World, 1961), 168. See also the discussion of "cultural radicalism" in Richard H. Pells, *Radical Visions and American Dreams: Culture and Social Thought in the Depression Years* (Middletown, CT: Wesleyan University Press, 1973); and see Cassidy, *J. Robert Oppenheimer,* 194.

34. Oppenheimer, quoted in "Physicist Oppenheimer," 76.

35. "Transcript," 8.

36. Kitty Oppenheimer, quoted in Goodchild, *J. Robert Oppenheimer,* 38. Her first marriage had lasted only a few months. See also Bird and Sherwin, *American Prometheus,* 155–61.

37. Goodchild, *J. Robert Oppenheimer,* 38–40, quoting Kitty Oppenheimer on 39; Stern, *The Oppenheimer Case,* 29; Serber, *Peace and War,* 59–60; "Transcript," 10.

38. Steve Nelson, James R. Barrett, and Rob Ruck, *Steve Nelson, American Radical* (Pittsburgh: University of Pittsburgh Press, 1981), 268–69; Bird and Sherwin, *American Prometheus,* 162. On Oppenheimer's reading of *Capital,* see also Chevalier, *Oppenheimer,* 16; and Kuznick, *Beyond the Laboratory,* 140.

39. Jackie Oppenheimer, quoted in Goodchild, *J. Robert Oppenheimer,* 40; Oppenheimer, quoted in "Physicist Oppenheimer," 76. See also Kitty Oppenheimer's testimony during the 1954 hearings, in "Transcript," 571–77.

40. Weisskopf, *The Joy of Insight,* 115. See also "Transcript," 10; and Cassidy, *J. Robert Oppenheimer,* 189.

41. Chevalier, *Oppenheimer,* 32. See also Stern, *The Oppenheimer Case,* 25; Herken, *Brotherhood of the Bomb,* 28–29; and Chevalier's fictionalized account, *The Man Who Would be God,* 66.

42. Segrè, *A Mind Always in Motion,* 138, 139; Chevalier, *Oppenheimer,* 36; Johnson, "Subject: Frank Friedman Oppenheimer," section titled "Acquaintance Check," July 22, 1943. Herken discusses the pamphlet in *Brotherhood of the Bomb,* 31. It seems likely that the pamphlet in question was *Report to Our Colleagues,* by the College Faculties Committee, Communist Party of California (copy held at Bancroft Library). Herken has made the document available on the Web, together with related documents: http://www.brotherhoodofthebomb.com/bhbsource/documents.html. See also Bird and Sherwin, *American Prometheus,* 144–45.

43. Morrison, interviewed by the author, December 19, 2001; Oppenheimer, quoted in Goodchild, *J. Robert Oppenheimer,* 41.

44. "Transcript," 327; Bethe's recollection of Oppenheimer's words, quoted in Schweber, *In the Shadow of the Bomb,* 108. See also Bethe's testimony at the security hearings, in "Transcript," 332; and Carey McWilliams, in Roy Hoopes, *Americans Remember the Home Front: An Oral Narrative of the World War II Years* (New York: Berkley Press, 2002), 22.

45. Herbert Childs, *An American Genius: The Life of Ernest Orlando Lawrence* (New York: E. P. Dutton, 1968), 172–73; Oppenheimer to Lawrence, August 30, 1945, Oppenheimer Papers, box 45, in Smith and Weiner, *Robert Oppenheimer,* 300–302.

46. The agents had asked Millikan for information on Frank Oppenheimer, but he steered the conversation toward Robert, stating that "Frank Oppenheimer was an appendage to his brother Robert." Johnson, "Subject: Frank Friedman Oppenheimer," section titled "Acquaintance Check," July 22, 1943.

47. Birge, "History of the Physics Department," vol. 5, "The Period 1942–1950," chap. 17, p. 6, quoting from notes made on August 18, 1945; Birge, quoted in Davis, *Lawrence and Oppenheimer,* 83.

48. Martin D. Kamen, *Radiant Science, Dark Politics: A Memoir of the Nuclear Age* (Berkeley and Los Angeles: University of California Press, 1985), 179. Birge said that "Lawrence didn't like bohemians" (quoted in Davis, *Lawrence and Oppenheimer,* 83). Kamen quotes this statement and says, "I may add that Birge didn't either" (*Radiant Science, Dark Politics,* 316 n. 2 for chap. 10). On Frank Oppenheimer as "bohemian," see comments by Millikan, William A. Fowler of Caltech, and Frank's landlady, in Johnson, "Subject: Frank Friedman Oppenheimer," sections titled "Acquaintance Check," "Reference," and "Residence Check," July 22, 1943. Physicist Robert Wilson, interviewed years later for Jon Else's film *The Day after Trinity,* referred to Robert Oppenheimer as having been "bohemian" in this Berkeley period.

49. Condon, interviewed by Weiner, September 11 and 12, 1973, 208. See also Bird and Sherwin, *American Prometheus,* 116–17. Oppenheimer's association with

Peters later became a subject of investigation by the House Un-American Activities Committee (HUAC) and the AEC. In 1949, Oppenheimer denounced Peters as a "Red" in front of HUAC. See also Schweber, *In the Shadow of the Bomb*, 115–30; and see the further discussion in chapter 7 of this book.

50. Oppenheimer to Birge, November 4, 1943, and further correspondence regarding Feynman, in Smith and Weiner, *Robert Oppenheimer*, 268–69, 275–77, 283–84.

51. Kuznick, *Beyond the Laboratory*, 40.

52. Herbert F. York, *Making Weapons, Talking Peace: A Physicist's Odyssey from Hiroshima to Geneva* (New York: Basic Books, 1987), 25; Childs, *An American Genius*, 266; Morrison, interviewed by the author, December 19, 2001; Sidney and Beatrice Webb, *Soviet Communism: A New Civilization?* vol. 1 (Worker's Education Association, 1935), chap. 11, "Science the Salvation of Mankind," 944–1016, quoting 948. On Bernal and other socialists, see Gary Werskey, *The Visible College: A Collective Biography of British Scientists and Socialists of the 1930s* (London: Free Association Books, 1988); on their influence in the United States, see Helena Sheehan, *Marxism and the Philosophy of Science: A Critical History* (Atlantic Highlands, NJ: Humanities Press International, 1993), 409.

53. On the OSRD and the Rad Lab's work in 1941 on electromagnetic separation of isotopes, see Hewlett and Anderson, *The New World*, 41, 56–60, 75.

54. Kamen, *Radiant Science, Dark Politics*, 183–85.

55. Oppenheimer to Lawrence, November 12, 1941, in Smith and Weiner, *Robert Oppenheimer*, 220. Oppenheimer further distanced himself from FAECT in his security-hearings testimony ("Transcript," 131).

56. Kamen, *Radiant Science, Dark Politics*, 186; see also 161–77.

57. Lawrence, quoted in Smith and Weiner, *Robert Oppenheimer*, 222–23; Hewlett and Anderson, *The New World*, 46, 56–60; Cassidy, *J. Robert Oppenheimer*, 203. Oppenheimer had initially greeted news of the discovery of fission with some skepticism. Rhodes, *Dark Sun*, 370; Herken, *Brotherhood of the Bomb*, 23; Oppenheimer to George Uhlenbeck, February 5 [1939], in Smith and Weiner, *Robert Oppenheimer*, 208–9.

58. Smith and Weiner, *Robert Oppenheimer*, 223–24; Lillian Hoddeson et al., *Critical Assembly: A Technical History of Los Alamos during the Oppenheimer Years, 1943–1945* (Cambridge: Cambridge University Press, 1993), 42; Arthur Holly Compton, *Atomic Quest: A Personal Narrative* (New York: Oxford University Press, 1956), 125–30.

59. Stern, *The Oppenheimer Case*, 33–35.

60. Nelson, Barrett, and Ruck, *Steve Nelson*, 269. See also "Transcript," 194–95.

61. Compton, quoted in Jungk, *Brighter Than a Thousand Suns*, 121; Hoddeson et al., *Critical Assembly*, 42–47; Jones, *Manhattan*, 73–77.

62. Jones, *Manhattan*, 83.

63. Ibid.; Robert F. Bacher, *Robert Oppenheimer, 1904–1967* (Los Alamos, NM: Los Alamos Historical Society, 1999), 10–11; "Transcript," 12, 28.

64. Jones, *Manhattan*, 83–85; Rhodes, *The Making of the Atomic Bomb*, 450–51; John H. Dudley, "Ranch School to Secret City," in *Reminiscences of Los Alamos, 1943–1945*, ed. Lawrence Badash, Joseph O. Hirschfelder, and Herbert P. Broida, 1–11 (Dordrecht, Holland: Reidel, 1980); Edwin McMillan, "Early Days at Los Alamos," ibid., 13–19, esp. 14–15.

65. James W. Kunetka, *City of Fire: Los Alamos and the Atomic Age, 1943–1945* (Englewood Cliffs, NJ: Prentice-Hall, 1978), 64–66; Smith and Weiner, *Robert Oppenheimer*, 238; Goodchild, *J. Robert Oppenheimer*, 72; Lawren, *The General and the Bomb*, 105; Brode, *Tales of Los Alamos*, 1; Oppenheimer, quoted in Rhodes, *The Making of the Atomic Bomb*, 452. See also Jungk, *Brighter Than a Thousand Suns*, 125.

66. Kunetka, *City of Fire*, 59; Smith and Weiner, *Robert Oppenheimer*, 229; Hewlett and Anderson, *The New World*, 228.

67. "Transcript," 561.

68. Jones, *Manhattan*, 87–88; Conant and Groves to Oppenheimer, February 25, 1943, LANL A-84-019 57-27, Los Alamos National Laboratory Archives, Los Alamos, NM.

69. Ellen Schrecker, *Many Are the Crimes: McCarthyism in America* (Boston: Little, Brown, 1998), 165; R. E. Mayer, "J. Robert Oppenheimer alias Robert J. Oppenheimer," March 28, 1941, in *Oppenheimer, FBI Security File*, 100-17828, sec. 1; Kunetka, *Oppenheimer*, 31.

70. "Transcript," 260. See also Groves's testimony in "Transcript," 165.

71. Smith and Weiner, *Robert Oppenheimer*, 222. See also Hewlett and Anderson, *The New World*, 230.

72. Conant to Groves, December 21, 1942, quoted in Hershberg, *James B. Conant*, 167. See also Lawren, *The General and the Bomb*, 106; Bird and Sherwin, *American Prometheus*, 186–87, 208; and Cassidy, *J. Robert Oppenheimer*, 217.

73. Kenneth D. Nichols, *The Road to Trinity* (New York: William Morrow, 1987), 72; Leslie R. Groves, *Now It Can Be Told: The Story of the Manhattan Project* (New York: Harper and Brothers, 1962), 61–63; Lawren, *The General and the Bomb*, 100; Jungk, *Brighter Than a Thousand Suns*, 124.

74. Luis Alvarez, *Alvarez: Adventures of a Physicist* (New York: Basic Books, 1987), 78; Herken, *Brotherhood of the Bomb*, 71; Rabi, quoted in Jeremy Bernstein, *Experiencing Science* (New York: Basic Books, 1978), 97; Eugene Wigner, *The Recollections of Eugene P. Wigner as Told to Andrew Szanton* (New York: Plenum Press, 1992), 243; Allison, quoted in Goodchild, *J. Robert Oppenheimer*, 71.

75. John Manley, "Assembling the Wartime Labs," *Bulletin of the Atomic Scientists* 30 (May 1974): 45; Manley, "A New Laboratory Is Born," in Badash, Hirschfelder, and Broida, *Reminiscences of Los Alamos*, 21–38, on 30.

76. Manley, "Assembling the Wartime Labs," 45; Wilson, "A Recruit for Los Alamos," 45.

77. Wilson, quoted in Goodchild, *J. Robert Oppenheimer*, 72.

78. Hawkins, interviewed by the author, November 16, 1998; Oppenheimer, quoted in Stern, *The Oppenheimer Case,* 40; Groves, *Now It Can Be Told,* 61–63; Lawren, *The General and the Bomb,* 102. See also Nichols, *The Road to Trinity,* 79, and Childs, *An American Genius,* 337.

79. Nichols, *The Road to Trinity,* 152; Oppenheimer, quoted in Lawren, *The General and the Bomb,* 102; Wilson, "A Recruit for Los Alamos," 42–43; Goodchild, *J. Robert Oppenheimer,* 73; Herken, *Brotherhood of the Bomb,* 74–75.

80. The length of Oppenheimer's hair in 1941 was discussed in Herbert S. Marks and Joseph Volpe, "Memorandum: Interview at Office of Assistant U. S. Attorney, William Hitz, Washington D.C., May 20, 1952," 16, Oppenheimer Papers, box 237. Bethe's description is given, retrospectively, in Herbert S. Marks to Oppenheimer, July 8, 1952, Oppenheimer Papers, box 237. Alvarez's description is in *Alvarez,* 128. Jane Wilson is quoted in Palevsky, *Atomic Fragments,* 139. See also Charles Thorpe and Steven Shapin, "Who Was J. Robert Oppenheimer?" 566.

81. Rabi, quoted in Lawren, *The General and the Bomb,* 101.

82. Nichols, *The Road to Trinity,* 152; Bacher, quoted in Lawren, *The General and the Bomb,* 103–4; Bacher and Rabi, quoted in Rigden, *Rabi,* 150 (emphasis in original); Alvarez, *Alvarez,* 128.

83. Oppenheimer to Conant, February 1, 1943, in Smith and Weiner, *Robert Oppenheimer,* 247–48, quoting 247; Conant and Groves to Oppenheimer, February 25, 1943; Stern, *The Oppenheimer Case,* 45.

84. War Department, "Complications of the Los Alamos Project," November 12, 1946, 5, file 322 (Los Alamos), Decimal Files, 1942–48, General Correspondence, 1942–48, General Administrative Files, Record Group 77, National Archives and Records Administration. See also Colonel K. E. Fields to General Groves, January 30, 1947, attached to this report.

85. Rhodes, *The Making of the Atomic Bomb,* 452; Robert F. Bacher, interviewed by Mary Terrall, June 9, 1981, 74, California Institute of Technology Archives, Pasadena.

86. Oppenheimer to Rabi, February 8, 1943, Oppenheimer Papers, box 59; Rabi, "Suggestions for Interim Organization and Procedure," with covering letter from Rabi to Oppenheimer, February 10, 1943, Oppenheimer Papers, box 59; Hans A. Bethe, interviewed by the author, May 12, 1998; Oppenheimer to Rabi [ca. December (19?), 1944], Oppenheimer Papers, box 59; Bethe, quoted in Rigden, *Rabi,* 149.

87. "Transcript," 174, 166; see also Groves, *Now It Can Be Told,* 154–55.

88. Condon to Oppenheimer, April 26, 1943, Oppenheimer Papers, box 27, in Groves, *Now It Can Be Told,* 429–32; "Transcript," 166, 173–74.

89. Stern, *The Oppenheimer Case,* 45.

90. The different accounts are neatly summarized in Stern, *The Oppenheimer Case,* 44; see also Chevalier, *Oppenheimer,* esp. 52–54. Much of the security-hearings transcript deals with this incident and the different accounts of it that Oppenheimer gave Manhattan Project intelligence officers. On Oppenheimer's recollection of Eltenton as having attended the meeting at his home, see "Transcript," 131.

91. "Transcript," 809, 811–12, 821–23; Stern, *The Oppenheimer Case*, 47. See also "Findings and Recommendation of the Personnel Security Board in the Matter of J. Robert Oppenheimer," in *Transcript and Texts*, 999–1021, esp. 1006.

92. On Weinberg and Bohm as FAECT members, see Russell Olwell, "Physical Isolation and Marginalization in Physics: David Bohm's Cold War Exile," *Isis* 90, no. 4 (1999): 740–41. Oppenheimer told Colonel Lansdale, "It was just generally obvious that [Lomanitz] was a member of the union" ("Transcript," 206). On the three men's FAECT membership, I am also indebted to Shawn Mullet, personal communication, March 19, 2002. See also Bird and Sherwin, *American Prometheus*, 175–76.

93. "Transcript," 811. According to Schrecker, Bohm was officially a Communist Party member between 1942 and 1943, but Lomanitz, although he attended party meetings, was not a member (*No Ivory Tower*, 135). Weinberg later denied the HUAC allegation that he had been a party member. He was charged with perjury on that count but was cleared in March 1953. Stern, *The Oppenheimer Case*, 57, 60; Herken, *Brotherhood of the Bomb*, 263–64.

94. Lomanitz, quoted in Schrecker, *No Ivory Tower*, 135.

95. Colonel K. D. Nichols to General Groves, "Communist Infiltration," June 30, 1943, Investigation Files, box 100, Records of the Office of the Commanding General, Manhattan Project. For more detail on the meeting between Nelson and "Joe," see Rhodes, *Dark Sun*, 122–23, 309. See also Olwell, "Physical Isolation," 741; Stern, *The Oppenheimer Case*, 48–49, 57; Herken, *Brotherhood of the Bomb*, 87–121; and Bird and Sherwin, *American Prometheus*, 188–92. Weinberg denied that he had ever met Nelson, and he was never charged with espionage. For further discussion of the Weinberg case, see chapter 7; and Shawn Mullet, "Strong Rope, Leaky Buckets, and Atomic Espionage" (manuscript, 2002). I am grateful to Shawn Mullet for helpful discussions on the Weinberg case.

96. Transcript of conversation between Steve Nelson and "Joe," attachment to John Lansdale to General Strong, "Subject: DSM Communist Infiltration," June 12, 1943, 2–6; and Colonel K. D. Nichols to General Groves, "Subject: Joseph Woodrow Weinberg," June 30, 1943, both in Investigation Files, box 100, Records of the Office of the Commanding General, Manhattan Project. In the conversation, Oppenheimer is mentioned by name once: "Joe" says, "Oppie, for instance, thinks that it [the experimental phase] might take as long as a year and a half, if handled . . ." (13; ellipsis points in original).

97. Transcript of conversation between Nelson and "Joe," 8, 19. What actually took place between Nelson and Oppenheimer is unknown. Oppenheimer testified at the 1954 hearings, "I never discussed anything . . . about the atomic bomb with Steve Nelson" ("Transcript," 13). In his autobiography, Nelson denied that he was involved in atomic espionage (*Steve Nelson*, 294–95), and it is important to note that he was never found guilty.

98. Transcript of conversation between Nelson and "Joe," 15, 24, 25. For the discussion of time estimates, see pp. 9 and 13. On the sensitivity of time schedules,

see Lieutenant Colonel Boris T. Pash, memorandum, "Subject: D.S.M. Project, Re: Transcript of Conversation between Dr. J. R. Oppenheimer, Lt. Col. Boris T. Pash and Lt. Lyall Johnson," August 27, 1943, in "Transcript," 863–71, on 867. On Pash's suspicions, see also Herken, *Brotherhood of the Bomb,* 98.

99. Boris T. Pash to Lieutenant Colonel Lansdale, "Subject: Julius Robert Oppenheimer," June 29, 1943, in "Transcript," 821–22. Lansdale thought it likely that Oppenheimer knew he was under surveillance ("Transcript," 264). On Hawkins, see Herken, *Brotherhood of the Bomb,* 101–2.

100. Special agent D. L. Johnson, memorandum, "Subject: Frank Friedman Oppenheimer, Re: Covering Memorandum," July 22, 1943, Investigation Files, box 100, Records of the Office of the Commanding General, Manhattan Project.

101. Stern, *The Oppenheimer Case,* 48–51, quoting 49. Lansdale later said that "our main row" about Lomanitz and other personnel matters was "with Lawrence" ("Transcript," 272).

102. "Transcript," 123, 133; Condon, interviewed by Weiner, September 11 and 12, 1973, 92–93. See also G. Rossi Lomanitz, interviewed by Shawn Mullet, July 26, 2001, Niels Bohr Library, AIP.

103. Colonel Lansdale to General Groves, "Subject: J. R. Oppenheimer," August 12, 1943, in "Transcript," 275–76, quoting 276. Lomanitz had complained to Oppenheimer that he was being targeted because of his union organizing (Herken, *Brotherhood of the Bomb,* 107, 110). In the spring of 1942, Lomanitz had done important theoretical work under Oppenheimer on designing a new electromagnetic separation method. Stern, *The Oppenheimer Case,* 49; Condon, interviewed by Weiner, September 11 and 12, 1973, 92.

104. Lansdale to Groves, "Subject: J. R. Oppenheimer," August 12, 1943, 276.

105. Pash, "Transcript of Conversation between Dr. J. R. Oppenheimer, Lt. Col. Boris T. Pash, and Lt. Lyall Johnson," August 27, 1943, 863–65.

106. Ibid., 864–65. Oppenheimer's naming of Chevalier was first reported in a wire dated December 13, 1943 ("Transcript," 6, 14, 168, and esp. 153).

107. "Transcript," 168.

108. Lieutenant Colonel Boris T. Pash to General Groves, "Re: DSM Project (J. R. Oppenheimer)," September 2, 1943, in "Transcript," 816.

109. Peer de Silva to Lieutenant Colonel Boris T. Pash, "Subject: J. R. Oppenheimer," September 2, 1943, in "Transcript," 273–75, quoting 274.

110. Bird and Sherwin, *American Prometheus,* 198–99, 237–41, 247–48, 356–59. In addition, there is the apparent fabrication in Oppenheimer's account that other people who were approached came to talk with him about these approaches. Bird and Sherwin speculated that Oppenheimer could have been recalling conversations with colleagues who may have been sympathetic to passing information and technology to the Russians (239–40). In his memoir, Chevalier described attending a party at Oppenheimer's Berkeley home after the war and mentioned Lawrence as one of the physicists among the guests "whom I knew" (*Oppenheimer,* 72). But it seems that this

was knowledge at a distance. Even if Chevalier and Lawrence were acquainted via Oppenheimer or the university, this was in no way sufficient to make it conceivable that Chevalier would have discussed Eltenton's proposal with Lawrence. Even if the Lawrence-Alvarez scenario had merely implanted in Oppenheimer's mind the general and vague idea of three approaches, his account to the security officers was nevertheless a fiction.

111. De Silva to Pash, September 2, 1943, 275.

112. Bacher, interviewed by Terrall, June 9, 1981, 80–81.

CHAPTER FOUR

1. Arno Roensch, interviewed by Los Alamos Historical Society (hereafter LAHS), March 21, 1992, LAHM-M1992-112-1-56, Los Alamos Historical Museum (hereafter LAHM), Los Alamos, NM.

2. Rabi, "Introduction," in Rabi et al., *Oppenheimer*, 8; unnamed source, quoted in Howard Gardner, "J. Robert Oppenheimer: The Teaching of Physics, the Lessons of Politics," in *Leading Minds: An Anatomy of Leadership* (New York: Basic Books, 1995), 89–109, on 99; Bethe, "Oppenheimer," 1082; Robert R. Wilson, "The Conscience of a Physicist," in *Alamogordo plus Twenty-five Years*, ed. Richard S. Lewis and Jane Wilson, with Eugene Rabinowitch (New York: Viking Press, 1971), 71; Bradbury, quoted and paraphrased in Goodchild, *J. Robert Oppenheimer*, 278; Bradbury, speech at Los Alamos, February 1971, 5, LAHM (M) 1856 (b)/72.208; Laura Fermi, *Atoms in the Family: My Life with Enrico Fermi* (1954; Albuquerque: University of New Mexico Press, 1982), 205; Enrico Fermi, quoted in Davis, *Lawrence and Oppenheimer*, 182; Teller, quoted in Rhodes, *The Making of the Atomic Bomb*, 570; Edward Teller, *The Legacy of Hiroshima*, with Allen Brown (1962; Westport, CT: Greenwood Press, 1975), 13.

3. Untitled address, "Los Alamos Correspondence, 1942–46," Oppenheimer Papers, quoted in Jon Hunner, *Inventing Los Alamos: The Growth of an Atomic Community* (Norman: University of Oklahoma Press, 2004), 94.

4. See W. I. Thomas, "Situational Analysis: The Behavior Pattern and the Situation" (1927), in Thomas, *On Social Organization and Social Personality*, ed. Morris Janowitz, 154–67 (Chicago: University of Chicago Press, 1966); Peter McHugh, *Defining the Situation: The Organization of Meaning in Social Interaction* (New York: Bobbs-Merrill Co., 1968); Thorpe and Shapin, "Who Was J. Robert Oppenheimer?" esp. 547–63; and Steven Shapin, "The House of Experiment in Seventeenth-Century England," *Isis* 79 (1988), esp. 390.

5. Alice Kimball Smith, "Los Alamos: Focus of an Age," in Lewis and Wilson, *Alamogordo plus Twenty-five Years*, 33–46, quoting 34.

6. Ruth Marshak, "Secret City," in Wilson and Serber, *Standing By and Making Do*, 1–19, quoting 5. See also Segrè, *A Mind Always in Motion*, 181; and War Department, "Complications of the Los Alamos Project," 8.

7. Dudley, "Ranch School to Secret City," in Badash, Hirschfelder, and Broida, *Reminiscences of Los Alamos, 3–4.*

8. Hales, *Atomic Spaces,* 57–60; Peter Bacon Hales, "Topographies of Power: The Forced Spaces of the Manhattan Project," in *Mapping American Culture,* ed. Wayne Franklin and Michael Steiner, 251–90 (Iowa City: University of Iowa Press, 1992); Jon Hunner, "Family Secrets: The Growth of Community at Los Alamos, New Mexico, 1943–1957" (PhD diss., University of New Mexico, 1996), 20–22; Hunner, *Inventing Los Alamos,* 16–17; Fermor S. Church and Peggy Pond Church, *When Los Alamos Was a Ranch School* (Los Alamos, NM: Los Alamos Historical Society, 1974); Hal K. Rothman, *On Rims and Ridges: The Los Alamos Area since 1880* (Lincoln: University of Nebraska Press, 1992), 209.

9. Françoise Ulam, interviewed in Jon Else's film *The Day after Trinity;* Otto Frisch, "Impressions of Los Alamos during the Manhattan Project," interview transcript, June 19, 1973, LAHM (M) 1845 (b)/68.5, 5.

10. Phyllis Fisher, *Los Alamos Experience* (New York: Japan Publications, 1985), 26. See also Marshak, "Secret City," 4; and Wilson, "Not Quite Eden," in Wilson and Serber, *Standing By and Making Do,* 43.

11. Smith and Weiner, *Robert Oppenheimer,* 253; Gregg Herken, "The University of California, the Federal Weapons Labs, and the Founding of the Atomic West," in *The Atomic West,* ed. Bruce Hevly and John M. Findlay (Seattle: University of Washington Press, 1998), 130 n. 1. See also Smith and Weiner, *Robert Oppenheimer,* 247.

12. Joseph P. Masco, "Nuclear Borderlands: The Legacy of the Manhattan Project in Post–Cold War New Mexico" (PhD diss., University of California, San Diego, 1999), 68, 97 n. 2, 146–47; Marshak, "Secret City," 3.

13. Rhodes, *The Making of the Atomic Bomb,* 444; Peggy Pond Church, *The House at Otowi Bridge: The Story of Edith Warner and Los Alamos* (Albuquerque: University of New Mexico Press, 1960), 86; Oppenheimer, quoted in Rhodes, *The Making of the Atomic Bomb,* 451.

14. Rothman, *On Rims and Ridges,* 209. See also Brode, *Tales of Los Alamos,* 15; Marshak, "Secret City," 4; and Canaday, *The Nuclear Muse,* 151–59.

15. Elsie Blumer McMillan, *The Atom and Eve* (New York: Vantage Press, 1995), 5.

16. Charles H. Bagley, interviewed by the author, June 22, 1998.

17. Jacob J. Wechsler, interviewed by the author, December 18, 1997.

18. Marshak, "Secret City," 2–3.

19. Dorothy McKibben, "109 East Palace," in Wilson and Serber, *Standing By and Making Do,* 21–27; Marshak, "Secret City," 8.

20. Marshak, "Secret City," 7–8. See also Rachel Fermi and Esther Samra, *Picturing the Bomb: Photographs from the Secret World of the Manhattan Project* (New York: Harry N. Abrams, 1995), 34. The procedure is also discussed in Governing Board Minutes, August 19, 1943, LANL A-83-013 1-22, Los Alamos National Laboratory Archives (hereafter LANL).

21. This policy was suggested to Groves by Oppenheimer. Oppenheimer to Groves, November 2, 1943, Decimal Files, 1942–48, Record Group (RG) 77, National Archives and Records Administration.

22. Serber, *Peace and War,* 79. See also Jette, *Inside Box 1663,* 34.

23. Alice Kimball Smith, "Law and Order," in Wilson and Serber, *Standing By and Making Do,* 73–87, quoting 74; Manhattan District History, bk. 8, Los Alamos Project (Y), vol. 1, General, 7.15; "Manhattan District History," 1942–46, Records of the Office of the Chief of Engineers, RG 77.

24. Governing Board Minutes, May 6, 1943, LANL A-83-013 1-3; Governing Board Minutes, September 30, 1943, LANL A-83-013 1-27; Mason, *Children of Los Alamos,* 11; Shirley B. Barnett, "Operation Los Alamos," in Wilson and Serber, *Standing By and Making Do,* 89–101, on 89.

25. Mason, *Children of Los Alamos,* 11; "Memorandum on the Los Alamos Project" [ca. 1943], in Jette, *Inside Box 1663,* 125–28; Hedy Dunn, Los Alamos Historical Society, personal communication to the author, May 6, 2003.

26. Governing Board Minutes, May 10, 1943, LANL A-83-013 1-4; Governing Board Minutes, October 21, 1943, LANL A-83-013 1-30.

27. Governing Board Minutes, October 28, 1943, LANL A-83-013 1-31. See also Governing Board Minutes, September 23, 1943, LANL A-83-013 1-26; "Censorship Regulations" [ca. 1943], in Jette, *Inside Box 1663,* 129–32; Mason, *Children of Los Alamos,* 10; Governing Board Minutes, May 8, 1943, LANL A-83-013 1-20; and Eleanor Roensch, interviewed by LAHS, March 21, 1992, LAHM-M1992-112-1-55.

28. McMillan, *The Atom and Eve,* 15; Wilson, "Not Quite Eden," 44–45; "Memorandum on the Los Alamos Project," 127. See also Hales, "Topographies of Power," 274; and Smith, "Los Alamos," 37.

29. Oppenheimer to Conant, November 30, 1942, in Smith and Weiner, *Robert Oppenheimer,* 240–42, quoting 242.

30. Bacher, interviewed by Mary Terrall, June 9, 1981, 76; Governing Board Minutes, May 3, 1943, LANL A-83-013 1-2; Governing Board Minutes, July 15, 1943, LANL A-83-013 1-17; Oppenheimer to Bacher, April 28, 1943, in Smith and Weiner, *Robert Oppenheimer,* 254–55, quoting 255.

31. Oppenheimer to Ernest Lawrence, May 19, 1942, quoted in Hunner, *Inventing Los Alamos,* 31; Dudley, "Ranch School to Secret City," 3; Goodchild, *J. Robert Oppenheimer,* 71; Hales, *Atomic Spaces,* 57; "Transcript," 12; Hugh Bradner, personal communication to the author, April 2000. See also Peter Goodchild, *Edward Teller: The Real Dr. Strangelove* (Cambridge, MA: Harvard University Press, 2004), 75.

32. Captain Musser to Colonel Tyler, "Post Population," February 5, 1945, file 091.4, Peoples (Census), Decimal Correspondence Relating to Los Alamos Laboratory, Records of the Santa Fe (NM) Engineer Office, Records of the Office of the Commanding General, Manhattan Project, RG 77.

33. Manhattan District History, bk. 8, Los Alamos Project (Y), vol. 1, General, 7.15, and Appendix B-1: "Hospital Statistics—Los Alamos, New Mexico." See also

Jones, *Manhattan,* 502; and see the rougher estimates in "Transcript," 12; and Fermi, *Atoms in the Family,* 206.

34. David Hawkins, "Toward Trinity," part 1 of *Project Y: The Los Alamos Story* (Los Angeles: Tomash Publishers, 1983), 483 (this official history was written in 1947 and declassified in 1961); Stafford L. Warren, Colonel, Medical Corps, to Major General L. R. Groves, June 22, 1944, "Subject: Hospital Requirements at 'Y,'" 2, Ex 31, bk. 2, Manhattan District History, bk. 8, Los Alamos Project (Y), series 1, vol. 9, bk. 2 (Exhibits), undated, LANL A-85-002 2-2.

35. Jean Bacher, "Fresh Air and Alcohol," in Wilson and Serber, *Standing By and Making Do,* 103–15, quoting 103, 112. See also John Balagna, interviewed by the author, January 13, 1998.

36. Quoting Marshak, "Secret City," 11. See also Barnett, "Operation Los Alamos," 91.

37. Governing Board Minutes, May 24, 1943, LANL A-83-013 1-8.

38. Eric Kent Clarke, MD, to Colonel S. L. Warren, "Mental Hygiene Survey at 'Y,' August 23–27, 1944," August 29, 1944, folder 330.11, Morale, Decimal Correspondence Relating to Los Alamos Laboratory, 1942–48. See also Hales, *Atomic Spaces,* 223–24.

39. Mason, *Children of Los Alamos,* 8. See also Hunner, "Family Secrets," xviii; and Hales, *Atomic Spaces,* 112.

40. Mason, *Children of Los Alamos,* 8.

41. Barnett, "Operation Los Alamos," 101; Fermi, *Atoms in the Family,* 229.

42. Tyler, "Resume of Instructions," October 1944, quoted in Lenore Fine and Jesse A. Remington, *The Corps of Engineers: Construction in the United States* (Washington, DC: Office of the Chief of Military History, United States Army, 1972), 699.

43. Smith, "Law and Order"; Weisskopf, *The Joy of Insight,* 141.

44. Governing Board Minutes, May 24, 1943, LANL A-83-013 1-8.

45. Ellen Herman, *The Romance of American Psychology: Political Culture in the Age of Experts* (Berkeley and Los Angeles: University of California Press, 1995), esp. 29, 69–72; Morris Janowitz, "Military Elites and the Study of War," in *War: Studies from Psychology, Sociology, Anthropology,* ed. Leon Bramson and George W. Goethals (New York: Basic Books, 1968), esp. 354. See also Samuel A. Stouffer et al., *The American Soldier,* vol. 1, *Adjustment during Army Life* (1949; New York: Science Editions, 1965).

46. War Department, "Complications of the Los Alamos Project," 4, 8.

47. Bradbury, speech at Los Alamos, February 1971, 3, 18; Eric Kent Clarke, MD, "Psychiatric Problems in the Community at Y," May 2, 1945, file 702, Medical Examinations, Decimal Correspondence Relating to Los Alamos Laboratory; Clarke, "Mental Hygiene Survey at 'Y.'"

48. See Albert B. Christman, *Target Hiroshima: Deak Parsons and the Creation of the Atomic Bomb* (Annapolis, MD: Naval Institute Press, 1998).

49. Jones, *Manhattan,* 485–86; Herken, *Brotherhood of the Bomb,* 76; War Department, "Complications of the Los Alamos Project," 3.

50. Jones, *Manhattan,* 502; Hawkins, "Toward Trinity," 484–86. See also Ruth H. Howes and Caroline L. Herzenberg, *Their Day in the Sun: Women of the Manhattan Project* (Philadelphia: Temple University Press, 1999), 86, 132, 148–51, 166.

51. For further discussion of implosion, see chapter 5.

52. Kistiakowsky to Oppenheimer, "Organization of Field Sites," July 28, 1944, LANL A-84-019 4-1.

53. Conference with Senior Staff Officers [ca. 1945], 3–4, file 337, Staff Conferences, Decimal Correspondence Relating to Los Alamos Laboratory; Clarke, "Mental Hygiene Survey at 'Y.'" See also Hoddeson et al., *Critical Assembly,* 98.

54. McAllister H. Hull Jr., interviewed by LAHS, March 24, 1992, LAHM-M1992-112-1-64; Bagley, interviewed by the author, June 22, 1998; Bagley, interviewed by LAHS, October 25, 1991, LAHM-M1992-112-1-1; Roger Rasmussen, interviewed by LAHS, January 11, 1992, LAHM-M1992-112-1-26; Roy Merryman, interviewed by LAHS, December 14, 1991, LAHM-M1992-112-1-18; Miriam White Campbell, interviewed by LAHS, December 9, 1991, LAHM-M1992-112-1-15.

55. Clarke, "Mental Hygiene Survey at 'Y.'"

56. Conant and Groves to Oppenheimer, February 25, 1943, in Hawkins, "Toward Trinity," 495–97, quoting 495.

57. Hales, *Atomic Spaces,* 119. See, for example, the organization chart printed in Fine and Remington, *The Corps of Engineers,* 678. See also Robert S. Norris, *Racing for the Bomb: General Leslie R. Groves, the Manhattan Project's Indispensable Man* (South Royalton, VT: Steerforth Press, 2002).

58. Colonel Nichols, memorandum [to key officers in the District], "Subject: Information concerning Compliance with Instructions," April 22, 1944, box 1, file 000, Decimal Correspondence Relating to Los Alamos Laboratory, 1942–48.

59. Groves, interviewed by J. J. Ermenc, November 7, 1967, in Ermenc, *Atomic Bomb Scientists: Memoirs, 1939–1945* (Westport, CT: Meckler, 1984), 15, 32. See also Norris, *Racing for the Bomb,* 278.

60. Groves, quoted in Ermenc, *Atomic Bomb Scientists,* 33–34; Hewlett and Anderson, *The New World,* 333–35; Norris, *Racing for the Bomb,* 334.

61. Groves, memorandum, "What Occurred during a Meeting at the White House," December 30, 1944; "Notes to Aid in the Preparation of the Agenda for the Meeting between the President and the Secretary of War, with Major General Groves," n.d., both in Correspondence ("Top Secret") of the Manhattan Engineer District, 1942–46, Records of the Office of the Chief of Engineers, National Archives Microfilm Publication M150 (hereafter MED Correspondence), microfilm roll 3. See also Fine and Remington, *The Corps of Engineers,* 661–63.

62. Nichols, *The Road to Trinity,* 60–61.

63. Groves and Lawrence, quoted in Edward Teller, *Memoirs: A Twentieth-Century Journey in Science and Politics,* with Judith L. Shoolery (Oxford: Perseus

Press, 2001), 204; Hughes, *American Genesis*, 409–10; Norris, *Racing for the Bomb*, 205.

64. Groves, *Now It Can Be Told*, 140. See also Stanley Goldberg, "Groves and the Scientists: Compartmentalization and the Building of the Bomb," *Physics Today* 48, no. 8 (August 1995): 38–43; Hewlett and Anderson, *The New World*, 227–28; and Norris, *Racing for the Bomb*, 254–60.

65. Marshall, quoted in Fine and Remington, *The Corps of Engineers*, 658; Groves, *Now It Can Be Told*, 20. See also Norris, *Racing for the Bomb*, 228, 256.

66. Groves, *Now It Can Be Told*, 140. See also "Transcript," 164.

67. Compton, *Atomic Quest*, 130; Hales, *Atomic Spaces*, 118.

68. Groves, quoted in Ermenc, *Atomic Bomb Scientists*, 41; Sherwin, *A World Destroyed*, 53–63, esp. 61–62.

69. Szilard to Compton, November 25, 1942, "[Re] Compartmentalization of Information and the Effect of Impurities of 49," in *Leo Szilard: His Version of the Facts; Selected Recollections and Correspondence*, ed. Spencer R. Weart and Gertrud Weiss Szilard (Cambridge, MA: MIT Press, 1978), 160–61, quoting 160.

70. Eugene Wigner, "Memoir of the Uranium Project," in *The Collected Works of Eugene Paul Wigner*, vol. 5, *Nuclear Energy, Part A: The Scientific Papers*, ed. Alvin M. Weinberg with Alfred M. Perry (Berlin: Springer-Verlag, 1992), 79.

71. Groves, Memorandum for the Secretary of War, "Subject: Leo Szilard," October 29, 1945, and attachment, MED Correspondence, microfilm roll 2; Rhodes, *The Making of the Atomic Bomb*, 506.

72. Stanley Goldberg, "General Groves and the Atomic West: The Making and Meaning of Hanford," in *The Atomic West*, ed. Bruce Hevly and John M. Findlay (Seattle: University of Washington Press, 1998), 46; Goldberg, "Groves and the Scientists," 38.

73. Groves, quoted in Ermenc, *Atomic Bomb Scientists*, 43.

74. Cf. Richard Whitley, *The Intellectual and Social Organization of the Sciences* (Oxford: Clarendon Press, 1984).

75. Nichols, *The Road to Trinity*, 104; "Transcript," 172. See also Norris, *Racing for the Bomb*, 244–45.

76. Hewlett and Anderson, *The New World*, 228–29.

77. See, for example, Oppenheimer to Bacher, June 10, 1942, in Smith and Weiner, *Robert Oppenheimer*, 225.

78. Sociologist Diane Vaughan calls this de facto compartmentalization "structural secrecy." *The Challenger Launch Decision: Risky Technology, Culture and Deviance at NASA* (Chicago: University of Chicago Press, 1996), 238–77; Diane Vaughan, "The Role of the Organization in the Production of Techno-scientific Knowledge," *Social Studies of Science* 29 (1999), esp. 916–18.

79. Governing Board Minutes, November 11, 1943, LANL A-83-013 1-33; Minutes of the Meetings of the Liaison Committee, January 11 and February 11, 1944, LANL A-83-013 3-25. On Teller's role in liaison, see also Teller, *Memoirs*, 178.

80. Governing Board Minutes, May 6, 1943, LANL A-83-013 1-3. See also Teller, *Memoirs*, 172.

81. Governing Board Minutes, August 5, 1943, LANL A-83-013 1-20.

82. Governing Board Minutes, October 28, 1943, LANL A-83-013 1-31.

83. Ibid.

84. See, for example, Governing Board Minutes, March 23, 1944, LANL A-83-013 1-45; and Oppenheimer to Groves (teletype), June 19, 1943, Decimal Files, 1942–48.

85. Governing Board Minutes, June 10, 1943, LANL A-83-013 1-12; Governing Board Minutes, October 23, 1943, LANL A-83-013 1-29; Governing Board Minutes, November 4, 1943, LANL A-83-013 1-32; Oppenheimer to Groves, "Liaison with Site X," October 4, 1943, file 001, Decimal Files, 1942–48.

86. Individual exceptions were occasionally made. Berlyn Brixner, interviewed by the author, January 14, 1998.

87. McAllister Hull Jr., interviewed by the author, January 16, 1998.

88. Wechsler, interviewed by the author, December 18, 1997.

89. Ibid.; Bagley, interviewed by the author, June 22, 1998; Hull, interviewed by the author, January 16, 1998.

90. Hawkins, interviewed by the author, November 16, 1998.

91. Governing Board Minutes, May 31, 1943, LANL A-83-013 1-10; Governing Board Minutes, August 5, 1943, LANL A-83-013 1-20; Hawkins, "Toward Trinity," 30; Groves, *Now It Can Be Told*, 167; Philip Morrison, interviewed by the author, December 19, 2001; Bethe, "Oppenheimer," 1082; Victor Weisskopf, interviewed by Lillian Hoddeson and Gordon Baym, March 10, 1978, 16–17, LANL TR-78-006.

92. Victor Kumin, interviewed by LAHS, July 26, 1995, LAHM-M1995-1-1; Weisskopf, "The Los Alamos Years," in Rabi et al., *Oppenheimer*, 24; Hawkins, interviewed by the author, November 16, 1998.

93. Bethe, "Oppenheimer," 1082; Alvarez, *Alvarez*, 128; Rudolf E. Peierls, *Atomic Histories* (Woodbury, NY: AIP Press, 1997), 51.

94. Donald E. Hirsch, interviewed by LAHS, May 7, 1997, LAHM-M1997-1-1-1; Morrison, interviewed by the author, December 19, 2001.

95. Hawkins, interviewed by the author, November 16, 1998; Joseph O. Hirschfelder, "The Scientific and Technological Miracle at Los Alamos," in Badash, Hirschfelder, and Broida, *Reminiscences of Los Alamos*, 67–88, quoting 78; Harold Agnew, interviewed by LAHS, November 20, 1991, LAHM-M1992-112-1-5.

96. Bethe, quoted in Rhodes, *The Making of the Atomic Bomb*, 570; Bethe, "Oppenheimer," 1082; Peierls, *Atomic Histories*, 51; Smith and Weiner, *Robert Oppenheimer*, 264.

97. Weisskopf, interviewed by Hoddeson and Baym, March 10, 1978, 2; Stanislaw M. Ulam, *Adventures of a Mathematician* (New York: Scribner, 1976), 147; Weisskopf, "The Los Alamos Years," 24.

98. Weisskopf, interviewed by Hoddeson and Baym, March 10, 1978, 2–3. Weisskopf said during this interview that in his later career as director of the European

nuclear physics laboratory, CERN, he tried to model his leadership style on Oppenheimer's example. See also Weisskopf, *The Joy of Insight*, 133, 230.

99. Wilson, "A Recruit for Los Alamos," 15

100. Wigner, *Recollections*, 245–46.

101. Tuck, quoted in Davis, *Lawrence and Oppenheimer*, 187; Peierls, *Atomic Histories*, 51; Bethe, "Oppenheimer," 1082.

102. Peierls, *Atomic Histories*, 59, 51; Hughes, quoted in Davis, *Lawrence and Oppenheimer*, 184; Jungk, *Brighter Than a Thousand Suns*, 125; Cyril Stanley Smith, interviewed by LAHS, May 29, 1992, LAHM-M 1992-112-1-7; Teller, quoted in Davis, *Lawrence and Oppenheimer*, 129; Hawkins, interviewed by the author, November 16, 1998.

103. Jungk, *Brighter Than a Thousand Suns*, 128; Ed Doty to his parents, August 7, 1945, LAHM.

104. Kunetka, *City of Fire*, 57–58; Hirschfelder, "Scientific and Technological Miracle," 78; Bernice Brode, "Tales of Los Alamos," in Badash, Hirschfelder, and Broida, *Reminiscences of Los Alamos*, 133–60, quoting 142. The baby was Katherine "Toni" Oppenheimer, born in the first year of the project. Goodchild, *J. Robert Oppenheimer*, 127.

105. Smith, "Los Alamos," 38; Brode, *Tales of Los Alamos*, 97. See also Jungk, *Brighter Than a Thousand Suns*, 132.

106. Hull, interviewed by the author, January 16, 1998.

107. Hawkins, interviewed by the author, November 16, 1998. On the history of these dualisms, see Steven Shapin and Barry Barnes, "Head and Hand: Rhetorical Resources in British Pedagogical Writing, 1770–1850," *Oxford Review of Education* 2, no. 3 (1976): 231–54.

108. Cf. Steven Shapin, "Invisible Technicians: Masters, Servants, and the Making of Experimental Knowledge," in *A Social History of Truth*, 355–407; Chandra Mukerji, *A Fragile Power: Scientists and the State* (Princeton, NJ: Princeton University Press, 1989), 125–45; Stephen Barley and Beth Bechky, "In the Backrooms of Science: The Work of Technicians in Science Labs," *Work and Occupations* 21 (1994): 85–126; Benjamin Sims, "Concrete Practices: Testing in an Earthquake-Engineering Laboratory," *Social Studies of Science* 29, no. 4 (August 1999), esp. 493.

109. See Hoddeson et al., *Critical Assembly*, 160; Jennifer Light, "When Computers Were Women," *Technology and Culture* 40 (1999): 455–83.

110. Nevertheless, Groves saw such visits as important for maintaining morale ("Transcript," 172–73).

111. Charles and Jean Critchfield, interviewed by LAHS, November 9, 1991, LAHM-M1992-112-1-2; Bagley, interviewed by the author, June 22, 1998.

112. Agnew, interviewed by LAHS, November 20, 1991; Eleanor Roensch, interviewed by LAHS, March 21, 1992; Bagley, interviewed by the author, June 22, 1998.

113. Agnew, interviewed by LAHS, November 20, 1991; Hull, interviewed by the author, January 16, 1998.

114. Groves, quoted in Ermenc, *Atomic Bomb Scientists,* 44; War Department, "Complications of the Los Alamos Project," 8–9.

115. George Kistiakowsky, interviewed by LANL, April 1981, LANL TR-81-010; Raemer Schreiber, interviewed by LAHS, March 21, 1992, LAHM-M1992-112-1-57.

116. Marjorie Ulam, interviewed by LAHS, November 9, 1991, LAHM-M1992-1-3; Robert Serber, *The Los Alamos Primer: The First Lectures on How to Build an Atomic Bomb,* ed. Richard Rhodes (Berkeley and Los Angeles: University of California Press, 1992), xxx; Teller, quoted in Stanley A. Blumberg and Gwinn Owens, *Energy and Conflict: The Life and Times of Edward Teller* (New York: G. P. Putnam's Sons, 1976), 126. See also Goodchild, *Edward Teller,* 88–91.

117. Teller to Mayer, n.d., Papers of Maria Goeppert Mayer, Mandeville Special Collections Library, University of California, San Diego. The use of the phrase "physicist-reservation" suggests that this letter was written after August 1945, since it would have been censored during the war. Some of Teller's postwar letters to Mayer are extracted in Teller, *Memoirs,* 258, 275–76, 279, 283–84, 290, 305–6, 355, 434–35.

118. Teller to Mayer, n.d., Mayer Papers.

119. Teller to Mayer, n.d., Mayer Papers.

120. Teller to Mayer [April 15, 1945], Mayer Papers.

121. Teller to Mayer, n.d., Mayer Papers.

122. Teller to Mayer [ca. 1945], Mayer Papers.

123. Teller and Bethe, quoted in Blumberg and Owens, *Energy and Conflict,* 128–29; Weisskopf, *The Joy of Insight,* 134–35.

124. Blumberg and Owens, *Energy and Conflict,* 128.

125. Bethe, quoted in Blumberg and Owens, *Energy and Conflict,* 128; see also Teller, *Memoirs,* 159–61.

126. Bethe, quoted in Rhodes, *The Making of the Atomic Bomb,* 453; Teller, quoted in Blumberg and Owens, *Energy and Conflict,* 126, 135, and in Davis, *Lawrence and Oppenheimer,* 129; Teller, interviewed by the author, June 20, 1998. See also Teller, *Memoirs,* esp. 163, 379.

127. Teller, quoted in Blumberg and Owens, *Energy and Conflict,* 136; see also Teller, *Memoirs,* 180–81.

128. Quoted in Blumberg and Owens, *Energy and Conflict,* 130.

129. Edward Teller, "Seven Hours of Reminiscences," *Los Alamos Science* 4, no. 7 (Winter/Spring 1983): 192; Teller to Szilard, July 2, 1945, in Weart and Szilard, *Leo Szilard,* 208–9; Teller, *Memoirs,* 204–8; Goodchild, *Edward Teller,* 102–3.

130. Teller, interviewed by the author, June 20, 1998.

131. Smith and Weiner, *Robert Oppenheimer,* 256; Teller to Mayer [ca. early 1947], Mayer Papers, printed in part in Teller, *Memoirs,* 264.

132. Hawkins, "Toward Trinity," 184–87; Hoddeson et al., *Critical Assembly,* 203–4.

133. Hawkins, interviewed by the author, November 16, 1998.

134. Parsons to Oppenheimer, "Organization," September 7, 1944, Oppenheimer Papers, box 56.

135. Oppenheimer to Parsons, "Organization," September 15, 1944, LANL A-84-019 34-12.

136. Manley, quoted in Smith and Weiner, *Robert Oppenheimer*, 263.

137. Hewlett and Anderson, *The New World*, 380–92; Lawrence Badash, *Scientists and the Development of Nuclear Weapons: From Fission to the Limited Test Ban Treaty, 1939–1963* (Atlantic Highlands, NJ: Humanities Press International, 1995), 53.

138. Groves, Memorandum for the Secretary of War, "Subject: The Test," July 18, 1945, MED Correspondence, microfilm roll 1; Brigadier General Thomas F. Farrell, quoted in Groves, Memorandum for the Secretary of War, July 18, 1945.

CHAPTER FIVE

1. Hawkins, *Project Y*, xviii; Thomas P. Hughes, "Technological Momentum in History: Hydrogenation in Germany, 1898–1933," *Past and Present* 44 (1969): 106–32; Hughes, *American Genesis*, esp. 381–442, 459–61.

2. Cf. E. P. Thompson, "Time, Work-Discipline and Industrial Capitalism," *Past and Present* 38 (1967): 56–97; Eviatar Zerubavel, *Hidden Rhythms: Schedules and Calendars in Social Life* (Chicago: University of Chicago Press, 1981).

3. Hales, *Atomic Spaces*.

4. Bohr, quoted in Rhodes, *The Making of the Atomic Bomb*, 500; and Teller, *Memoirs*, 186.

5. Stanislaw Ulam, H. W. Kuhn, A. W. Tucker, and Claude E. Shannon, "John von Neumann, 1903–1957," in Fleming and Bailyn, *The Intellectual Migration*, 264.

6. Roensch, interviewed by LAHS, March 21, 1992.

7. Groves, *Now It Can Be Told*, 167. See also Groves to Tolman, April 16, 1943, MED Correspondence, microfilm roll 5.

8. Hawkins, *Project Y*, 17.

9. Hoddeson et al., *Critical Assembly*, 251.

10. However, the pile was shut down almost immediately because of a problem with fission product contamination, which delayed full operation until December. Plutonium from Hanford began to arrive at Los Alamos in February 1945. See Rhodes, *The Making of the Atomic Bomb*, 557–60; and Hoddeson et al., *Critical Assembly*, 290–91. On Hanford, see also Harry Thayer, *Management of the Hanford Engineer Works in World War Two: How the Corps, DuPont and the Metallurgical Laboratory Fast Tracked the Original Plutonium Works* (New York: ASCE Press, 1996).

11. In July 1944, the work of the Governing Board was divided between an Administrative Board and a Technical Board. The work of the Technical Board was quickly taken over by other senior committees, and the Administrative Board continued the

Governing Board's functions of organizational planning and workforce management. Hoddeson et al., *Critical Assembly,* 247.

12. Administrative Board Minutes, August 17, 1944, LANL A-83-013 1-54.

13. Administrative Board Minutes, September 28, 1944, LANL A-83-013 1-56. (The minutes refer, in error, to "Colonel Kilpatrick.") Kirkpatrick's visit of September 20–October 1, 1944, was also recorded in the post's log of arrivals and departures (folder 280.2, Visits, Decimal Files, 1942–48, General Correspondence, 1942–48, General Administrative Files, Records of the Office of the Commanding General, Manhattan Project, RG 77). See also Kunetka, *City of Fire,* 114; Jones, *Manhattan,* 526–27; and Norris, *Racing for the Bomb,* 322–24. The Pacific Mariana island of Tinian was chosen in February 1945 as the base from which the atomic bomb attack against Japan would be launched (Jones, *Manhattan,* 524).

14. Brode, *Tales of Los Alamos,* 33.

15. Fermi, *Atoms in the Family,* 229; Brode, *Tales of Los Alamos,* 37; Groves, in "Transcript," 167.

16. Serber, *Peace and War,* 83; Marshak, "Secret City," in Wilson and Serber, *Standing By and Making Do,* 10–11.

17. Marshak, "Secret City," 10; Mitchell, quoted in Davis, *Lawrence and Oppenheimer,* 183; Lillian Hoddeson, "Mission Change in the Large Laboratory: The Los Alamos Implosion Program, 1943–1945," in *Big Science: The Growth of Large-Scale Research,* ed. Peter Galison and Bruce Hevly (Stanford, CA: Stanford University Press, 1992), 286.

18. Davis, *Lawrence and Oppenheimer,* 169; Rhodes, *The Making of the Atomic Bomb,* 466; Hawkins, *Project Y,* 10; Hoddeson et al., *Critical Assembly,* 69–75, 268; Serber, *The Los Alamos Primer,* esp. 59.

19. Hoddeson et al., *Critical Assembly,* 87; Hawkins, *Project Y,* 22; John H. Manley, "A New Laboratory Is Born," in Badash, Hirschfelder, and Broida, *Reminiscences of Los Alamos,* 21–38, on 34; Parsons, quoted in Rhodes, *The Making of the Atomic Bomb,* 479.

20. Thayer, *Hanford Engineer Works,* 87; Hawkins, *Project Y,* 14; Hoddeson et al., *Critical Assembly,* 87; McMillan, "Early Days at Los Alamos," in Badash, Hirschfelder, and Broida, *Reminiscences of Los Alamos,* 13–19, on 16.

21. Thompson to Oppenheimer, June 25, 1943, and Thompson to Neddermeyer, June 23, 1943, LANL A-84-019 4-1.

22. George Kistiakowsky, quoted in Goodchild, *J. Robert Oppenheimer,* 113; William Higinbotham and Neddermeyer, quoted in Davis, *Lawrence and Oppenheimer,* 169, 172, 219. See also Rhodes, *The Making of the Atomic Bomb,* 467. See also Hoddeson et al., *Critical Assembly,* 245–46.

23. Hawkins, *Project Y,* 125–26; Hoddeson et al., *Critical Assembly,* 130–31; Rhodes, *The Making of the Atomic Bomb,* 480; Jones, *Manhattan,* 506.

24. Rhodes, *The Making of the Atomic Bomb,* 479, 542; Hoddeson et al., *Critical Assembly,* 131, 445 n. 2; Hawkins, *Project Y,* 125.

25. Critchfield, quoted in Hoddeson et al., *Critical Assembly,* 131–32; Neddermeyer, quoted in Davis, *Lawrence and Oppenheimer,* 217. See also Hawkins, *Project Y,* 124–25.

26. Governing Board Minutes, October 28, 1943, LANL A-83-013 1-31; Governing Board Minutes, November 4, 1943, LANL A-83-013 1-32; Hoddeson et al., *Critical Assembly,* 134–35; Hawkins, *Project Y,* 69; Kunetka, *City of Fire,* 86–87. See also "H.E. Program Discussed at Meeting," October 25, 1943, LANL A-84-019 4-1.

27. Governing Board Minutes, September 23, 1943, LANL A-83-013 1-26; Governing Board Minutes, October 28, 1943. See also Hoddeson et al., *Critical Assembly,* 136; and Kunetka, *City of Fire,* 86.

28. Oppenheimer to Conant, November 1, 1943, quoted in Hoddeson et al., *Critical Assembly,* 137; Kistiakowsky to Oppenheimer, "E-5 Group," June 3, 1944, Oppenheimer Papers, box 43; Kistiakowsky, "Reminiscences of Wartime Los Alamos," in Badash, Hirschfelder, and Broida, *Reminiscences of Los Alamos,* 49–65, quoting 49–50. See also Davis, *Lawrence and Oppenheimer,* 218.

29. Kistiakowsky, "Reminiscences of Wartime Los Alamos," 50. Such conflicts are often present in a field's transition to "big science"; see Harry Collins, "LIGO Becomes Big Science," *Historical Studies in the Physical and Biological Sciences* 32, no. 2 (2003): 235–96; and Collins, *Gravity's Shadow: The Search for Gravitational Waves* (Chicago: University of Chicago Press, 2004).

30. Kistiakowsky to Parsons, November 24, 1943, quoted in Hoddeson et al., *Critical Assembly,* 139; Hoddeson et al., *Critical Assembly,* 139–40.

31. In early 1944, an implosion group within the Theoretical Division was established, headed by Teller (Hoddeson et al., *Critical Assembly,* 157). See also Hoddeson, "Mission Change," 273.

32. Hans Bethe, interviewed by Lillian Hoddeson, October 1986, LANL TR-86-0787.

33. Kistiakowsky to Oppenheimer, "E-5 Group," June 3, 1944.

34. Hawkins, *Project Y,* 70; Kunetka, *City of Fire,* 116.

35. Kistiakowsky to Oppenheimer, "E-5 Group," June 3, 1944.

36. Ibid.; Hugh Bradner, interviewed by the author, August 9, 2001.

37. Neddermeyer, quoted in Davis, *Lawrence and Oppenheimer,* 219 (see also 230–31); Kistiakowsky to Oppenheimer, "E-5 Group," June 3, 1944; Oppenheimer to Neddermeyer, quoted in Rhodes, *The Making of the Atomic Bomb,* 547 (quoted also in Kunetka, *City of Fire,* 88); Charles Critchfield, "First Implosion at Los Alamos," in *Behind Tall Fences,* 101–2.

38. Kistiakowsky to Parsons, "Organization of the H.E. Project," June 21, 1944, LANL A-84-019 4-1.

39. Kistiakowsky to Oppenheimer, "Delegation of Authority within the H.E. Project," May 30, 1944, LANL A-84-019 4-1. Group E-5 was designated Implosion Experimentation. Group E-9, under the leadership of physicist Kenneth Bainbridge, was named High Explosive Development. The latter worked on the design and testing

of full-scale high-explosive assemblies (see Hoddeson et al., *Critical Assembly,* 139, 245–46). It is worth noting that Kistiakowsky tended to refer to the entire implosion experimentation and engineering program (including the work of E-5) as the HE Project or HE Program. I follow Kistiakowsky in this, and my references in the text to the HE program or HE research relate to this broader program of activity including, especially, the work of E-5.

40. Hoddeson, "Mission Change," 274–82; Hoddeson et al., *Critical Assembly,* 228–44; Rhodes, *The Making of the Atomic Bomb,* 540–41.

41. Hoddeson, "Mission Change," 279–85.

42. Administrative Board Minutes, July 20, 1944, quoted in Hoddeson et al., *Critical Assembly,* 243.

43. Hawkins, *Project Y,* 208–9, 488.

44. Kistiakowsky to Bainbridge, Bradbury, and Stevens, "Interim Reports in the Explosives Division," August 21, 1944, LANL A-84-019 40-18.

45. "Hyman Rudoff's Memoirs of his Life in Los Alamos during the Manhattan Project" (manuscript, 1991, LAHM M1991-41-1-1).

46. Hawkins, *Project Y,* 157; Christman, *Target Hiroshima,* 85–101, esp. 94.

47. Parsons to Oppenheimer, "Organization," September 7, 1944, Oppenheimer Papers, box 56.

48. Hoddeson et al., *Critical Assembly,* 168–69, 294–99; Davis, *Lawrence and Oppenheimer,* 218; Hull, interviewed by the author, January 16, 1998.

49. Kistiakowsky said that Parsons was not alone in his pessimism about lenses by late 1944 ("Reminiscences of Wartime Los Alamos," 54).

50. Parsons to Groves via Oppenheimer, "Special Report of Ordnance and Engineering Activities of Project Y," September 25, 1944, Papers of Admiral William S. Parsons, Manuscript Division, Library of Congress. I am grateful to Al Christman for providing me with this document.

51. Parsons to Oppenheimer, "Design of Non-Lens Assembly," October 16, 1944, and Oppenheimer, quoted in Hoddeson et al., *Critical Assembly,* 300.

52. Parsons to Oppenheimer, "'Home Stretch' Measures," February 19, 1945, LANL A-84-019 13-3; Hoddeson et al., *Critical Assembly,* 300.

53. Hoddeson et al., *Critical Assembly,* 300. See also John Russell, interviewed by Hoddeson, July 16, 1986, LANL OH-133; and Bethe, interviewed by Hoddeson, October 1986.

54. Kistiakowsky, "Reminiscences of Wartime Los Alamos," 54; Bethe, interviewed by the author, May 12, 1998.

55. Parsons to Oppenheimer, "'Home Stretch' Measures," February 19, 1945, LANL A-84-019 13-3.

56. Hawkins, *Project Y,* 157–58; Hoddeson et al., *Critical Assembly,* 316; Parsons, quoted in Christman, *Target Hiroshima,* 169. The minutes of the Cowpuncher Committee remain classified.

57. Kistiakowsky, quoted in Hoddeson et al., *Critical Assembly,* 315, and in Hoddeson, "Mission Change," 283.

58. Peggy Corbett, "29 Years Ago Los Alamos, New Mexico," part 9, "Pace at Post, Lab Quickens," *Santa Fe New Mexican,* October 16, 1973; Hull, interviewed by the author, January 16, 1998.

59. Groves, *Now It Can Be Told,* 256 n. 2; see also Jones, *Manhattan,* 519–20.

60. Parsons, "Description of Work Performed at Y," June 18, 1945, written in longhand in answer to questionnaire, Manhattan District, Scientific Research and Development Personnel, Captain William Sterling Parsons, June 18, 1945, Parsons Papers; Parsons to Oppenheimer, "Organization," September 7, 1944.

61. Parsons to Groves, "Special Report of Ordnance and Engineering Activities of Project Y," September 25, 1944; Christman, *Target Hiroshima,* 158; Al Christman, personal communication to the author, April 3, 2001. The discussions at Chicago were formalized in the Committee on Political and Social Problems, chaired by the physicist James Franck. Scientist Ralph Lapp noted, "Although historians may date the inception of the Franck Committee as June 2, 1945, it was already in existence as a loosely organized discussion group at the Metallurgical Laboratory and dated back to the previous summer." Ralph E. Lapp, *The New Priesthood: The Scientific Elite and the Uses of Power* (New York: Harper and Row, 1965), 76.

62. Oppenheimer to Groves, October 6, 1944, LANL A-84-019 34-11.

63. Parsons, "Description of Work Performed at Y," June 18, 1945; Parsons to Groves, "Special Report of Ordnance and Engineering Activities of Project Y," September 25, 1944; Christman, *Target Hiroshima,* 167.

64. Parsons to Groves, "Special Report of Ordnance and Engineering Activities of Project Y," September 25, 1944; Christman, *Target Hiroshima,* 158; Parsons, quoted in Christman, *Target Hiroshima,* 164. See also Jones, *Manhattan,* 496; and Hawkins, *Project Y,* 247–58.

65. Parsons to Mr. Carter T. Barron, October 21, 1946, Parsons Papers. Parsons was here complaining about the script for the postwar movie *The Beginning or the End,* which portrayed a fictional accident on Tinian. Work at Tinian is documented in letters from physicist Norman Ramsey to Oppenheimer, in Oppenheimer Papers, box 60. See also Jones, *Manhattan,* 536.

66. General Groves, [undated] notes on meeting with the Under Secretary of War, March 27, 1945, MED Correspondence, microfilm roll 5. On the Manhattan Project mission code-named "Alsos," to gather intelligence on the German atomic bomb effort, see Mark Walker, *German National Socialism and the Quest for Nuclear Power, 1939–1949* (Cambridge: Cambridge University Press, 1989), 153–60; and Samuel A. Goudsmit, *Alsos: The Failure in German Science* (London: Sigma Books, 1947). For the case that Japan had been the official target from early on in the Manhattan Project, see Arjun Makhijani, "'Always' the Target?" *Bulletin of the Atomic Scientists* 31, no. 3 (May–June 1995): 23–27.

67. This is Joseph Rotblat's recollection of Groves's comment. Rotblat said, "Whatever his exact words, his real meaning was clear." "Leaving the Bomb Project," *Bulletin of the Atomic Scientists* 41 (August 1985): 15–19, on 18. See also Norris, *Racing for the Bomb,* 331. Rotblat gave the precise date of the dinner party as March 4, 1944. Rotblat, "We Are on a Slippery Slope, Heading for Disaster," *Guardian,* January 8, 2002, http://www.guardian.co.uk/nuclear/article/0,2763,870939,00.html.

68. Rotblat, "Leaving the Bomb Project," 18–19. Rotblat's discounting of the possibility of a German bomb was based on his view that "the war in Europe would be over before the bomb project was completed . . . If it took the Americans such a long time, then my fear of the Germans being first was groundless" (ibid.).

69. Jones, *Manhattan,* 528–29; Major J. A. Derry and Dr. N. F. Ramsey, Memorandum for General L. R. Groves, "Summary of Target Committee Meetings on 10 and 11 May 1945," May 12, 1945, MED Correspondence, microfilm roll 5. See also Cassidy, *J. Robert Oppenheimer,* 241, 243–44.

70. Derry and Ramsey, Memorandum for General L. R. Groves, May 12, 1945. When Groves presented the secretary of war with the list of targets, Stimson objected to the choice of Kyoto, because of its religious and cultural significance for the Japanese. By July 15, Kyoto had been replaced with Nagasaki on the target list (Jones, *Manhattan,* 529–30).

71. Lapp, *The New Priesthood,* 68–86, quoting 83; Hewlett and Anderson, *The New World* (AEC official history), quoting 199 (see also 200–201, 365–67, 399–400); Matt Price, "Roots of Dissent: The Chicago Met Lab and the Origins of the Franck Report," *Isis* 86 (1995): 222–44; Compton, *Atomic Quest,* 233–36; Jones, *Manhattan,* 532–33; Robert R. Wilson, "Hiroshima: The Scientists' Social and Political Reaction," *Proceedings of the American Philosophical Society* 140, no. 3 (September 1996): 350–57; Smith, *A Peril and a Hope,* esp. 15–16, 41–59.

72. Weisskopf, *The Joy of Insight,* 127–28, 137; Richard Feynman, "The Pleasure of Finding Things Out," *The Listener,* November 26, 1981, 635–36, on 635, quoted also in Brian Easlea, *Fathering the Unthinkable: Masculinity, Scientists and the Nuclear Arms Race* (London: Pluto Press, 1983), 83–84; Rotblat, "Leaving the Bomb Project," 18. See also Richard Feynman, "Los Alamos from Below," in Badash, Hirschfelder, and Broida, *Reminiscences of Los Alamos,* 105–32, esp. 132.

73. Rabi, "Suggestions for Interim Organization and Procedure," with covering letter from Rabi to Oppenheimer, February 10, 1943, Oppenheimer Papers, box 59.

74. Arney, *Experts,* 113; Segrè, *A Mind Always in Motion,* 190; Teller to Mayer [ca. 1945], Mayer Papers.

75. Norris Bradbury, speech at Los Alamos, February 1971, 8–9, LAHM (M) 1856 (b)/72.208.

76. Robert R. Wilson, "Niels Bohr and the Young Scientists," in *Assessing the Nuclear Age: Selections from the "Bulletin of the Atomic Scientists,"* ed. Len Acklund and Steve McGuire (Chicago: Educational Foundation for Nuclear Science, 1986), 38. Wilson said elsewhere that the meeting was held "something like a year after

Los Alamos had started," which would put it in mid- to late 1944. He also admitted to being "hazy now as to who came or what was said." The assembled group, he said, filled a small meeting room ("Conscience of a Physicist," 72; see also Wilson, interview in Palevsky, *Atomic Fragments,* 125–50, on 135–36). Weisskopf referred to such a meeting having taken place in March 1945 (*The Joy of Insight,* 145–46). See also Bird and Sherwin, *American Prometheus,* 287–89.

77. Wilson, "Conscience of a Physicist," 72; Wilson, "Niels Bohr and the Young Scientists," 38–39.

78. Morrison, quoted in Lynn Margulis, "Sunday with J. Robert Oppenheimer," in Lynn Margulis and Dorion Sagan, *Slanted Truths: Essays on Gaia, Symbiosis, and Evolution,* 5–28 (New York: Springer-Verlag, 1997), 6–7.

79. Oppenheimer's Remarks at Memorial Services for President Roosevelt, Los Alamos, April 15, 1945, Oppenheimer Papers, box 262; reprinted also in Smith and Weiner, *Robert Oppenheimer,* 288.

80. "A Petition to the President of the United States," July 17, 1945, Oppenheimer Papers, box 70; Teller to Szilard, July 2, 1945, Oppenheimer Papers, box 71. See also Teller to Oppenheimer, attached to Teller to Szilard, July 2, 1945; Szilard to Oppenheimer, July 10, 1945; Szilard to Oppenheimer, July 23, 1945, all in Oppenheimer Papers, box 70; and Smith, *A Peril and a Hope,* 55–56.

81. Edward Hammel, interviewed by the author, December 8, 1997.

82. Wilson, "Conscience of a Physicist," 71.

83. Ibid., 72–73.

84. Bernard Feld, interviewed by Ian Low, in "Science for Peace," *New Scientist,* July 24, 1975, 208–9; quoted also in Easlea, *Fathering the Unthinkable,* 84.

85. Jette, *Inside Box 1663,* 42. Cf. Erving Goffman, *Asylums: Essays on the Social Situation of Mental Patients and Other Inmates* (New York: Doubleday, 1961).

86. Szilard to Oppenheimer, May 16, 1945, Oppenheimer Papers, box 70.

87. Emilio Segrè, *Enrico Fermi, Physicist* (Berkeley and Los Angeles: University of California Press, 1970), 145; Oppenheimer, in "Transcript," 32–33; both also quoted in Easlea, *Fathering the Unthinkable,* 84–85.

88. Wilson, "Niels Bohr and the Young Scientists," 39.

CHAPTER SIX

1. Harold Cherniss, interviewed by Alice Kimball Smith, April 21, 1976, 6, Oppenheimer Oral History Collection, MIT Archives.

2. "Atomic Doldrums," *Time,* February 25, 1946, 88.

3. Oppenheimer, "The Atom Bomb and College Education," *General Magazine and Historical Chronicle,* University of Pennsylvania General Alumni Society, 1946, quoted in Rhodes, *The Making of the Atomic Bomb,* 676.

4. "Physicist Oppenheimer," *Time,* November 8, 1948, 77. On the history of Oppenheimer's use of the quotation, I am indebted to Hijiya, "The *Gita* of J. Robert

Oppenheimer," esp. 123 n. 3 and 132; and Hijiya, personal communication to the author, December 14, 2002.

5. Jungk, *Brighter Than a Thousand Suns,* 183–84. The book was first published in German in 1956. Oppenheimer repeated the quotation for a 1965 NBC television documentary, *The Decision to Drop the Bomb,* produced by Fred Freed. It was printed in the accompanying book: Len Giovanitti and Fred Freed, *The Decision to Drop the Bomb* (New York: Coward McCann, 1965), 197. The NBC clip was also used in Jon Else's documentary *The Day after Trinity.*

6. Laurence, *Men and Atoms,* 118; William L. Laurence, for the *New York Times,* September 27, 1945, reprinted in Laurence, *The Story of the Atomic Bomb* (Washington, DC: War Department, 1945), 15–17, on 17.

7. William L. Laurence, *Dawn over Zero: The Story of the Atomic Bomb* (London: Museum Press, 1947), 153.

8. Frank Oppenheimer, quoted in Rhodes, *The Making of the Atomic Bomb,* 675.

9. Hewlett and Anderson, *The New World,* 344–45.

10. Ibid., 345; Lapp, *The New Priesthood,* 73.

11. Alperovitz, *The Decision to Use the Atomic Bomb,* 185. See also Makhijani, "'Always' The Target?"

12. Oppenheimer, interviewed by Giovanitti and Freed, October 27, 1964, in *The Decision to Drop the Bomb,* 8, 328; quoted also in Ian Clark, *Limited Nuclear War: Political Theory and War Conventions* (Oxford: Martin Robertson, 1982), 235.

13. See Interim Committee Minutes, May 31, 1945, in Robert C. Williams and Philip L. Cantelon, eds., *The American Atom: A Documentary History of Nuclear Policies from the Discovery of Fission to the Present, 1939–1984* (Philadelphia: University of Pennsylvania Press, 1984), 58–63.

14. Hewlett and Anderson, *The New World,* 367–68; Alperovitz, *The Decision to Use the Atomic Bomb,* 17–46.

15. "Transcript," 34.

16. Alperovitz, *The Decision to Use the Atomic Bomb,* 164–65, quoting 164; and 448–71, esp. 461–62. See also Bird and Sherwin, *American Prometheus,* 293–97; and Cassidy, *J. Robert Oppenheimer,* 249–50.

17. Notes of the Interim Committee Meeting, May 31, 1945, quoted in Alperovitz, *The Decision to Use the Atomic Bomb,* 170.

18. Hewlett and Anderson, *The New World,* 367. On the Scientific Panel's deliberations, see also Sherwin, *A World Destroyed,* 204–6; and Herken, *Brotherhood of the Bomb,* 131–34.

19. Truman-Stalin conversation, Potsdam, July 24, 1945, in Cantelon, Hewlett, and Williams, *The American Atom,* 61–62. Stalin, of course, already had information from spies at Los Alamos—Klaus Fuchs, David Greenglass, and Theodore Hall. Norman Moss, *Klaus Fuchs: The Man Who Stole the Atom Bomb* (London: Grafton Books, 1987); Joseph Albright, *Bombshell: The Secret Story of America's Unknown Atomic Spy Conspiracy* (New York: Times Books, 1997).

20. Alperovitz, *The Decision to Use the Atomic Bomb,* esp. 221–317; Gar Alperowitz, *Atomic Diplomacy: Hiroshima and Potsdam; The Use of the Atomic Bomb and the American Confrontation with Soviet Power* (1965; New York: Penguin Books, 1985); Sherwin, *A World Destroyed;* Barton J. Bernstein, ed., *The Atomic Bomb: The Critical Issues* (Boston: Little, Brown, 1976); Gregg Herken, *The Winning Weapon: The Atomic Bomb in the Cold War, 1945–1950* (Princeton, NJ: Princeton University Press, 1988); J. S. Walker, "The Decision to Use the Atomic Bomb: A Historiographical Update," *Diplomatic History* 14 (1990): 97–114.

21. Wallace, diary, October 19, 1945, quoted in Alperovitz, *The Decision to Use the Atomic Bomb,* 429. See also Bird and Sherwin, *American Prometheus,* 330–31.

22. Wilson, "Hiroshima," 351. See also Goodchild, *J. Robert Oppenheimer,* 169–73; and Palevsky, *Atomic Fragments,* 125–50.

23. Freeman Dyson, *Disturbing the Universe* (New York: Harper Colophon, 1979), 53. See also Goodchild, *J. Robert Oppenheimer,* 180; and Donald Strickland, *Scientists in Politics: The Atomic Scientists Movement, 1945–46* (Lafayette, IN: Purdue University Studies, 1968), 8–9.

24. Brian Balogh, *Chain Reaction: Expert Debate and Public Participation in American Commercial Nuclear Power, 1945–1975* (Cambridge: Cambridge University Press, 1991), 38. See also Lilienthal, *The Atomic Energy Years,* 127; Steven M. Neuse, *David E. Lilienthal: The Journey of an American Liberal* (Knoxville: University of Tennessee Press, 1996), 168; and Paul Boyer, *By the Bomb's Early Light: American Thought and Culture at the Dawn of the Atomic Age* (New York: Pantheon, 1985), 109–21.

25. See Felix Frankfurter–Niels Bohr papers, in Oppenheimer Papers, box 34; Sherwin, *A World Destroyed,* 91–114; Sherwin, "Niels Bohr"; Gowing, "Niels Bohr and Nuclear Weapons"; Aage Bohr, "The War Years and the Prospects Raised by the Atomic Weapons," in *Niels Bohr: His Life and Work as Seen by His Friends and Colleagues,* ed. S. Rozenthal, 191–214 (Amsterdam: North-Holland Publishing Co., 1967); and Rhodes, *The Making of the Atomic Bomb,* 524–38.

26. Niels Bohr, "Science and Civilization," *Times* (London), August 11, 1945, reprint in Oppenheimer Papers, box 21; Bohr, "A Challenge to Civilization," *Science* 102 (1945): 363–64.

27. Oppenheimer, unedited version of 1963 lecture "Niels Bohr and His Times." The finished version read, "He made the enterprise seem hopeful, when many were not free of misgiving." Both versions are quoted in Rhodes, *The Making of the Atomic Bomb,* 524.

28. Strickland, *Scientists in Politics,* 38; Smith, *A Peril and a Hope,* 75–122, 128, 279–81; Alice Kimball Smith, "Scientists and the Public Interest, 1945–46," *Newsletter on Science, Technology, and Human Values* 24 (June 1978): 24–32; Paul J. Piccard, "Scientists and Public Policy: Los Alamos, August–November, 1945," *Western Political Quarterly* 18, no. 2 (June 1965): 251–62. See also Edward Shils, "Freedom and

Influence: Observations on the Scientists' Movement in the United States," *Bulletin of the Atomic Scientists* 8 (January 1957): 13–18.

29. Smith, *A Peril and a Hope*, 115–16; Wilson, "Hiroshima," 353.

30. Smith, *A Peril and a Hope*, 117; William Higinbotham, notes on a conversation with Oppenheimer, quoted in Piccard, "Scientists and Public Policy," 255; Strickland, *Scientists in Politics*, 38–43.

31. Smith, *A Peril and a Hope*, 116–19.

32. Oppenheimer to Bradbury, September 28, 1945, quoted in Smith, *A Peril and a Hope*, 118–19.

33. Ibid., 119.

34. Higinbotham's meeting notes, September 28, 1945, quoted in Smith, *A Peril and a Hope*, 119–20; ALAS Papers, quoted in Strickland, *Scientists in Politics*, 42.

35. Wilson, "Hiroshima," 353. The statement, released to the press on October 14, 1945, is in Oppenheimer Papers, box 171.

36. Smith, *A Peril and a Hope*, 128.

37. Badash, *Scientists and the Development of Nuclear Weapons*, 64; Hewlett and Anderson, *The New World*, 428–55.

38. Smith, *A Peril and a Hope*, 140–41.

39. Anderson to Higinbotham, October 11, 1945, quoted in Smith, *A Peril and a Hope*, 139–40. See also Higinbotham to Anderson, October 16, 1945, Records of the Association of Los Alamos Scientists, Department of Special Collections, Joseph Regenstein Library, University of Chicago.

40. Oppenheimer, Fermi, and Lawrence to Patterson, October 11, 1945, quoted in Smith, *A Peril and a Hope*, 142–43.

41. Smith and Weiner, *Robert Oppenheimer*, 310–11.

42. Strickland, *Scientists in Politics*, 46–47.

43. Read into the testimony by Senator Fulbright, *Hearings on Science Legislation (S. 1297 and Related Bills), Hearings before a Subcommittee of the Committee on Military Affairs*, U.S. Senate, 79th Congress, 1st session, pt. 2, October 15, 16, 17, 18, and 19, 1945 (Washington, DC: U.S. Government Printing Office, 1945), 297–318, quoting 318; see also Smith and Weiner, *Robert Oppenheimer*, 310–11.

44. Jette, *Inside Box 1663*, 123.

45. *Hearings on Science Legislation*, 301. See also Smith, *A Peril and a Hope*, 152–55. On the history of postwar science legislation, see Kevles, *The Physicists*, 343–64; Daniel J. Kevles, "The National Science Foundation and the Debate over Postwar Research Policy, 1942–1945: A Political Interpretation of *Science—the Endless Frontier*," *Isis* 68 (1977): 5–26; Jessica Wang, "Liberals, the Progressive Left, and the Political Economy of Post-war American Science: The National Science Foundation Debate Revisited," *Historical Studies in the Physical and Biological Sciences* 26 (1995): 139–66; Nathan Reingold, "Vannevar Bush's New Deal for Research: Or the Triumph of the Old Order," *Historical Studies in the Physical and Biological Sciences* 17 (1987): 299–344; and Greenberg, *Science, Money, and Politics*, 41–58.

46. *Hearings on Science Legislation,* 302.

47. Ibid., 308.

48. *Hearings before the Committee on Military Affairs, House of Representatives, Seventy-ninth Congress, First Session, on H.R. 4280, an Act for the Development and Control of Atomic Energy,* October 9, 18, 1945 (Washington, DC: U.S. Government Printing Office, 1945), 126–29, on 127.

49. *Hearings on Science Legislation,* 309–10.

50. *Hearings before the Committee on Military Affairs, House of Representatives,* 128–29.

51. Quoted in Strickland, *Scientists in Politics,* 48. See also Pais, *A Tale of Two Continents,* 241; Chevalier, *Oppenheimer,* 79–80; Bird and Sherwin, *American Prometheus,* 356; and Balogh, *Chain Reaction,* 40.

52. Cf. Peter Galison, "Physics between War and Peace," in *Science, Technology, and the Military,* vol. 1, ed. Everett Mendelsohn, Merritt Roe Smith, and Peter Weingart, 47–86 (Dordrecht: Kluwer Academic Publishers, 1988).

53. *Hearings on Science Legislation,* 300, 301.

54. Ibid., 300; *Hearings before the Committee on Military Affairs, House of Representatives,* 126.

55. *Hearings on Science Legislation,* 305.

56. Ibid., 318–33, quoting 321. Robert Wilson from ALAS also testified at this session (330–33).

57. Ibid., 322, 325–26.

58. Ibid., 322, 325–29.

59. See Smith, *A Peril and a Hope,* 150–56, quoting 155.

60. Quoted in Smith, *A Peril and a Hope,* 154.

61. Quoted in Strickland, *Scientists in Politics,* 46–47.

62. Badash, *Scientists and the Development of Nuclear Weapons,* 66. See also Kevles, *The Physicists,* 351–52; and Hewlett and Anderson, *The New World,* 482–530.

63. Michael S. Sherry, *In the Shadow of War: The United States since the 1930s* (New Haven, CT: Yale University Press, 1995), 137; Cassidy, *J. Robert Oppenheimer,* 264.

64. Oppenheimer, "Speech to the Association of Los Alamos Scientists," November 2, 1945, in Smith and Weiner, *Robert Oppenheimer,* 315–25, quoting 317.

65. Ibid., 319. See also J. Robert Oppenheimer, "Atomic Weapons and the Crisis in Science," *Saturday Review of Literature,* November 24, 1945, 9–11.

66. Oppenheimer, "Speech to the Association of Los Alamos Scientists," 319–20.

67. Ibid., 323.

68. Robert Gilpin, *American Scientists and Nuclear Weapons Policy* (Princeton, NJ: Princeton University Press, 1962), 53. The report was published as Chester I. Barnard et al., *A Report on the International Control of Atomic Energy, Prepared for the Secretary of State's Committee on Atomic Energy by a Board of Consultants* (Washington, DC: U.S. Government Printing Office, 1946). The report was dated March 16,

1946, and was made public by the State Department on March 28. Oppenheimer set out his individual views on international control in Oppenheimer to David Lilienthal, February 2, 1946, Oppenheimer Papers, box 46.

69. Barnard et al., *Report,* 31, 61. For background on the report, see Hewlett and Anderson, *The New World,* 531–34.

70. Bohr to Oppenheimer, April 17, 1946, Oppenheimer Papers, box 21.

71. See, for example, Federation of Atomic Scientists, "Resolution on Acheson Report," Council Meeting, April 10–21 [1946], Oppenheimer Papers, box 195.

72. Lilienthal, *The Atomic Energy Years,* 13.

73. Ibid., 16.

74. Lilienthal to Herbert Marks, January 14, 1948, Oppenheimer Papers, box 46; quoted also in Neuse, *David E. Lilienthal,* 170. See also Dean Acheson, *Present at the Creation: My Years in the State Department* (New York: W. W. Norton, 1969), 153.

75. David E. Lilienthal, "How Can Atomic Energy Be Controlled?" *Bulletin of the Atomic Scientists* 2 (October 1947): 14–15. See also the journal entry for July 14, 1948, in Lilienthal, *The Atomic Energy Years,* 67; Lilienthal, *TVA: Democracy on the March* (New York: Pocket Books, 1945); Lilienthal, *This I Do Believe* (New York: Harper and Brothers, 1949); Neuse, *David E. Lilienthal,* esp. 134–38, 177; and Gilpin, *American Scientists and Nuclear Weapons Policy,* 58.

76. J. Robert Oppenheimer, "The New Weapon: The Turn of the Screw," in *One World or None,* ed. Dexter Masters and Katharine Way (1945; London: Purnell and Sons, 1946), 56.

77. J. Robert Oppenheimer, "The International Control of Atomic Energy," Messenger Lecture at Cornell University, May 1946, esp. F-21; "The Atomic Potential," editorial in *Cornell Bulletin,* May 17, 1946, both in Oppenheimer Papers, box 254. A revised version of the lecture was published in *Bulletin of the Atomic Scientists* 1, no. 12 (June 1, 1946): 1–5, and reprinted as "International Control of Atomic Energy," in *The Atomic Age: Scientists in National and World Affairs,* ed. Morton Grodzins and Eugene Rabinowitch, 53–63 (New York: Basic Books, 1963). In the transcript of the original lecture, but not in the published version, Oppenheimer follows a discussion of the possibilities of cooperation of scientists and engineers of different nationalities by saying, "I have seen this work, in little part, at Los Alamos."

78. J. Robert Oppenheimer, "Atomic Physics in Civilization," Messenger Lecture at Cornell University, April 30, 1946, 2, Oppenheimer Papers, box 254.

79. J. Robert Oppenheimer, "Atomic Explosives," in *The Open Mind,* 3–17, quoting 12.

80. *Hearings on Science Legislation,* 314–15. This justification was essentially the same as that which Oppenheimer had used at Los Alamos in dissuading Teller from circulating Szilard's petition. Cassidy, *J. Robert Oppenheimer,* 245–46.

81. David S. McLellan, *Dean Acheson: The State Department Years* (New York: Dodd, Mead, 1976), 80–81; Jordan A. Schwarz, *The Speculator: Bernard M. Baruch*

in Washington, 1917–1965 (Chapel Hill: University of North Carolina Press, 1981), 5, 490.

82. Schwarz, *The Speculator*, 493. See also Goodchild, *J. Robert Oppenheimer*, 180; John Newhouse, *War and Peace in the Nuclear Age* (New York: Alfred A. Knopf, 1989), 63; Herken, *Brotherhood of the Bomb*, 165; and Norris, *Racing for the Bomb*, 482.

83. Swope, quoted in W. L. White, *Bernard Baruch: Portrait of a Citizen* (New York: Harcourt, Brace, 1950), 111.

84. Bernard M. Baruch, *Baruch: The Public Years* (New York: Holt, Rinehart and Winston, 1960), 364–65, quoting 364.

85. Baruch, *Baruch*, 361; Neuse, *David E. Lilienthal*, 175; Badash, *Scientists and the Development of Nuclear Weapons*, 69; McLellan, *Dean Acheson*, 82; Bird and Sherwin, *American Prometheus*, 342–49.

86. Hewlett and Anderson, *The New World*, 581–82; Johnathan M. Weisgall, *Operation Crossroads: The Atomic Tests at Bikini Atoll* (Annapolis, MD: Naval Institute Press, 1994), esp. 252–55; Alperovitz, *The Decision to Use the Atomic Bomb*, 443; Badash, *Scientists and the Development of Nuclear Weapons*, 69–70; Williams and Cantelon, *The American Atom*, 176. See also W. A. Shurcliff, *Bombs at Bikini: The Official Report of Operation Crossroads* (New York: William H. Wise, 1947).

87. Norman Cousins, "The Standardization of Catastrophe," *Saturday Review of Literature*, August 10, 1946, 18. See also William L. Laurence, "The Bikini Tests and Public Opinion," *Bulletin of the Atomic Scientists* 2 (1946): 2, 17; and Weisgall, *Operation Crossroads*, 246.

88. Joseph I. Lieberman, *The Scorpion and the Tarantula: The Struggle to Control Atomic Weapons, 1945–1949* (Boston: Houghton Mifflin, 1970), 227–412; Cassidy, *J. Robert Oppenheimer*, 248. See also Oppenheimer to Bernard Baruch, January 10, 1947, in Oppenheimer Papers, box 195.

89. Journal entry for July 24, 1946, in Lilienthal, *The Atomic Energy Years*, 69–70. See also Oppenheimer to Lincoln Gordon, April 21, 1947, Oppenheimer Papers, box 195; and Herken, *Brotherhood of the Bomb*, 180.

90. Hewlett and Anderson, *The New World*, 562; John W. Spanier and Joseph L. Nogee, *The Politics of Disarmament: A Study in Soviet-American Gamesmanship* (New York: Praeger, 1962), 58; Bird and Sherwin, *American Prometheus*, 343.

91. Bernard Lovell, "Patrick Maynard Stuart Blackett, Baron Blackett of Chelsea, 18 November 1897–13 July 1974," *Biographical Memoirs of Fellows of the Royal Society* 21 (1975): 1–115; P. M. S. Blackett, *Fear, War, and the Bomb: Military and Political Consequences of Atomic Energy* (New York: Whittlesey House, 1949); Nye, *Blackett*, esp. 89–92. The title of the 1948 British edition of Blackett's book was *Military and Political Consequences of Atomic Energy*.

92. See Paul Boyer's essays "How Americans Imagined the Bomb They Dropped," "President Truman, the American People, and the Atomic Bomb," and "Hiroshima in American Memory," in *Fallout: A Historian Reflects on America's Half-Century*

Encounter with Nuclear Weapons (Columbus: Ohio State University Press, 1998), 9–16, 17–40, 226–45; Alperovitz, *The Decision to Use the Atomic Bomb,* 423–641; and Barton J. Bernstein, "A Post-War Myth: 500,000 U.S. Lives Saved," in *Hiroshima's Shadow,* ed. Kai Bird and Lawrence Lifschultz, 130–34 (Stony Creek, CT: Pamphleteers Press, 1998).

93. Blackett, *Fear, War, and the Bomb,* 131–43, quoting 139. Blackett's account is now supported by revisionist historians; see, above all, Alperovitz, *The Decision to Use the Atomic Bomb,* esp. 128, 573–74.

94. "Some Observations and Comments on the Book, by Professor P. M. S. Blackett," and Albin E. Johnson to Oppenheimer, January 28, 1949, both in Oppenheimer Papers, box 195.

95. Blackett, *Fear, War, and the Bomb,* 139.

96. Ibid., 142–43.

97. Ibid., 140–41, 143. The Oppenheimer quotation is from Oppenheimer, "International Control of Atomic Energy," 53–54. Bird and Sherwin emphasized the critical tone and import of Oppenheimer's statement (*American Prometheus,* 389). However, it is worth noting that the notion that Japan was "essentially defeated" is still compatible with the idea that without the atomic bombings, the Japanese could have mounted resistance that (though ultimately futile) would have been costly in terms of lives and could have significantly delayed the achievement of unconditional surrender. As I see it, even here Oppenheimer did not directly contradict the official account.

98. Blackett, *Fear, War, and the Bomb,* 143.

99. Ibid., 143–44, 157. In a later book, Blackett referred to the Acheson-Lilienthal report as the "Lilienthal-Oppenheimer Plan" and noted disparagingly its "high-flown and idealistic phraseology." P. M. S. Blackett, *Atomic Weapons and East-West Relations* (Cambridge: Cambridge University Press, 1956). For Blackett, with his background in operational research and his overtly hard-boiled realist style of analysis, questions of arms and military strategy were not appropriate subjects for idealism. Nevertheless, this later book, which was written after Oppenheimer's downfall in the security hearings and expressed sympathy with Oppenheimer's' "ordeal" (73), reserved its main criticisms for Baruch's adaptations.

100. Blackett, *Fear, War, and the Bomb,* 158.

101. I. I. Rabi, "Playing Down the Bomb: Blackett versus the Atom," *Atlantic Monthly* 183 (April 1949): 23. See also Mary Jo Nye, "What Price Politics? Scientists and Political Controversy," *Endeavour* 23 (1999): 148–54; Nye, *Blackett,* 88–92; Boyer, *By the Bomb's Early Light,* 192; and Bird and Sherwin, *American Prometheus,* 389. Philip Morrison was the exception: he wrote an article in the *Bulletin of the Atomic Scientists* supporting Blackett (see Nye, *Blackett,* 91; and Bird and Sherwin, *American Prometheus,* 657).

102. Lilienthal, however, noted that at a reunion of the members of the Board of Consultants, "Oppenheimer got off a rather subtle and fine-spun discussion, the

purport of which was that the 'just another bomb' argument is meaningless" (journal entry for February 3, 1949, in Lilienthal, *The Atomic Energy Years,* 454). Blackett's argument was often attacked for denying the significance of the atomic bomb. Bird and Sherwin suggested that Oppenheimer may have been sympathetic to Blackett's analysis; they noted that Oppenheimer wrote to Blackett in 1956 expressing agreement with the "major thesis" (if not all the details) of Blackett's later (1956) book, *Atomic Weapons and East-West Relations* (see Bird and Sherwin, *American Prometheus,* 389, 578). However, it is not clear that this does imply sympathy with Blackett's specific criticisms and analysis of the Hiroshima and Nagasaki bombings. Blackett's 1956 book was primarily an exploration of the implications of nuclear parity (following the development of the Soviet A-bomb and H-bomb) for military strategic thought. It also included a discussion of the H-bomb controversy and the Oppenheimer case. The book's major thesis concerned the fallacy of seeing security as resting on Western technological superiority. Blackett's analysis here fit closely with Oppenheimer's critical stance toward the doctrines of the Strategic Air Command. One can easily imagine Oppenheimer agreeing with Blackett that "the goal of permanent Western technological superiority is neither attainable nor is it a necessary condition for peace. What is a reasonable and attainable goal is approximate technological parity in certain essential weapons and weapon systems" (Blackett, *Atomic Weapons,* 96–97). But agreement on that cannot be taken to imply assent, in 1956 or in 1948–49, to Blackett's criticisms of the bombing of Hiroshima or to his views on the origins of the Cold War.

103. For Oppenheimer's view on the Bikini tests, see Bird and Sherwin, *American Prometheus,* 349–50.

104. Frances Stonor Saunders, *Who Paid the Piper? The CIA and the Cultural Cold War* (London: Granta Books, 1999), 24–25, quoting 25. See also Bird and Sherwin, *American Prometheus,* 354.

105. J. Robert Oppenheimer, "Atomic Energy as a Contemporary Problem," in *The Open Mind,* 21–41, on 25–26. George F. Kennan's anonymous article, "The Sources of Soviet Conduct," appeared in *Foreign Affairs* 25, no. 4 (July 1947), esp. 572. On Kennan's role in shaping American views of the Soviet Union, see Charles E. Nathanson, "The Social Construction of the Soviet Threat: A Study in the Politics of Representation," *Alternatives* 8 (1988), esp. 454–63.

106. Oppenheimer, "The Open Mind," 49–51.

107. Cf. Paul K. Hoch, "The Crystallization of a Strategic Alliance: The American Physics Elite and the Military in the 1940s," in Mendelsohn, Smith, and Weingart, *Science, Technology and the Military,* 87–116.

108. Oppenheimer, "Atomic Energy as a Contemporary Problem," 34.

109. Ibid., 32, 36, 34.

110. Jessica Wang, *American Science in an Age of Anxiety: Scientists, Anticommunism, and the Cold War* (Chapel Hill: University of North Carolina Press, 1999), 234–35; Badash, *Scientists and the Development of Nuclear Weapons,* 67; Hoch,

"The Crystallization of a Strategic Alliance," esp. 93–100; Silvan S. Schweber, "The Mutual Embrace of Science and the Military: ONR and the Growth of Physics in the United States after World War Two," in Mendelsohn, Smith, and Weingart, *Science, Technology and the Military,* 1:3–45; Mukerji, *A Fragile Power;* Naomi Oreskes and Paul Rainger, "Science and Security before the Atomic Bomb: The Loyalty Case of Harold U. Sverdrup," *Studies in History and Philosophy of Modern Physics* 31 (2000): 309–69; Daniel S. Greenberg, *The Politics of Pure Science* (1967; Chicago: University of Chicago Press, 1999).

111. See Paul Forman, "Behind Quantum Electronics: National Security as Basis for Physical Research in the United States, 1940–1960," *Historical Studies in the Physical and Biological Sciences* 18 (1987): 149–229; Stuart Leslie, *The Cold War and American Science: The Military-Industrial-Academic Complex at MIT and Stanford* (New York: Columbia University Press, 1993); Robert Seidel, "The Postwar Political Economy of High-Energy Physics," in *Pions to Quarks: Particle Physics in the 1950s,* ed. Laurie Brown, Max Dresden, and Lillian Hoddeson, 497–507 (Cambridge: Cambridge University Press, 1989); Daniel J. Kevles, "Cold War and Hot Physics: Science, Security, and the American State, 1945–1956," *Historical Studies in the Physical and Biological Sciences* 20 (1990): 239–64.

112. Karl Cohen, "A Re-examination of the McMahon Act," *Bulletin of the Atomic Scientists* 4, no. 1 (January 1948): 7–10.

113. "Transcript," 45; see also Gilpin, *American Scientists and Nuclear Weapons Policy,* 71.

114. Oppenheimer, quoted in Smith, *A Peril and a Hope,* 350. On the decline of the Federation of Atomic Scientists, see also Coser, *Men of Ideas,* 308–12.

115. Oppenheimer, "Physics in the Contemporary World," 92, 81, 91.

116. Hilbert, quoted in Oppenheimer, "Physics in the Contemporary World," 89. See also J. Robert Oppenheimer, "The Need for New Knowledge," in *Symposium on Basic Research,* ed. Dael Wolfle (Washington, DC: American Association for the Advancement of Science, 1959), 9.

117. Oppenheimer, "Physics in the Contemporary World," 90–91.

118. Ibid., 88, 83.

119. "Physicist Oppenheimer," 77.

120. Oppenheimer, "Physics in the Contemporary World," 83.

121. Neuse, *David E. Lilienthal,* 168–69, 177–200; Hershberg, *James B. Conant,* 332; Lapp, *The New Priesthood,* 103.

122. Journal entry for July 29, 1947, in Lilienthal, *The Atomic Energy Years,* 229; Oppenheimer, in "Transcript," 69; Oppenheimer, quoted in Hershberg, *James B. Conant,* 313. See also Gilpin, *American Scientists and Nuclear Weapons Policy,* 70; Balogh, *Chain Reaction,* 60–90; and Lilienthal's speeches in Oppenheimer Papers, box 46.

123. Harrison S. Brown to Oppenheimer, February 26, 1948, and Oppenheimer to Brown, March 5, 1948, Oppenheimer Papers, box 32.

124. Gilpin, *American Scientists and Nuclear Weapons Policy,* 64–161. See also Samuel R. Williamson Jr. and Steven L. Rearden, *The Origins of U.S. Nuclear Strategy, 1945–1953* (New York: St. Martin's Press, 1993).

125. York, *The Advisors,* 46–56; Richard T. Sylves, *The Nuclear Oracles: A Political History of the General Advisory Committee of the Atomic Energy Commission, 1947–1977* (Ames: Iowa State University Press, 1987), esp. 142–70; Rhodes, *Dark Sun,* 377–81; Cassidy, *J. Robert Oppenheimer,* 265, 294.

126. The comparison of "super" and atomic bombs is from the GAC's report of October 30, 1949, in York, *The Advisors,* 156, 158.

127. Journal entry for October 30, 1949, in Lilienthal, *The Atomic Energy Years,* 582.

128. Hershberg, *James B. Conant,* 470–75; Rhodes, *Dark Sun,* 388–91. See also James B. Hershberg, " 'Over My Dead Body': James Bryant Conant and the Hydrogen Bomb," in Mendelsohn, Smith, and Weingart, *Science, Technology and the Military,* 2:379–430.

129. Journal entry for October 29, 1949, in Lilienthal, *The Atomic Energy Years,* 580–81, quoting 581.

130. See York, *The Advisors,* 46–56; the GAC report is on 150–59. See also Peter Galison and Barton Bernstein, " 'In Any Light': Scientists and the Decision to Build the Hydrogen Bomb," *Historical Studies in the Physical and Biological Sciences* 19 (1989): 267–347.

131. Rhodes, *Dark Sun,* 406–7, quoting 406; York, *The Advisors,* 65–74; Badash, *Scientists and the Development of Nuclear Weapons,* 84–85.

132. Neuse, *David E. Lilienthal,* 208–31.

133. Lilienthal to President Truman, November 21, 1949; "Lilienthal Seen Ready to Quit A Commission," *Rochester Democrat and Chronicle,* November 12, 1949, both in Oppenheimer Papers, box 46.

134. President Truman to Lilienthal, November 23, 1949, Oppenheimer Papers, box 46. See also Richard G. Hewlett and Francis Duncan, *Atomic Shield, 1947/1950,* vol. 2 of *A History of the United States Atomic Energy Commission* (Berkeley and Los Angeles: University of California Press, 1991), 444–45.

135. Gilpin, *American Scientists and Nuclear Weapons Policy,* 73.

136. Lilienthal, quoted in Wang, *American Science in an Age of Anxiety,* 237. See also David E. Lilienthal, *Change, Hope, and the Bomb* (Princeton, NJ: Princeton University Press, 1963), 115.

137. "Transcript," 305.

138. "Transcript," 251. See also York, *The Advisors,* 127.

139. See Schweber, *In the Shadow of the Bomb,* esp. 156–68. See also Hans A. Bethe, "The Technological Imperative," in Acklund and McGuire, *Assessing the Nuclear Age,* 73–78.

140. Williamson and Rearden, *Origins of U.S. Nuclear Strategy,* 168–69; Gilpin, *American Scientists and Nuclear Weapons Policy,* 112–21. Gilpin pointed out that

Oppenheimer had argued as early as 1948 for research on tactical nuclear weapons (114). See also W. Patrick McCray, "Project Vista, Caltech and the Dilemmas of Lee DuBridge," *Historical Studies in the Physical and Biological Sciences* 34, no. 2 (2000): 339–70; and Bird and Sherwin, *American Prometheus*, 444–45.

141. J. Robert Oppenheimer, "Comments on the Military Value of the Atom," *Bulletin of the Atomic Scientists* 7, no. 2 (February 1951): 43 (my emphasis); quoted also in Gilpin, *American Scientists and Nuclear Weapons Policy*, 118.

142. J. Robert Oppenheimer, "The Atom Bomb as a Great Force for Peace," *New York Times Magazine*, June 9, 1946, 59; Oppenheimer, "Comments on the Military Value of the Atom," 44–45. These passages are quoted also in Gilpin, *American Scientists and Nuclear Weapons Policy*, 77 n. 16, 118. There is a minor discrepancy in that Gilpin transcribed the first quotation as referring to "policy weapons." In the original, it is "police weapons."

143. "Transcript," 470. In a talk at the Council on Foreign Relations in 1953, Oppenheimer spoke of the need to look "the tiger in the eye," but added, "That does not mean that I think [preventive war] is a good idea" (quoted in Bird and Sherwin, *American Prometheus*, 448).

144. "Scientists Tell SHAPE of Nevada Atom Tests," *New York Herald Tribune*, December 5, 1951. See also "U.S. Atom Experts Visit West Germany," *New York Herald Tribune*, December 6, 1951; "Oppenheimer Visits Eisenhower for Hour," *New York Times*, December 5, 1951; and "Oppenheimer Lands in Paris," *New York Times*, December 4, 1951.

145. Williamson and Rearden, *Origins of U.S. Nuclear Strategy*, 169. See also John P. Rose, *The Evolution of U.S. Army Nuclear Doctrine, 1945–1980* (Boulder, CO: Westview Press, 1980); and Steven Leonard Newman, "The Oppenheimer Case: A Reconsideration of the Role of the Defense Department and National Security" (PhD diss., New York University, 1977), 34–51. Blackett stated that by 1956, Oppenheimer's contributions to Vista had become military orthodoxy (*Atomic Weapons*, 74).

146. Freeman Dyson, "The Scholar-Soldier," in *Weapons and Hope* (New York: Harper and Row, 1984), 137.

147. Ibid., 144–45. Conant had thought it important that he and Oppenheimer act as "good soldiers" by not resigning from the GAC after their defeat on the H-bomb (Conant, quoted in Bird and Sherwin, *American Prometheus*, 429).

148. Teller, "The Role of the Scientist," 234.

149. York, *The Advisors*, 56.

CHAPTER SEVEN

1. Robert Coughlan, "The Equivocal Hero of Science: Robert Oppenheimer," *Life*, February 1967, 34–34A. As well as the GAC, Oppenheimer served on, for example, the Joint Research and Development Board to the Departments of the Army and Navy (which in 1947 became the Research and Development Board of the newly

established Department of Defense) and on the Scientific Advisory Committee to the White House Office of Defense Mobilization (York, *The Advisors,* 17, 47).

2. The agent was referring in particular to Oppenheimer's influence over AEC chairman Lilienthal and commissioner Robert F. Bacher. "Re: J. Robert Oppenheimer—Justification for Continuation of Technical or Microphone Surveillance," November 5, 1946, in *Oppenheimer, FBI Security File,* 100-17828, sec. 6.

3. Cf. Steven Shapin, foreword to Greenberg, *The Politics of Pure Science,* xvii.

4. On the notion of a "file person," see Rom Harré, *Personal Being* (Cambridge, MA: Harvard University Press, 1984). See also Spencer Cahill, "Toward a Sociology of the Person," *Sociological Theory* 16 (1998), esp. 143.

5. See the materials in *Transcript and Texts,* including "Transcript of Hearing before Personnel Security Board," 1–992 (most citations of this transcript in this chapter will be given in the text); "Findings and Recommendation of the Personnel Security Board in the Matter of Dr. J. Robert Oppenheimer" (hereafter "Findings"), 999–1021; "Recommendations of the General Manager to the United States Atomic Energy Commission in the Matter of Dr. J. Robert Oppenheimer" (hereafter "Recommendations of the General Manager"), 1041–46; and "Decision and Opinions of the United States Atomic Energy Commission in the Matter of Dr. J. Robert Oppenheimer" (hereafter "Decision and Opinions"), 1049–65. Secondary literature includes Joseph Alsop and Stewart Alsop, *We Accuse! The Story of the Miscarriage of Justice in the Case of J. Robert Oppenheimer* (New York: Simon and Schuster, 1954); Barton J. Bernstein, "In the Matter of J. Robert Oppenheimer," *Historical Studies in the Physical Sciences* 12 (1982): 195–252; Bernstein, "The Oppenheimer Loyalty-Security Case Reconsidered," *Stanford Law Review* 42 (July 1990): 1383–1484; Curtis, *The Oppenheimer Case;* Galison and Bernstein, "'In Any Light'"; Harold P. Green, "The Oppenheimer Case: A Study in the Abuse of Law," *Bulletin of the Atomic Scientists* 33 (1977): 12–16, 56–61; Rachael L. Holloway, *In the Matter of J. Robert Oppenheimer: Politics, Rhetoric and Self-Defense* (Westport, CT: Praeger, 1993); Lakoff, "The Trial of Dr. Oppenheimer"; John Major, *The Oppenheimer Hearing* (1971; New York: Scarborough Books, 1983); Michelmore, *The Swift Years;* Richard Polenberg, ed., *In the Matter of J. Robert Oppenheimer: The Security Clearance Hearing* (Ithaca, NY: Cornell University Press, 2002); Rhodes, *Dark Sun;* Stern, *The Oppenheimer Case;* Charles Thorpe, "Disciplining Experts: Scientific Authority and Liberal Democracy in the Oppenheimer Case," *Social Studies of Science* 32, no. 4 (August 2002): 525–62; and York, *The Advisors.*

6. Schrecker, *Many Are the Crimes,* 165; Kunetka, *Oppenheimer,* 31, 203.

7. Chevalier to Oppenheimer [ca. November 1943], and Chevalier to Oppenheimer, December 3, 1943, Oppenheimer Papers, box 26.

8. J. Edgar Hoover to Major General Leslie R. Groves, June 13, 1946, and Groves to Hoover, June 21, 1946, box 99, General Correspondence, 1943–47, Investigation Files, Records of the Office of the Commanding General, Manhattan Project, RG 77.

9. Rhodes, *Dark Sun,* 125–26; Chevalier, *Oppenheimer,* 61–68; Bird and Sherwin, *American Prometheus,* 356–59.

10. Rhodes, *Dark Sun,* 309; Bird and Sherwin, *American Prometheus,* 352; Cassidy, *J. Robert Oppenheimer,* 265, 279.

11. H. B. Fletcher to Director, FBI, "Re: Julius Robert Oppenheimer," May 14, 1946, in *Oppenheimer, FBI Security File,* 100-17828, sec. 1; Lapp, quoted in Stern, *The Oppenheimer Case,* 112. See also Pais, *A Tale of Two Continents,* 240. Oppenheimer's comment to Lapp was made in 1948. Although the FBI had begun to gather information about Oppenheimer from 1941, and although it was indeed tapping his phone at Berkeley after the war and was keeping files on his activities at Princeton, the bureau did not install phone taps in his home and office at Princeton until the beginning of 1954 (Bird and Sherwin, *American Prometheus,* 344–45, 405, 483). On Oppenheimer's postwar job offers and academic appointments, see Smith and Weiner, *Robert Oppenheimer,* 298–99, 307–9; and Bird and Sherwin, *American Prometheus,* 320, 329–30, 351, 369–71.

12. Stern, *The Oppenheimer Case,* 107–10, quoting 107.

13. Herken, *Brotherhood of the Bomb,* 189–90, Thomas quoted on 189; "Red Spy's Plot on U.S. Atomic Secrets Bared," *New York World Telegram,* October 30, 1947; Oppenheimer's Statement to Associated Press, October 30, 1947, Oppenheimer Papers, box 236. On HUAC's interest in the Chevalier incident, see also Schrecker, *No Ivory Tower,* 139, 141; Chevalier, *Oppenheimer,* 76–78; and Cassidy, *J. Robert Oppenheimer,* 279–80. A few months after the end of the war, Hoover had forwarded to HUAC a report detailing the FBI's information on atomic espionage (Herken, *Brotherhood of the Bomb,* 145).

14. Wang, *American Science in an Age of Anxiety,* 130–47.

15. Quoted in Wang, *American Science in an Age of Anxiety,* 132.

16. Harold C. Urey to Oppenheimer, March 15, 1948, and Oppenheimer to Urey, March 15, 1948, Oppenheimer Papers, box 32.

17. "'Civil Liberties': from stenotyped report of NAS Business Session, April 27, 1948 (page 54)," Oppenheimer Papers, box 134. See also Wang, *American Science in an Age of Anxiety,* 184–92, esp. 190.

18. Wang, *American Science in an Age of Anxiety,* 192.

19. E. U. Condon to Martin Kamen, May 24, 1948, and Kamen to Condon, May 26, 1948, Papers of Martin David Kamen, Mandeville Special Collections Library, University of California, San Diego.

20. Alfred N. Richards to Oppenheimer, November 2, 1948, and Richards to J. B. Conant, December 8, 1948, Oppenheimer Papers, box 134; Wang, *American Science in an Age of Anxiety,* 193–94.

21. National Academy of Sciences, "The Problem of Security in Government Service: A Statement Adopted by the Council of the National Academy of Sciences Authorized for Transmission to the President of the United States," February 3, 1949, Oppenheimer Papers, box 134.

22. Wang, *American Science in an Age of Anxiety,* 194–96, Oppenheimer quoted on 194.

23. Herken, *Brotherhood of the Bomb,* 192–93.

24. House Committee on Un-American Activities, 80th Cong., 2d sess., Public Law 601, *Interim Report on Hearings regarding Communist Espionage in the United States* (Washington, DC: U.S. Government Printing Office, August 28, 1948); House Committee on Un-American Activities, 80th Cong., 2d sess., Public Law 601, *Report on Soviet Espionage Activities in Connection with the Atomic Bomb* (Washington, DC: U.S. Government Printing Office, September 28, 1948), esp. 182–83; newspaper clippings in Oppenheimer Papers, box 236: "Text of Report by House Committee on Un-American Activities Relating to Atomic Espionage," *New York Times,* September 28, 1948; "Scientist X Sought," *New York Sun,* September 28, 1948; William S. White, "Indictment of Five Is Urged in Report on Atomic Spying," *New York Times,* September 28, 1948; James E. Warner, "Atom Inquiry Accuses Five of Espionage: Says Two Scientists Sent Data to Russia," *New York Herald Tribune,* September 28, 1948; Ruth Montgomery, "Atom Probers Demand Indictment for Four," *New York Daily News,* September 28, 1948; Oliver Pilat, "Prosecution of Four in Atom Plot Asked," *New York Post,* September 28, 1948; "Indictment of 4 as Soviet Spies Demanded by Atom Bomb Inquiry," *New York Sun,* September 28, 1948. See also Schrecker, *No Ivory Tower,* 130–42; Rhodes, *Dark Sun,* 122–23, 309; and Olwell, "Physical Isolation," 741–44.

25. Wang, *American Science in an Age of Anxiety,* 211–12.

26. Stern, *The Oppenheimer Case,* 118–19; Herken, *Brotherhood of the Bomb,* 196. For Nixon's comments, see *New York Times,* May 11, 1950.

27. "Dr. Oppenheimer Once Termed Peters 'Quite Red,'" *Rochester Times Union,* June 15, 1949. I am grateful to Shawn Mullet for providing me with a copy of the article. See also Stern, *The Oppenheimer Case,* 119–20, 124–26; Schweber, *In the Shadow of the Bomb,* 119–29; and Bird and Sherwin, *American Prometheus,* 394–400.

28. Weisskopf to Oppenheimer, June 27 [1949], Oppenheimer Papers, box 77 (emphasis in original); quoted also in Schweber, *In the Shadow of the Bomb,* 123–24. See also "Dr. Peters Denies He Is Communist," *New York Times,* April 23, 1954.

29. Oppenheimer to the Editor, *Rochester Democrat and Chronicle,* June 30, 1949, published July 6, 1949. I am very grateful to Shawn Mullet for providing me with the letter, as well as for comments on its significance. The letter is also quoted in "J. Robert Oppenheimer: His Life and Times," *Time,* April 26, 1954, 19–26, on 21. See also Cassidy, *J. Robert Oppenheimer,* 281–84.

30. Condon, quoted in Stern, *The Oppenheimer Case,* 125.

31. Condon, interviewed by Weiner, September 11 and 12, 1973, esp. 209–14, quoting 210, 211, AIP.

32. Edward Condon to Emilie Condon, June 23, 1949, attached to D. M. Ladd to J. Edgar Hoover, December 15, 1953, and [Hoover?] to Admiral Lewis L. Strauss, January 14, 1954, in *Oppenheimer, FBI Security File,* 100-17828, sec. 16. Condon's

letter to his wife was publicized during the 1954 Oppenheimer hearings, and Condon then wrote, "Under stress of difficult conditions, I was led privately to consider doubts about Oppenheimer which I never stated publicly, which I soon learned were wholly unjustified, and which I now publicly repudiate" (Condon to Editor, *Time,* published May 17, 1954, 8).

33. Stern, *The Oppenheimer Case,* 128–30.

34. Ibid., 107–8, 130–31; Wang, *American Science in an Age of Anxiety,* 212, 287; Schrecker, *No Ivory Tower,* 147–48, 272; Herken, *Brotherhood of the Bomb,* 196–97; Goodchild, *J. Robert Oppenheimer,* 196.

35. Chevalier, *Oppenheimer,* 78–79; Jungk, *Brighter Than a Thousand Suns,* 144.

36. Chevalier to Oppenheimer, February 21, 1950; Oppenheimer to Chevalier (two letters), February 24, 1950; Chevalier to Oppenheimer, February 28, 1950; Chevalier to Oppenheimer, August 10, 1950; and Chevalier to Oppenheimer, September 8, 1950, Oppenheimer Papers, box 26; Chevalier, *Oppenheimer,* 71–84.

37. Stern, *The Oppenheimer Case,* 115; Hershberg, *James B. Conant,* 425–26; Sherry, *In the Shadow of War,* 172.

38. C. A. Rolander, Memo to file, "J. R. Oppenheimer—Interview with Col. Boris T. Pash," March 15, 1954, in *Oppenheimer, FBI Security File,* 100-17828, sec. 23.

39. Quoted in Stern, *The Oppenheimer Case,* 162. See also Ben C. Duniway, "Summary of Testimony Given before the California Senate Committee on Unamerican Activities, Monday, May 8, Tuesday, May 9, and Wednesday, May 10, 1950," Oppenheimer Papers, box 235.

40. Schrecker, *No Ivory Tower,* 142; Dan Gillmor, *Fear, the Accuser* (New York: Abelard-Schuman, 1954), 180; Stern, *The Oppenheimer Case,* 427–28.

41. Statement by J. Robert Oppenheimer, May 9, 1950, Oppenheimer Papers, box 237; Stern, *The Oppenheimer Case,* 161–63.

42. *Oakland Tribune,* May 9, 1950; *New York World-Telegram,* May 10, 1950; and other clippings, in Oppenheimer Papers, box 236.

43. Nixon, quoted in "Nixon Defends the Loyalty of Dr. Oppenheimer," *San Francisco Chronicle,* May 11, 1950, and "Nixon Champions Dr. Oppenheimer," *New York Times,* May 11, 1950, both in Oppenheimer Papers, box 236. See also Stern, *The Oppenheimer Case,* 163. Oppenheimer wrote to thank Nixon for his "good words" (Oppenheimer to Richard Nixon, May 15, 1950, Oppenheimer Papers, box 235).

44. Robert R. Brunn, "Californians Rally to Back Oppenheimer," *Christian Science Monitor,* May 16, 1950; "Robert Oppenheimer, Patriot," editorial, *Santa Fe New Mexican,* May 10, 1950; "The Magic of Ex-Communists," editorial, *Washington Star,* May 11, 1950; "Sabotage by Slander," editorial, *Washington Post,* May 11, 1950; "On Growing Up," editorial, *Baltimore Sun,* May 11, 1950; "The Oppenheimer Story," editorial, *San Francisco Chronicle,* May 11, 1950; "And Speaking of Propriety," *San Francisco Chronicle,* May 12, 1950; "AEC Chiefs Unmoved by Charge against Robert Oppenheimer," *Washington Star,* May 12, 1950; "An Unjustified Smear of a Loyal

Scientist," editorial, *San Francisco News,* May 12, 1950; Tony Smith, "Red Unwittingly Verified Oppenheimer's Loyalty," *Washington Daily News,* May 12, 1950; "A Man to Trust," editorial, *Washington Daily News,* May 12, 1950; "A Defense, as Well as an Attack, Is News," editorial, *Baltimore Sun,* May 13, 1950, clippings in Oppenheimer Papers, box 236.

45. Lilienthal to Oppenheimer, May 10, 1950, Oppenheimer Papers, box 46; Raymond Lawrence, "At Home and Abroad—the Oppenheimer Case," *Oakland Tribune,* May 10, 1950; "West Coast," *San Francisco Chronicle, This World,* May 14, 1950; Oppenheimer to Robert F. Bacher, May 23, 1950, Oppenheimer Papers, box 18.

46. See, for example, "Oppenheimer's Home Scene of Red Meeting, Court Told," *Trenton Evening Times,* December 2, 1952; Note by Oppenheimer's secretary on telephone call from G. S. Gould of Associated Press, December 2 [1952], Oppenheimer Papers, box 237.

47. Quoting "Summary of Interview between Hitz, Cunningham, Roney, Volpe, Marks, Oppenheimer, and, at times, Crouch," May 21, 1952, 10–11; Herbert S. Marks and Joseph Volpe, "Memorandum: Interview at Office of Assistant U. S. Attorney, William Hitz, Washington D. C., May 20, 1952," esp. 11–18, Oppenheimer Papers, box 237.

48. Herbert S. Marks and Joseph Volpe Jr. to William Hitz, "Re: United States v. Weinberg," December 1, 1952, Oppenheimer Papers, box 237. See also correspondence in folders labeled "Weinberg Perjury Trial—1953—Letters to and from JR Oppenheimer," "Weinberg Perjury Trial—1952—Puening Correspondence," and "Weinberg Perjury Trial—1953—Marks and Volpe Memos to RO—1953," Oppenheimer Papers, box 237. The considerable personal attention that Oppenheimer gave to the Weinberg case is suggested by his handwritten notes on the trial, collected in the folder labeled "Weinberg, Perjury Trial—1953—RO/KO Hand Notes," Oppenheimer Papers, box 237. Oppenheimer was subpoenaed to appear as a witness (see documents in folder labeled "Weinberg Perjury Trial—1953—Subpoena and Government Paper," Oppenheimer Papers, box 237). As it happened, however, the Crouches' testimony was too unreliable for the prosecution to risk using their story, so Oppenheimer never had to appear. See Herken, *Brotherhood of the Bomb,* 251, 263–64, 398 nn. 87 and 88; and Bird and Sherwin, *American Prometheus,* 438–43, 456–60.

49. Quoted in "Jurors Acquit 'Scientist X' of Perjury Charge," *New York Herald Tribune,* March 6, 1953. See also "Weinberg Case Nears Jurors," *Trenton Evening Times,* March 4, 1953, and "'Scientist X' Trial Jury Deadlocks, Tries Again," *New York Herald Tribune,* March 5, 1953, clippings in Oppenheimer Papers, box 237. See also *United States v. Weinberg,* United States District Court, D. Columbia, November 21, 1952, *Federal Supplement* 108 (1953): 567–71.

50. York, *The Advisors,* 127–43, quoting 134.

51. W. A. Branigan to A. H. Belmont, "Dr. J. Robert Oppenheimer," May 19, 1952, in *Oppenheimer, FBI Security File,* 100-17828, sec. 13. See also Bird and Sherwin, *American Prometheus,* 443–34.

52. SAC, Albuquerque to Director FBI, "Dr. J. Robert Oppenheimer," May 27, 1952, in *Oppenheimer, FBI Security File,* 100-17828, sec. 13. For Teller's own exculpatory account of his role in the case, see Teller, *Memoirs,* 369–84, esp. 372. On Teller's motives and self-justification, see also Goodchild, *Edward Teller,* 242–43.

53. Quoted in James R. Shepley and Clay Blair Jr., *The Hydrogen Bomb: The Men, the Menace, the Mechanism* (New York: David McKay, 1954), 111; Newman, "The Oppenheimer Case," 25.

54. Rhodes, *Dark Sun,* 537; Herken, *Brotherhood of the Bomb,* 249, 283, 292.

55. Souers, paraphrased in Hoover to Mrs. Tolson, Ladd, and Nichols, July 10, 1952, in *Oppenheimer, FBI Security File,* 100-17828, sec. 13. See also Rhodes, *Dark Sun,* 496–97, 530; and Hewlett and Duncan, *Atomic Shield,* 518.

56. Robert Oppenheimer to Frank Oppenheimer, July 12, 1952, Oppenheimer Papers, box 237. Truman noted Oppenheimer's "strong desire, which you expressed to me last month, to complete your service [on the GAC] . . . with the expiration of your present term" (Harry S. Truman to J. Robert Oppenheimer, September 27, 1952, Oppenheimer Papers, box 73). Herken suggested that Oppenheimer's fear that he would be drawn into the Weinberg trial was a factor in his decision not to seek reappointment (*Brotherhood of the Bomb,* 251). I believe the overriding factor was the H-bomb controversy. Bird and Sherwin wrote that Oppenheimer was "generally fed up with Washington" (*American Prometheus,* 448).

57. See the comments by Secretary of Defense Charles Wilson in J. Edgar Hoover to Tolson, Ladd, and Nichols, December 2, 1953, in *Oppenheimer, FBI Security File,* 100-17828, sec. 15.

58. Strauss, quoted in A. H. Belmont to D. M. Ladd, "Julius Robert Oppenheimer," June 5, 1953, in *Oppenheimer, FBI Security File,* 100-17828, sec. 14; Rhodes, *Dark Sun,* 309–10, 530; Stern, *The Oppenheimer Case,* 114, 128–29.

59. See, for example, D. M. Ladd to the Director, "J. Robert Oppenheimer," May 25, 1953, in *Oppenheimer, FBI Security File,* 100-17828, sec. 14. Strauss, alongside Teller, was a key proponent of the idea that Oppenheimer was deliberately hindering the H-bomb program (Bird and Sherwin, *American Prometheus,* 433–34, 467, 471).

60. William Borden, interviewed by C. A. Rolander and Roger Robb, February 20, 1954 (and attachments), in *Oppenheimer, FBI Security File,* 100-17828, sec. 27; Wilson, quoted in "Regardless of Outcome: Oppenheimer's Services Ended, Wilson Indicates," *New York Herald Tribune,* April 15, 1954; Rhodes, *Dark Sun,* 530–31; York, *The Advisors,* 139.

61. "The Hidden Struggle for the H-Bomb," *Fortune,* May 1953, 109, 110, 230, quoting 230.

62. Stern, *The Oppenheimer Case,* 201–4. According to Bird and Sherwin, Strauss was Murphy's "unacknowledged collaborator" (*American Prometheus,* 468).

63. "U.S. Atom Bomb Boss Lewis Strauss: The Bomb Race Runs on Moscow Time," *Time*, September 21, 1953, 25–27, quoting 26.

64. J. Robert Oppenheimer, "Atomic Weapons and American Policy," *Foreign Affairs* 31, no. 4 (July 1953): 528–29. The article was reprinted in the *Bulletin of the Atomic Scientists* 9, no. 6 (July 1953): 202–5, and in Oppenheimer, *The Open Mind*, 61–77. On Oppenheimer's campaign for "candor," see Newman, "The Oppenheimer Case," 63–70. At the time of Truman's H-bomb decision in late January 1950, Conant had passed on to Oppenheimer Dean Acheson's warning that public debate about the H-bomb would be "contrary to the national interest." As early as February, Oppenheimer made public remarks critical of the secrecy surrounding the H-bomb decision. On this, and on Oppenheimer's work for the State Department disarmament panel as background for his call for "candor," see Bird and Sherwin, *American Prometheus*, 429–30, 448–52.

65. Stern, *The Oppenheimer Case*, 204.

66. [Hoover] to Tolson, Ladd, Belmont, and Nichols, May 19, 1953, in *Oppenheimer, FBI Security File*, 100-17828, sec. 14; Hoover to Tolson, Ladd, and Nichols, December 3, 1953, ibid., sec. 15; Hoover to Tolson and Ladd, June 24, 1953, ibid., sec. 14; Bernstein, "The Oppenheimer Loyalty-Security Case Reconsidered," 1431–38; Hoover to Tolson, Ladd, and Nichols, December 2, 1953. Hoover had long felt stymied by the protection afforded Oppenheimer because of his perceived importance to national security (Herken, *Brotherhood of the Bomb*, 144–45). See also Bird and Sherwin, *American Prometheus*, 471–72; and Goodchild, *Edward Teller*, 224.

67. Ibid., 1–7; Rhodes, *Dark Sun*, 534–35.

68. Oppenheimer to Strauss, December 22, 1953, quoted in Newman, "The Oppenheimer Case," 108. See also Bird and Sherwin, *American Prometheus*, 479–84.

69. Green, "The Oppenheimer Case," 61. See also Bird and Sherwin, *American Prometheus*, 537.

70. Bernstein, "In the Matter of J. Robert Oppenheimer," 217; Stern, *The Oppenheimer Case*, 259–62.

71. Curtis, *The Oppenheimer Case*, 2. See also Bernstein, "The Oppenheimer Loyalty-Security Case Reconsidered," 1387, 1461; Bernstein, "In the Matter of J. Robert Oppenheimer," 221; Holloway, *In the Matter of J. Robert Oppenheimer*, 111; Lakoff, "The Trial of Dr. Oppenheimer," 70–72; and "Transcript," 201.

72. Curtis, *The Oppenheimer Case*, 2, 42. See, for example, Robb's interrogation of Bethe, in "Transcript," 331.

73. Green, "The Oppenheimer Case," 59, 60; Rhodes, *Dark Sun*, 536. Before the start of the hearings, Evans remarked that from his experience on security boards, he believed that nearly all subversives were Jews (Bernstein, "In the Matter of J. Robert Oppenheimer," 217–18; Bird and Sherwin, *American Prometheus*, 539).

74. Oppenheimer's attorney, Lloyd Garrison, complained about his lack of access to these files (Garrison to Nichols, June 1, 1954, in *Oppenheimer, FBI Security File*,

100-17828, sec. 44). Rhodes argued that Garrison made an important mistake in not applying for security clearance for himself (*Dark Sun,* 540).

75. Green, "The Oppenheimer Case," 59.

76. McCarthy, quoted in "McCarthy vs. Murrow," *New York Herald Tribune,* April 11, 1954. See also "McCarthy Says He'll Show Oppenheimer Was a Red," *New York Herald Tribune,* April 14, 1954; "Delay and the H-Bomb," editorial, *New York Herald Tribune,* April 10, 1954; "H-Bomb Held Back, M'Carthy Asserts," *New York Times,* April 7, 1954; "Eisenhower Sees No Need to Build a Larger H-Bomb," *New York Times,* April 8, 1954; "McCarthy and the H-Bomb," *New York Times,* editorial, April 8, 1954; "The H-Bomb Decision," *New York Times,* April 9, 1954; Thomas C. Reeves, *The Life and Times of Joe McCarthy: A Biography* (London: Madison Books, 1997), 589–90; and Bernstein, "In the Matter of J. Robert Oppenheimer," 219–21. For press clippings on the security hearings, see Oppenheimer Papers, boxes 208–16.

77. Bernstein, "In the Matter of J. Robert Oppenheimer," 219–21; Bird and Sherwin, *American Prometheus,* 503, 546.

78. Curtis, *The Oppenheimer Case,* 10–16, 57; Green, "The Oppenheimer Case," 14–15; Bernstein, "The Oppenheimer Loyalty-Security Case Reconsidered," 1465–67; Eleanor Bontecou, "President Eisenhower's 'Security' Program," *Bulletin of the Atomic Scientists* 9, no. 6 (July 1953): 215–17, 220. See also Forman, "Behind Quantum Electronics"; Oreskes and Rainger, "Science and Security"; and Lawrence Badash, "Science and McCarthyism," *Minerva* 38 (2000): 53–80.

79. Oppenheimer, quoted in Rigden, *Rabi,* 228.

80. Malraux, quoted in Chevalier, *Oppenheimer,* 115.

81. Cf. the discussion of the "documentary method of interrogation" in Michael Lynch and David Bogen, *The Spectacle of History: Speech, Text, and Memory at the Iran-Contra Hearings* (Durham, NC: Duke University Press, 1996), esp. 201–35.

82. However, after Garrison protested, the transcripts of Oppenheimer's 1943 interrogations by Pash and Lansdale were made available to Oppenheimer and his attorneys. See "Transcript," 201, 203, 281, 285–93, 871–86; and Curtis, *The Oppenheimer Case,* 213–18.

83. For examples, see "Transcript," 123, 144.

84. Groves testified that when he had ordered Oppenheimer to tell him the name of the intermediary, "I got what to me was the final story" ("Transcript," 168).

85. "Transcript," 152–53. See also Chevalier, *Oppenheimer,* 165–66. Oppenheimer contested the fixing of the date of his revelation to Groves by reference to this telegram. He said, "I thought it was earlier. It could have been that late. I thought it was earlier" ("Transcript," 153).

86. "Recommendations of the General Manager," 1043.

87. "Decision and Opinions," 1050–51.

88. Ibid., 1049.

89. The reality was, of course, different. Lilienthal said that Strauss's role in the Oppenheimer case showed "how much the course of events is affected by wholly

personal quirks, by what seem at the time little personal things" (quoted in Stern, *The Oppenheimer Case,* 130).

90. "Findings," 1013.

91. Oppenheimer's wartime appeal concerning Giovanni Rossi Lomanitz's draft deferment was also regarded as fitting this pattern.

92. "Findings," 1018–19. See also "Transcript," 134, 210–15, 252–53.

93. "Findings," 1015.

94. See Rieff, "The Case of Dr. Oppenheimer," in *On Intellectuals,* 363–64.

95. Lakoff, "The Trial of Dr. Oppenheimer," 66. An FBI agent described Oppenheimer as "a master at inuendo [*sic*] and evasive answers" (W. A. Branigan to A. H. Belmont, "Dr. J. Robert Oppenheimer," March 31, 1954, in *Oppenheimer, FBI Security File,* 100-17828, sec. 29).

96. For a general discussion of "denigration through contextualization," see Adi Ophir and Steven Shapin, "The Place of Knowledge: A Methodological Survey," *Science in Context* 4 (1991): 4.

97. "Transcript," 517. On these accounts of Oppenheimer's role in the GAC, see also Wilson, *The Great Weapons Heresy,* 138–39.

98. "Transcript," 461. See also the similar testimony of Walter G. Whitman (496).

99. See, for example, the testimony of Thomas Keith Glennan, in "Transcript," 256.

100. "Transcript," 566–67, 565. See also Vannevar Bush, "If We Alienate Our Scientists," *New York Times Magazine,* June 13, 1954, 9, 60, 62, 64–65, 67.

101. Major, *The Oppenheimer Hearing,* 144–45.

102. "Transcript," 385; quoted in part also in Major, *The Oppenheimer Hearing,* 145.

103. Garrison, quoted in Major, *The Oppenheimer Hearing,* 145. See also Stern, *The Oppenheimer Case,* 510.

104. "Findings," 1019 (hereafter cited in text).

105. Curtis, *The Oppenheimer Case,* 152.

106. *New York Times* columnist James Reston wrote of the PSB opinion, "They suggest that a scientist, like a soldier, is expected not only to obey but also to show 'enthusiasm' for Government policies, regardless of his own convictions" ("How Expendable Are the Atomic Scientists?" *New York Times,* June 6, 1954).

107. "Findings," 1020–21; Major, *The Oppenheimer Hearing,* 194. Bizarrely, it seems that Evans's dissent was actually penned by Robb. Evans's original version was apparently so poorly constructed that Gray feared embarrassment to the board (Bernstein, "In the Matter of J. Robert Oppenheimer," 239; Bird and Sherwin, *American Prometheus,* 540).

108. "McCarthy Skeptical," *New York Herald Tribune,* April 10, 1954, 10. Eisenhower's statement is discussed in A. H. Belmont to L. V. Boardman, "J. Robert Oppenheimer," April 11, 1954, in *Oppenheimer, FBI Security File,* 100-17828, sec. 31.

109. "Recommendations of the General Manager," 1045.

110. "Decision and Opinions," 1051 (hereafter cited in text).

111. Karl K. Darrow to H. A. Bethe and R. T. Birge, June 5, 1954, Birge Papers.

112. *New York Journal-American,* April 17, 1954.

113. Major, *The Oppenheimer Hearing,* 276–97.

114. Columns by David Lawrence in the *New York Herald Tribune* during this period included "Oppenheimer's Side Gave Publicity to His Suspension" (April 14), "Drive to Discredit A.E.C. Is Laid to 'Left Wingers'" (June 17), "Oppenheimer Verdict Held Devastating Condemnation" (June 30), "Security Task Complicated by Solidarity of Scientists" (June 22), "Oppenheimer Gave to Reds after Hitler-Stalin Alliance" (June 24), and "'Leftist' Drive to Sabotage Security System Is Seen" (December 30). Columns by Joseph and Stewart Alsop in the *Herald Tribune* included "What Is Security?" (June 4), "Super and Security" (June 7), "The Scientific Hornets" (June 18), and "Operation Spill-the-Beans" (July 12).

115. "The Oppenheimer Case," editorial, *New York Herald Tribune,* April 14, 1954.

116. "The Gray Report: Findings and Recommendations," editorial, *New York Herald Tribune,* June 3, 1954.

117. "The Essence of Law," editorial, *New York Herald Tribune,* June 8, 1954. See also "The Oppenheimer Documents," editorial, *New York Herald Tribune,* June 17, 1954.

118. "The A.E.C. Verdict," editorial, *New York Herald Tribune,* July 1, 1954.

119. Ibid.

120. Arthur Krock, "Verdict on the Criterion of 'Trustworthiness,'" *New York Times,* July 1, 1954.

121. Extracts from editorials, reprinted in "Editorial Views on the Oppenheimer Decision," *New York Times,* July 2, 1954. See also "Press Comment on Oppenheimer," *New York Herald Tribune,* July 1, 1954.

122. "Science Schism Widens: Oppenheimer Debate Impedes Progress in United States Military Technology," *New York Times,* July 1, 1954.

123. William L. Laurence, "Oppenheimer Case Recalls Momentous H-Bomb Debate," *New York Times,* June 3, 1954. See also E. W. Kenworthy, "The Drama of the Hydrogen Bomb—and Dr. Oppenheimer's Key Role," *New York Times,* April 18, 1954.

124. Waldemar Kaempffert, "X-Ray of the Scientific Mind," *New York Times Magazine,* April 25, 1954, 7, 54–56.

125. "Scientist's Views Stir Panel Worry," *New York Times,* June 2, 1954.

126. *New York Times,* September 7, 1954.

127. "214 Atomic Scientists in Protest," *New York Herald Tribune,* June 22, 1954; Joseph and Stewart Alsop, "Operation Don't Argue!" *New York Herald Tribune,* July 14, 1954.

128. Killian, quoted in Newman, "The Oppenheimer Case," 181.

129. Goudsmit, quoted in William L. Laurence, "Oppenheimer Case Divides Physicists: Subjecting of Atomic Expert to 'Security Risk' Hearing Debated by Colleagues," *New York Times*, April 30, 1954. See also Cassidy, *J. Robert Oppenheimer*, 328–30.

130. Joseph and Stewart Alsop, "Gamy Stuff," *New York Herald Tribune*, October 3, 1954; "Shepley and Blair Discuss Their Book on the H-Bomb," *New York Herald Tribune*, October 14, 1954; "Science Fiction," *Santa Fe New Mexican*, September 26, 1954. See also Goodchild, *Edward Teller*, 253–55.

131. "Teller Denies Responsibility for 'Any Part' of H-Bomb Book," *Santa Fe New Mexican*, September 26, 1954; Edward Teller, "The Work of Many People," *Science*, February 25, 1955, 267–75; Teller, *Memoirs*, 401–4. See also "Strauss Tried to Block Hydrogen Bomb Book," *New York Herald Tribune*, September 27, 1954.

132. Nat S. Finney, "The Paradox of the H-bomb and the Power Vacuum in Washington," *New York Herald Tribune*, October 3, 1954.

133. "Division in the A.E.C.," editorial, *New York Herald Tribune*, June 7, 1954. See also Walter Kerr, "Split over Oppenheimer in A.E.C. Called Likely," *New York Herald Tribune*, June 3, 1954; and Cassidy, *J. Robert Oppenheimer*, 337.

134. "Critic of Strauss to Quit Atom Post," *New York Times*, May 22, 1954.

135. "Smyth Quits Atom Post, Voted for Oppenheimer," *New York Herald Tribune*, September 21, 1954.

136. "Nichols to Quit A.E.C., Open Office," *New York Herald Tribune*, January 30, 1955.

137. This comment was recorded in a memo to Strauss by Charter Heslep, an AEC liaison officer with the U.S. Information Agency. Heslep added in parentheses, "This last phrase is mine and he [Teller] agrees it is apt" (quoted in Blumberg and Owens, *Energy and Conflict*, 359; Bird and Sherwin, *American Prometheus*, 532; Goodchild, *Edward Teller*, 240–41).

138. Alistair Cooke, "Suspension of Dr. Oppenheimer a Test Case," *Guardian* (Manchester), April 22, 1954.

139. Max Ascoli, "Editorial: The Jurisprudence of Security," *The Reporter*, July 6, 1954, 8–9.

140. "Paper, Oppenheimer Continue to Differ," *New York Times*, June 5, 1954.

141. Michael Amrine, "A Generation on Trial," *The Progressive*, June 1954, 5–7, esp. 5.

142. Bell, *Post-industrial Society*, 400.

CHAPTER EIGHT

1. Coughlan, "The Equivocal Hero of Science: Robert Oppenheimer," *Life*, February 1967, 34A; Alfred Friendly, "Oppenheimer: Scientist Beset by Political Ills," *Washington Post*, February 20, 1967.

2. Victor Weisskopf, interviewed by Los Alamos National Laboratory, ca. 1981, LANL TR-81-013 4/81.

3. Pais, *A Tale of Two Continents,* 331.

4. Quoted in Jungk, *Brighter Than a Thousand Suns,* 298.

5. Rabi, quoted in Rigden, *Rabi,* 243.

6. Jackson and Murray, quoted in Cushing Strout, *Conscience, Science, and Security: The Case of Dr. J. Robert Oppenheimer* (Chicago: Rand McNally, 1963), 57–58, on 58; Oppenheimer, quoted in Goodchild, *J. Robert Oppenheimer,* 275; Coughlan, "The Equivocal Hero of Science," 34A; Friendly, "Oppenheimer."

7. "Oppenheimer Kept in Institute Post," *New York Times,* April 14, 1954; Friendly, "Oppenheimer." Bird and Sherwin noted that Oppenheimer had turned in his resignation the previous year, citing as reasons his poor relations with the faculty and his wife's health problems (*American Prometheus,* 580).

8. See, for example, Walter Kerr, "Several Unwitting Distortions Noted in Reporting the Oppenheimer Case," *New York Herald Tribune,* Weekly Review, June 20, 1954; "Keeping the Record Straight," editorial, *New York Herald Tribune,* June 22, 1954; and Roscoe Drummond, "Weakness of the Gray Report," *New York Herald Tribune,* June 27, 1954. Press clippings relating to the security hearings are in Oppenheimer Papers, boxes 208–16; clippings from the *New York Herald Tribune* and *New York Times* are in boxes 211–13.

9. Walter Lippmann, "Disorderly Government," *New York Herald Tribune,* June 3, 1954; "J. Robert Oppenheimer: His Life and Times," *Time,* April 26, 1954, 19–21, quoting 21. See also Walter Lippmann, "The Oppenheimer Case," *New York Herald Tribune,* June 7, 1954, in Strout, *Conscience, Science, and Security,* 47.

10. Joseph and Stewart Alsop, *We Accuse!* 13, 18–25, 59; Waldo Frank, "Oppenheimer's 'Folly': An Alsop Fable," *The Nation,* March 5, 1955, 193–95.

11. Arthur M. Schlesinger Jr., "The Oppenheimer Case," *Atlantic Monthly,* October 1954, in Strout, *Conscience, Science, and Security,* 47–48; Tolman, quoted in Murray Illson, "Anti-Intellectual Fever Impeding U.S. Science, Psychologists Told," *New York Times,* June 8, 1954; Illson, "Tolman Assails Fear of Thinking," *New York Times,* June 12, 1954.

12. Cf. Steven Shapin, "Who Is the Industrial Scientist? Commentary from Academic Sociology and from the Shop-Floor in the United States, ca. 1900–ca. 1960," paper presented to the 123rd Nobel Symposium, "Science and Industry in the 20th Century," Royal Swedish Academy of Sciences, November 21–23, 2002.

13. Rieff, "The Case of Dr. Oppenheimer," in *On Intellectuals,* esp. 368. The essay was first published in *The Twentieth Century* 156, nos. 930–31 (1954).

14. Trilling, quoted in Richard H. Pells, *The Liberal Mind in a Conservative Age: American Intellectuals in the 1940s and 1950s* (New York: Harper and Row, 1985), 273, 284; Diana Trilling, "The Oppenheimer Case: A Reading of the Testimony," *Partisan Review* 21, no. 6 (November–December 1954), 615, 633.

15. See Hans Meyerhoff, "Through the Liberal Looking Glass—Darkly," *Partisan Review* 22, no. 2 (Spring 1955): 238–48.

16. Congress for Cultural Freedom, *Science and Freedom: The Proceedings of a Conference Convened by the Congress for Cultural Freedom and Held in Hamburg on July 23rd–26th, 1953* (Boston: Beacon Press, 1955); American Committee for Cultural Freedom, press release, November 12, 1954, Oppenheimer Papers, box 110.

17. Sol Stein to Oppenheimer, March 5, 1954; memo on telephone conversation between Oppenheimer and Sol Stein, March 6, 1954; Oppenheimer, in transcript of telephone conversation between Dr. Oppenheimer and Mrs. van Gelder of Mr. Leidersdorf's office, March 5, 1954, Oppenheimer Papers, box 32. See also "2 Scientists Refuse Bid to Einstein Fete," *New York Post*, March 11, 1954; Frederick Woltman, "Einstein in Middle of New Commie Hassle," *New York World-Telegram and The Sun*, March 11, 1954; and Clark Foreman to Oppenheimer, April 7, May 5, and May 28, 1954, Oppenheimer Papers, box 32.

18. "'Refuse to Testify,' Einstein Advises Intellectuals Called In by Congress," *New York Times*, June 12, 1953; Abraham Pais, *Einstein Lived Here* (Oxford: Clarendon Press, 1994), 236–42, esp. 241; Pais, *A Tale of Two Continents*, 326; and esp. Pais, *'Subtle is the Lord . . . : The Science and Life of Albert Einstein* (Oxford: Clarendon Press, 1982), 10–11. See also "Einstein Backs Oppenheimer," *New York Times*, April 13, 1954.

19. Oppenheimer, in transcript of telephone conversation between Dr. Oppenheimer and Mrs. van Gelder of Mr. Leidersdorf's office, March 5, 1954; memo on telephone conversation between Oppenheimer and Sol Stein, March 6, 1954; memo on telephone conversation with Miss Helen Dukas, March 9 [1954], Oppenheimer Papers, box 32.

20. American Committee for Cultural Freedom, press release, November 12, 1954.

21. Pells, *Liberal Mind*, 343–44, quoting 344.

22. Robert Gorham Davis to Oppenheimer, July 20, 1954 (Oppenheimer's secretary wrote on this document that he sent in his membership card on November 9, 1954); American Committee for Cultural Freedom, press release, November 12, 1954, Oppenheimer Papers, box 110. See also Michael A. Day, "Oppenheimer on the Nature of Science," *Centaurus* 43 (2001), esp. 74, 109 n. 3.

23. Jane A. Sanders, "The University of Washington and the Controversy over J. Robert Oppenheimer," *Pacific Northwest Quarterly* 70, no. 1 (January 1979): 8–19; Schrecker, *No Ivory Tower*, 94–112, 267–68, 310–37; Wang, *American Science in an Age of Anxiety*, 271–74; Jane A. Sanders, *Cold War on Campus: Academic Freedom at the University of Washington, 1946–1964* (Seattle: University of Washington Press, 1979); Pells, *Liberal Mind*, 287–94; Sigmund Diamond, *Compromised Campus: The Collaboration of Universities with the Intelligence Community, 1945–1955* (New York: Oxford University Press, 1992).

24. Sanders, "University of Washington," 15.

25. C. A. G. Wiersma to Arthur W. Martin Jr., March 8, 1955, Papers of Lee A. DuBridge, box 94, California Institute of Technology Archives, Pasadena. See also "University Cancels Conference: Scientists Shun Meeting that Barred Oppenheimer," *New York Herald Tribune*, March 24, 1955.

26. Oppenheimer, quoted in Sanders, "University of Washington," 15. The proposal to invite Oppenheimer as a visiting professor was originally made by the physics department chairman, Oppenheimer's former Los Alamos colleague John Manley, whom Uehling was replacing while Manley was on leave.

27. Ibid., 19.

28. Weisskopf to Oppenheimer, quoted in David Halberstam, *The Fifties* (New York: Villard Books, 1993), 349.

29. "Scientists Express Confidence in Oppenheimer," *Bulletin of the Atomic Scientists* 10, no. 7 (September 1954): 283, 286. See also Alan Simpson, "The Re-trial of the Oppenheimer Case," *Bulletin of the Atomic Scientists* 10, no. 10 (December 1954): 387–88; "Oppenheimer Suspension Stirs Scientists," *F.A.S. [Federation of American Scientists] Newsletter*, April 26, 1954, DuBridge Papers, box 33; "Trust in Oppenheimer Expressed by Scientists," *New York Herald Tribune*, April 14, 1954; "Scientists Appeal to Eisenhower: Oppenheimer's 'Purge' is Hit," *New York Herald Tribune*, June 7, 1954; "Einstein among Signers: Oppenheimer Called Loyal by 26 Princeton Colleagues," *New York Herald Tribune*, July 1, 1954; and "Atom Scientists Charge U.S. Broke Faith on Oppenheimer," *New York Herald Tribune*, May 21, 1954.

30. Editorial, *Los Angeles Examiner*, June 24, 1954; Arthur Calvin Little to Lee DuBridge, June 24, 1954; DuBridge to Little, June 28, 1954, DuBridge Papers, box 33.

31. Christy Walsh to Lee DuBridge, July 3, 1954, DuBridge Papers, box 33. This, together with other items of public correspondence to DuBridge, is also quoted in Cassidy, *J. Robert Oppenheimer*, 337.

32. On these cultural conflicts, see David A. Hollinger, "Jewish Intellectuals and the De-Christianization of American Public Culture in the Twentieth Century," in *Science, Jews, and Secular Culture*, 17–41; and Hollinger, "Science as a Weapon in *Kulturkämpfe* in the United States during and after World War II," ibid., 155–74.

33. "Oppenheimer a Sermon Topic: President Hears Prayer for a Worthy Congress," *New York Herald Tribune*, November 1, 1954.

34. Frances P. Bartlett, "The Atomic Brain Trust," *American Mercury* 81 (July 1955): 75, 78.

35. DuBridge to Mrs. Gladys G. Dickey, July 16, 1954, DuBridge Papers, box 33.

36. DuBridge to Edward U. Condon, July 14, 1954, DuBridge Papers, box 33.

37. DuBridge to the Board of Trustees, California Institute of Technology, July 12, 1954, DuBridge Papers, box 33.

38. Blumberg and Owens, *Energy and Conflict*, 364–80; Teller, *Memoirs*, 401.

39. Teller, quoted in Blumberg and Owens, *Energy and Conflict,* 380. See also H. L. Nieburg, *In the Name of Science* (Chicago: Quadrangle Books, 1966), 131–32, 135 57.

40. Quoted in Blumberg and Owens, *Energy and Conflict,* 364.

41. Ed Regis, *Who Got Einstein's Office? Eccentricity and Genius at the Institute for Advanced Study* (Reading, MA: Addison-Wesley, 1987), 139–52; George Johnson, *Strange Beauty: Murray Gell-Mann and the Revolution in Twentieth-Century Physics* (New York: Alfred A. Knopf, 1999), 80. See also Cassidy, *J. Robert Oppenheimer,* 268–72.

42. Johnson, *Strange Beauty,* 82.

43. Pais, *A Tale of Two Continents,* 221–22; Oppenheimer to David E. Lilienthal, October 22, 1947, Oppenheimer Papers, box 46.

44. Bernstein, *The Life It Brings,* 99; Brown, *Through These Men,* 290–91.

45. Robert Oppenheimer to Frank Oppenheimer, July 12, 1952, Oppenheimer Papers, box 237.

46. Oppenheimer, interviewed by Kuhn, November 20, 1963, 32, Archives for the History of Quantum Physics, AIP.

47. James B. Conant, *Science and Common Sense* (1951; New Haven, CT: Yale University Press, 1963), 1. See also Steve Fuller, *Thomas Kuhn: A Philosophical History for Our Times* (Chicago: University of Chicago Press, 2000), 150–78, 179–226, esp. 151–54, 222–23; Hershberg, *James B. Conant;* David Hollinger, "Free Enterprise and Free Inquiry: The Emergence of Laissez-Faire Communitarianism in the Ideology of Science in the U.S.," in *Science, Jews, and Secular Culture,* 97–120; and Stephen Turner, *Liberal Democracy 3.0: Civil Society in an Age of Experts* (London: Sage, 2003), esp. 15–16, 117–28.

48. Oppenheimer to James B. Conant, May 13, 1947, Oppenheimer Papers, box 27. See also James B. Conant, *On Understanding Science: An Historical Approach* (New Haven, CT: Yale University Press, 1947).

49. Oppenheimer, "Physics in the Contemporary World," in *The Open Mind,* 97, 98.

50. J. Robert Oppenheimer, "Science as Action: Rutherford's World," in *Science and the Common Understanding,* 20–36.

51. Yaron Ezrahi, "Einstein and the Light of Reason," in *Albert Einstein: Historical and Cultural Perspectives,* ed. Gerald Holton and Yehuda Elkana (Mineola, NY: Dover Publications, 1982), esp. 259–73.

52. Oppenheimer, "Science as Action," 36.

53. Jeffries Wyman, interviewed by Charles Weiner, May 28, 1975, 29, Oppenheimer Oral History Collection, MIT Archives.

54. J. Robert Oppenheimer, "Uncommon Sense," in *Science and the Common Understanding,* 74–91, esp. 75, 91. On the philosophical background of complementarity, see Lewis Feuer, *Einstein and the Generations of Science* (New York: Basic Books, 1974), 121–26; Gerald Holton, "The Roots of Complementarity," in *Thematic Origins*

of Scientific Thought (Cambridge, MA: Harvard University Press, 1973), 115–61; and David Kaiser, "More Roots of Complementarity: Kantian Aspects and Influences," *Studies in History and Philosophy of Science and Technology* 23, no. 2 (1992): 213–39.

55. Oppenheimer, "Physics in the Contemporary World," 92.

56. J. Robert Oppenheimer, "The Sciences and Man's Community," in *Science and the Common Understanding*, 92–110, on 98, 97.

57. Ibid., 101–2.

58. Ibid., 102–3, quoting 103.

59. "The Message of the New Physics," review of Oppenheimer, *Science and the Common Understanding*, in *Times* (London), October 22, 1954.

60. Brown, *Through These Men*, 290.

61. "London Letter," *Manchester Guardian Weekly*, April 22, 1954.

62. Letters to Oppenheimer from BBC listeners, quoted in Ferenc Morton Szasz, "Great Britain and the Saga of J. Robert Oppenheimer," *War in History* 2, no. 3 (1995): 324.

63. J. L. Synge, review of Oppenheimer, *Science and the Common Understanding*, in *Universities Quarterly*, February 1955, 206, 208. See also Szasz, "Great Britain," 325.

64. Lewis Gannet, review of Oppenheimer, *Science and the Common Understanding*, in *New York Herald Tribune*, June 22, 1954.

65. J. Robert Oppenheimer, "Prospects in the Arts and Sciences," in *The Open Mind*, 133–46, quoting 144, 145.

66. Ibid., 140, 138, 142–43.

67. Cf. José Ortega y Gasset, *The Revolt of the Masses* (New York: W. W. Norton, 1932); Clement Greenberg, "Avant Garde and Kitsch," *Partisan Review* 6 (1939): 34–49; and George Cotkin, "The Tragic Predicament: Post-war American Intellectuals, Acceptance, and Mass Culture," in *Intellectuals in Politics: From the Dreyfus Affair to Salman Rushdie*, ed. Jeremy Jennings and Anthony Kemp-Welch, 284–70 (London: Routledge, 1997).

68. Alistair Cooke, "Isolation of the Specialist in America," *Guardian* (Manchester), December 30, 1954.

69. "The Scotsman's Log," *Scotsman*, December 30, 1954.

70. Transcript of conversation between J. Robert Oppenheimer and Edward R. Murrow, recorded December 16, 1954, for CBS television program *See It Now*, broadcast on January 4, 1955, in *Oppenheimer, FBI Security File*, 100-17828, sec. 55, quoting 7. For positive press reaction to the Murrow interview, see, for example, John Crosby, "Portrait by Television," *New York Herald Tribune*, January 7, 1955.

71. Transcript of conversation between Oppenheimer and Murrow, December 16, 1954, 3, 2. On Einstein's politics, see also Jerome, *The Einstein File*. On Oppenheimer's relationship with Einstein, see also Bird and Sherwin, *American Prometheus*, 379–82.

72. However, during the interview, Oppenheimer did criticize immigration restrictions that had prevented many prominent European scientists from visiting the United States. "Barring Foreign Scientists Absurd, Oppenheimer Says," *New York Herald Tribune,* January 5, 1955; "Oppenheimer Assailed for Alien Law Criticism," *New York Herald Tribune,* January 16, 1955; see also "Bricker Decries Scientists' Talk: Objects to Some Who Seek to Influence Public on Topics outside Their Fields," *New York Times,* February 6, 1955.

73. Victor Cohn, "Can a Security Risk Survive?" *Minneapolis Sunday Tribune,* June 16, 1957; Cohn, "Oppenheimer Survives 'Security Risk' Branding," *Detroit News,* June 24, 1957; clippings, and correspondence with Cohn, in Oppenheimer Papers, box 208.

74. John W. Finney, "Hagen Estimates Orbit," *New York Times,* October 12, 1957.

75. Vannevar Bush, "The Gentleman of Culture," talk delivered at Pingry School, Hillside, New Jersey, December 6, 1958, printed (with modifications) in Bush, *Science Is Not Enough* (New York: William Morrow, 1967), 31–49, quoting 6.

76. Ibid., 17, 9, 10, 1, 6.

77. J. Robert Oppenheimer, "Knowledge and the Structure of Culture," Helen Kenyon Lecture, Vassar College, Poughkeepsie, NY, October 29, 1958 (manuscript copy), 3.

78. Ibid., 14.

79. J. Robert Oppenheimer, *Some Reflections on Science and Culture,* 1959 lecture (Chapel Hill: University of North Carolina, 1960), 5.

80. Ibid., 8–9.

81. Ibid., 22–23.

82. C. P. Snow, *The Two Cultures* (1959; Cambridge: Cambridge University Press, 1993). Snow first set out his ideas in an October 1956 *New Statesman* article. While in Japan in 1960, Oppenheimer dismissed Snow with the remark, "Everything he says is of a degree triviality and childishness" ("An Afternoon with Professor Oppenheimer," *Science and Man,* September 23, 1960, 2, Oppenheimer Papers, box 261). On the "two cultures" debate in America and its context, see Michael A. Day, " I.I. Rabi: The Two Cultures and the Universal Culture of Science," *Physics in Perspective* 6, no. 4 (2004): 428–76.

83. Jacques Barzun, *The House of Intellect* (New York: Harper and Row, 1959), 11.

84. Clark Kerr, *The Uses of the University* (1963; Cambridge, MA: Harvard University Press, 1972), 101. These were Kerr's Godkin Lectures at Harvard University, on April 23, 24, and 25, 1963. Kerr was quoting from J. Robert Oppenheimer, "Science and the Human Community," in *Issues in University Education: Essays by Ten American Scholars,* ed. Charles Frankel (New York: Harper and Brothers, 1959), 56, 58; and referring to David Riesman, Reuel Denney, and Nathan Glazer, *The Lonely Crowd: A Study of the Changing American Character* (New Haven, CT: Yale University Press, 1950).

85. David Kaiser, "The Postwar Suburbanization of American Physics," *American Quarterly* 56 (2004): 851–88.

86. Shils, quoted in Peter Coleman, *The Liberal Conspiracy: The Congress for Cultural Freedom and the Struggle for the Mind of Postwar Europe* (New York: Free Press, 1989), 121.

87. Oppenheimer to Nicolas Nabokov, June 3, 1959, Oppenheimer Papers, box 253.

88. J. Robert Oppenheimer, "In the Keeping of Unreason," *Bulletin of the Atomic Scientists* 16, no. 1 (January 1960): 22.

89. Ibid.

90. Karl Jaspers, *The Atom Bomb and the Future of Man* (Chicago: University of Chicago Press, 1961), 202.

91. Mary McCarthy to Hannah Arendt, September 28, 1962, quoted in Saunders, *Who Paid the Piper?* 332.

92. Aron, *Memoirs*, 175.

93. "Intellectuals," *Time*, July 4, 1960, 17. See also Coleman, *The Liberal Conspiracy*, 174–76.

94. Journal entry for March 27, 1955, in David E. Lilienthal, *Venturesome Years, 1950–1955*, vol. 3 of *The Journals of David E. Lilienthal* (New York: Harper and Row, 1966), 618.

95. Ibid.

96. Bertrand Russell to Oppenheimer, February 8, 1957; Oppenheimer to Russell, February 18, 1957; Russell to Oppenheimer, March 11, 1957, Oppenheimer Papers, box 59. See also Joseph Rotblat, *Scientists in the Quest for Peace: A History of the Pugwash Conferences* (Cambridge, MA: MIT Press, 1972); and Rotblat, *Pugwash: The First Ten Years; History of the Conferences on Science and World Affairs* (London: Heinemann, 1967).

97. FBI Director to Legal Attache, London, August 27, 1954; W. A. Branigan to A. H. Belmont, "Dr. J. Robert Oppenheimer, Internal Security," August 27, 1954, in *Oppenheimer, FBI Security File*, 100-17828, sec. 49; "Oppenheimer to Paris: Will Lecture at Sorbonne as Exchange Professor," *New York Times*, February 8, 1958; "Oppenheimers Return from Europe; Toni Rides Belgian Prince's Horse," *Princeton Packet*, July 17, 1958. On FBI surveillance of Oppenheimer after the hearing and the bureau's fear that he might defect, see also Bird and Sherwin, *American Prometheus*, 554–55.

98. Giles Scott-Smith, *The Politics of Apolitical Culture: The Congress for Cultural Freedom, the CIA and Post-war American Hegemony* (London: Routledge, 2002), 105.

99. *Time*, September 19, 1960, 47; quoted also in Goodchild, *J. Robert Oppenheimer*, 274. The Japanese press reported his comments with a somewhat different emphasis: "I feel just as sorry tonight as I felt last night" ("Oppenheimer to Give Series of Talks Here," *Asahi Evening News*, September 6, 1960). Clippings and

correspondence relating to Oppenheimer's trip to Japan are in Oppenheimer Papers, box 41.

100. "Oppenheimer to Give Series of Talks Here."

101. Nicholas Evan Sarantakes, "Alliance in Doubt: American Reaction to the 1960 US-Japanese Security Treaty Crisis," *American Diplomacy* 4, no. 3 (Autumn 1999), http://www.unc.edu/depts/diplomat/AD_Issues/amdipl-13/sarantakes1.html; George R. Packard III, *Protest in Tokyo: The Security Treaty Crisis of 1960* (Princeton, NJ: Princeton University Press, 1966), esp. 3–17, 47–52, 258, 288–89, 291, 300–303.

102. John W. Dower, *Embracing Defeat: Japan in the Wake of World War Two* (London: Allen Lane, 1999), 487–95; Lise Yoneyama, *Hiroshima Traces: Time, Space, and the Dialectics of Memory* (Berkeley and Los Angeles: University of California Press, 1999), esp. 14; James J. Orr, *The Victim as Hero: Ideologies of Peace and National Identity in Postwar Japan* (Honolulu: University of Hawaii Press, 2001), esp. 36–71; John W. Dower, "The Bombed: Hiroshimas and Nagasakis in Japanese Memory," *Diplomatic History* 19, no. 2 (Spring 1995): 275–95.

103. "Oppenheimer Met by American Youth," *Mainichi Daily News,* September 18, 1960; Earle Reynolds to Oppenheimer, September 16, 1960, Oppenheimer Papers, box 41.

104. Dower, *Embracing Defeat,* 494.

105. On the role of Marxist-oriented "progressive intellectuals" in the popular movement against the security treaty, see Packard, *Protest in Tokyo,* 26–31, 138–41, 270–78.

106. "Dr. Oppenheimer Gives Lecture In Kyoto Univ.," *Mainichi Daily News,* September 15, 1960; "Oppenheimer in Kyoto," *Asahi Picture News,* September 1960; "1,000 Hear Oppenheimer Give Talk," *Asahi Evening News,* September 19, 1960; "Oppenheimer to Talk in Tokyo on Sept. 21," *Asahi Evening News,* September 15, 1960; "Oppenheimer Stresses Meaning of 'Science,'" *Asahi Evening News,* September 22, 1960; J. Robert Oppenheimer, "The Future of Civilization in the Scientific Age," *France-Asie,* March–April 1961, 1807–15.

107. Oppenheimer, "Science and Culture," *International House of Japan Bulletin* 6 (October 1960): 2–10, 21, quoting 10; Oppenheimer to Yasaka Takagi, October 27, 1960; Yasaka Takagi to Oppenheimer, November 26, 1960, Oppenheimer Papers, box 41.

108. Johnson, quoted in Goodchild, *J. Robert Oppenheimer,* 275. See also Major, *The Oppenheimer Hearing,* 293; and Pais, *A Tale of Two Continents,* 331.

109. *New York Herald Tribune,* April 8, 1963, quoted in Major, *The Oppenheimer Hearing,* 292.

110. *New York Times,* April 4, 1963, quoted in Stern, *The Oppenheimer Case,* 455.

111. Thomas B. Morgan, "A Visit with J. Robert Oppenheimer," *Look,* April 1, 1958, 35–38, quoting 38.

112. J. Robert Moskin, "Morality USA: Have Bigness, the Bomb, and the Buck Destroyed Our Old Morality?" *Look,* September 24, 1963, 74–88, quoting 74, 75; see also Moskin to Oppenheimer, October 7, 1963, Oppenheimer Papers, box 47.

113. Abraham Pais, *Niels Bohr's Times, in Physics, Philosophy, and Polity* (Oxford: Clarendon Press, 1991).

114. J. Robert Oppenheimer, "Niels Bohr and Atomic Weapons," *New York Review of Books,* December 17, 1964, 6–8, quoting 6. See also Oppenheimer's Pegram Lectures, delivered at Brookhaven National Laboratory in August 1963, Oppenheimer Papers, box 247.

115. Oppenheimer, "Niels Bohr and Atomic Weapons," 8.

116. See Margaret Gowing, *Britain and Atomic Energy, 1939–1945* (New York: St. Martin's Press, 1964), 358–59.

117. See, for example, Howard Sims, "Bohr's Motives Misunderstood, Oppenheimer Says. Danish Physicist's Wartime Aims Defended," *Washington Post,* December 7, 1969.

118. Oppenheimer to R. B. Silvers, December 5, 1964; Silvers to Oppenheimer, December 8, 1964, Oppenheimer Papers, box 262.

119. J. Robert Oppenheimer, "War and the Nations," in *The Flying Trapeze: Three Crises for Physicists* (London: Oxford University Press, 1964), 54–65, on 64.

120. Oppenheimer said that "the letter had very little effect, and that Einstein himself is not really answerable for all that came later." J. Robert Oppenheimer, "On Albert Einstein," *New York Review of Books,* March 17, 1966, 4–5.

121. Oppenheimer also had an article published in the CCF journal: "The Added Cubit," *Encounter* 21, no. 2 (August 1963): 43–47.

122. Oppenheimer to Nicolas Nabokov, April 13, 1962, Oppenheimer Papers, box 66. Correspondence between Oppenheimer and Agnes Meyer, dating from the late 1950s, is in Oppenheimer Papers, box 51. In 1955, Agnes Meyer had made a speech in which she asserted that the Oppenheimer case was indicative of "a rising 'anti-intellectualism' in this country [which] was driving brains out of Washington." "Security Program Is Called Vicious," *New York Times,* January 21, 1955, clipping in Oppenheimer Papers, box 213.

123. Statement by Oppenheimer attached to draft letter of invitation by Nicolas Nabokov, October 12, 1962; and (on Kennan) Oppenheimer to Nabokov, May 14, 1963, Oppenheimer Papers, box 66. A "trial run" meeting was held at the estate the previous summer (in September 1962), with guests including Oppenheimer, Meyer, Boyd, and Nabokov, among others, in order to "discuss whether the idea [of such meetings] is practical." See Agnes Meyer to Robert and Kitty Oppenheimer, July 3, 1962, Oppenheimer Papers, box 66.

124. Mrs. Eugene Meyer, letters of invitation, February 27, 1963, Oppenheimer Papers, box 66.

125. Statement by Oppenheimer attached to draft letters of invitation by Nicolas Nabokov, October 12, 1962, and by Mrs. Eugene Meyer, January 1963, Oppenheimer Papers, box 66.

126. Jeanne Hersch, "Colloquy on Mount Kisco," *Congress News,* Winter 1964, 1–2, quoting 1; Mrs. Eugene Meyer to Lester B. Pearson, January 28, 1967, Oppenheimer Papers, box 66; Meyer's comments transcribed within the original transcript of Oppenheimer's Seven Springs Farm talk, 1963, 1–3, Oppenheimer Papers, box 282; Mrs. Eugene Meyer to Oppenheimer, July 17, 1963; Mrs. Eugene Meyer to Kitty Oppenheimer [ca. March 29, 1966], Oppenheimer Papers, box 66.

127. Oppenheimer, "On Albert Einstein," 4–5. The UNESCO talk was delivered in Paris on December 13, 1965. See also Day, "Oppenheimer on the Nature of Science," 102–3.

128. Oppenheimer, "On Albert Einstein," 5; Abraham Pais to Oppenheimer, December 20 [1965]; Oppenheimer, cover letter, n.d., attached to transcript of UNESCO talk; Otto Nathan to Oppenheimer, January 14, 1966; Oppenheimer to Helen Dukas, January 14, 1966, Oppenheimer Papers, box 285. See also Henry Kamm, "Oppenheimer View of Einstein Warm but Not Uncritical," *New York Times,* December 14, 1965.

129. Oppenheimer, "On Albert Einstein," 5.

130. Michael Paul Rogin, *The Intellectuals and McCarthy: The Radical Specter* (Cambridge, MA: MIT Press, 1967), 10–11.

131. Ibid., 9.

132. Ibid., 19–20; Richard Hofstadter, *Anti-intellectualism in American Life* (London: Jonathan Cape, 1964); David Riesman and Nathan Glazer, "The Intellectuals and the Discontented Classes," *Partisan Review* 22, no. 1 (Winter 1955): 47–72.

133. See also Saunders, *Who Paid the Piper?* 359–62.

134. "C.I.A. Spies from 100 Miles Up; Satellites Probe Secrets of Soviet Union," *New York Times,* April 27, 1966; J. K. Galbraith, George Kennan, Robert Oppenheimer, and Arthur Schlesinger Jr., "Record of Congress for Cultural Freedom," letter to editor, *New York Times,* May 9, 1966. See also "Triple Pass: How C.I.A. Shifts Funds," *New York Times,* February 19, 1967; and "Units Linked with C.I.A.," *New York Times,* February 19, 1967.

135. MacDonald and Hampshire, quoted in Saunders, *Who Paid the Piper?* 378–79; on CCF members' knowledge of CIA funding, see, 394–95.

136. Mrs. Eugene Meyer to Oppenheimer, May 26, 1966; Oppenheimer to Mrs. Eugene Meyer, June 1, 1966; Mrs. Eugene Meyer to Oppenheimer, May 2, 1966, Oppenheimer Papers, box 66. On the topic of the meeting, see Mrs. Eugene Meyer, draft letter of invitation, October 1965, Oppenheimer Papers, box 66.

137. J. Robert Oppenheimer, "The Forbearance of Nations," *New York Herald Tribune–Washington Post,* Paris edition, December 6, 1966.

138. Ibid.

139. Oppenheimer to Josselson, February 13, 1967, Oppenheimer Papers, box 42; "Exploratory Memorandum. Subject: The Seminar Program of the Congress," enclosed with Josselson to Oppenheimer, January 17, 1967, Oppenheimer Papers, box 119. See also Josselson to Oppenheimer, February 16, 1967; Josselson to Oppenheimer, December 13, 1966; Josselson to Oppenheimer, December 23, 1966; "Ford Fund Helps Paris Art Group," *New York Times,* November 2, 1966; and Oppenheimer to Minoo Masani, December 7, 1966, Oppenheimer Papers, box 119.

140. Oppenheimer to Raymond Aron (chairman of the board of directors, CCF), February 13, 1967, Oppenheimer Papers, box 119.

141. Ellen Wilder, quoted in Mason, *Children of Los Alamos,* 148.

142. Jungk, *Brighter Than a Thousand Suns,* 298. The book first appeared in German in 1956.

143. Jungk, *Brighter Than a Thousand Suns,* 83–102, quoting 102. For recent historical debate on Heisenberg, see, for example, Thomas Powers, *Heisenberg's War: The Secret History of the German Bomb* (London: Cape, 1993); Paul Lawrence Rose, *Heisenberg and the Nazi Atomic Bomb Project: A Study in German Culture* (Berkeley and Los Angeles: University of California Press, 1998); Michael Frayn, *Copenhagen* (London: Methuen, 1998); Thomas Powers, "What Bohr Remembered," *New York Review of Books,* March 28, 2002, http://www.nybooks.com/articles/15226; and Gerald Holton et al., "Copenhagen: An Exchange," *New York Review of Books,* April 11, 2002, http://www.nybooks.com/articles/15264.

144. Wyman, interviewed by Weiner, May 28, 1975, 31–37, quoting 34; Chevalier, *Oppenheimer,* 88. See also Cassidy, *J. Robert Oppenheimer,* 318–19.

145. See Chevalier, *Oppenheimer,* 176–77; and Jungk, *Brighter Than a Thousand Suns,* 144.

146. Chevalier, *Oppenheimer,* 176. See also Chevalier correspondence in Oppenheimer Papers, box 26.

147. Kipphardt, *In the Matter of J. Robert Oppenheimer,* 17. See also Knust, "From Faust to Oppenheimer," esp. 135–36; and Haynes, *From Faust to Strangelove,* esp. 285–88.

148. Kipphardt, *In the Matter of J. Robert Oppenheimer,* 127. One British newspaper reported on the play under the headline "Scientist Who Had Scruples" (*Daily Telegraph,* November 19, 1966, clipping in Oppenheimer Papers, box 42).

149. Kipphardt made this point repeatedly. See Kipphardt to Oppenheimer, August 31, 1964; Kipphardt's notes on telephone conversation with Oppenheimer, September 8, 1964; and Kipphardt to Oppenheimer, October 24, 1964, in Heinar Kipphardt, *In der Sache J. Robert Oppenheimer: Ein Stück und seine Geschichte* (Reinbek bei Hamburg: Rowohlt, 1987), 159, 161–63, 167–70.

150. Oppenheimer to Kipphardt, October 12, 1964, in Kipphardt, *Ein Stück und seine Geschichte,* 164–65, quoting 165. See also further correspondence printed in Kipphardt, *Ein Stück und seine Geschichte,* 159–79. On possible legal action, see also Robert H. Montgomery Jr. to John Roberts, March 11, 1965, Oppenheimer Papers,

box 207. See also Kipphardt's notes on conversation with Professor Nabukow [*sic*], September 5, 1964, in Kipphardt, *Ein Stück und seine Geschichte,* 160–61, on 160; and Giovannitti and Freed, *The Decision to Drop the Bomb,* 348.

151. Oppenheimer to Françoise Spira, October 12, 1964, and Oppenheimer to Spira, November 13, 1964, both in Oppenheimer Papers, box 206; Oppenheimer to Josselson, October 12, 1964, Oppenheimer Papers, box 42; Oppenheimer to John Roberts, February 22, 1965, Oppenheimer Papers, box 207; Oppenheimer to Father Dominique Pire, February 8, 1965, Oppenheimer Papers, box 206. See also Spira to Oppenheimer, September 28, 1964; Spira to Oppenheimer, October 19, 1964; Oppenheimer to Spira, October 27, 1964, all in Oppenheimer Papers, box 206; Oppenheimer to Vilar, November 13, 1964, and Vilar to Oppenheimer, November 14, 1964, Oppenheimer Papers, boxes 206. See also Jean Vilar, *Le Dossier Oppenheimer* (Geneva: Éditions Gonthier, 1965).

152. Oppenheimer, quoted in Szasz, "Great Britain," 329; Kipphardt, *Ein Stück und seine Geschichte,* 171.

153. Victor Weisskopf and Jeanne Hersch to Oppenheimer, February 17, 1965. Weisskopf and Hersch warned Oppenheimer that he was wrong to appear to endorse Vilar's version of the play. According to them, Vilar had "badly distorted" the Chevalier incident. See also "Dans une lettre ouverte, Viktor Weisskopf, directeur général du CERN et Jeanne Hersch professeur de philosophie à l'Université de Genève prennent la défense de Robert Oppenheimer contre Jean Vilar," *Le Tribune de Genève,* April 23, 1965; "Jean Vilar a-t-il trahi Oppenheimer? Deux savants genevois accusent le théâtre—document," *Le Figaro Littéraire,* April 22–28, 1965; and "Autour du Dossier Oppenheimer," *Le Figaro Littéraire,* May 6–12, 1965, Oppenheimer Papers, box 207. Michael Josselson also intervened regarding the plays; see Josselson to Oppenheimer, September 18, 1964; Josselson to Oppenheimer, October 12, 1964; Peter Härtling to Josselson, October 2, 1964; Josselson to Oppenheimer, October 5, 1964; Josselson to Oppenheimer, November 3, 1966; Josselson to Oppenheimer, November 24, 1966; Josselson, letter to editor, *Times* (London), December 7, 1966, Oppenheimer Papers, box 42.

154. Oppenheimer to Kipphardt, December 16, 1965, in Kipphardt, *Ein Stück und seine Geschichte,* 177. See also Kipphardt to Oppenheimer, February 15, 1966, ibid., 178–79.

155. David Bohm to Oppenheimer, November 29, 1966, Oppenheimer Papers, box 20.

156. Oppenheimer to David Bohm, December 2, 1966, Oppenheimer Papers, box 20.

157. Original transcript of Oppenheimer's Seven Springs Farm talk, 1963, 9–10, Oppenheimer Papers, box 282; quoted also in Goodchild, *Shatterer of Worlds,* 278.

158. Oppenheimer statement, 1961, quoted in "Dr. J. Robert Oppenheimer, 'Father of the Atomic Bomb,' Dies in Princeton," *New York Times,* February 19, 1967.

159. Wyman, interviewed by Weiner, May 28, 1975, 41. Wyman was describing his impressions from a visit with Oppenheimer in 1958 or 1959.

160. Von Neumann, quoted in Schweber, *In the Shadow of the Bomb,* 196 n. 23; see also Steven Shapin, "Don't Let That Crybaby in Here Again," *London Review of Books,* September 7, 2000, http://www.lrb.co.uk/v22/n17/shap01_.html.

161. Martin Agronsky, Dr. Oppenheimer interview, excerpt from the *CBS Evening News,* August 5, 1965.

162. John R. Crowley to Oppenheimer, August 5, 1965, Oppenheimer Papers, box 285.

163. Thomas B. Morgan, "With Oppenheimer, on an Autumn Day: A Thoughtful Man Talks Searchingly about Science, Ethics and Nuclear War on a Quiet Afternoon during a Bad Time," *Look,* December 27, 1966, 61–63, quoting 61, 63.

164. Ibid., 63.

165. "World Scientists Mourn Death of J. Robert Oppenheimer," *Washington Post,* February 20, 1967; Friendly, "Oppenheimer: Scientist Beset by Political Ills"; Yukawa, quoted in "World Scientists Mourn Death"; Kennan, quoted in Leroy F. Aarons, "Affection, Not Tears, Marks Oppenheimer Tributes," *Washington Post,* February 26, 1967; Smyth, quoted in Aarons, "Affection, Not Tears, Marks Oppenheimer Tributes."

166. See, for example, *New York Times,* February 19, 1967, 1; *New York Herald Tribune–Washington Post,* Paris edition, February 20, 1967, 1, 2; *Washington Post,* February 20, 1967, 1.

167. Leslie, *The Cold War and American Science,* 233–49, esp. 241–43.

168. For an example from a few years later, see Bill Zimmerman et al., *Towards a Science for the People* (Brookline, MA: A People's Press, 1972), esp. 11: "The attitude of Oppenheimer and others, justified by the slogan of truth for truth's sake . . . becomes a rationalization for the maintenance of repressive or destructive institutions."

169. Bush, "The Gentleman of Culture," 32.

170. Cf. G. Pascal Zachary, *Endless Frontier: Vannevar Bush, Engineer of the American Century* (New York: Free Press, 1997), 402–4.

171. Rieff, preface to *On Intellectuals,* x.

172. Snow, *The Two Cultures,* 11. Snow also presented a more ambivalent picture in his 1960 Godkin Lectures: Snow, *Science and Government* (Cambridge, MA: Harvard University Press, 1961). See also Jessica Wang, "Scientists and the Problem of the Public in Cold War, 1945–1960," in "Science and Civil Society," ed. Lynn K. Nyhart and Thomas H. Broman, *Osiris* 17 (2002), esp. 344–45; and Welsh, *Mobilising Modernity,* chap. 1.

BIBLIOGRAPHY

ARCHIVES

Bancroft Library, University of California, Berkeley
· Papers of Raymond Thayer Birge
· University of California Oral Histories
California Institute of Technology Archives, Pasadena
· Papers of Lee A. DuBridge
Department of Special Collections, Joseph Regenstein Library, University of Chicago
· Records of the Association of Los Alamos Scientists
Library of Congress, Manuscript Division, Washington, DC
· Papers of Vannevar Bush
· Papers of J. Robert Oppenheimer
· Papers of Admiral William S. Parsons
Los Alamos Historical Museum, Los Alamos, NM
Los Alamos National Laboratory Archives, Los Alamos, NM
· Oral History Collection
· Wartime Committee Records
Mandeville Special Collections Library, University of California, San Diego
· Papers of Martin Kamen
· Papers of Maria Goeppert Mayer
Massachusetts Institute of Technology, Archives and Special Collections, Cambridge, MA
· Papers of Vannevar Bush
· J. Robert Oppenheimer Oral History Collection
National Archives and Records Administration, College Park, MD
· Records of the Office of the Commanding General, Manhattan Project,

Records of the Office of the Chief of Engineers, Record Group 77
· Papers of General Leslie R. Groves, Record Group 200
Niels Bohr Library, American Institute of Physics, College Park, MD
· Archives for the History of Quantum Physics
· Emilio Segrè Visual Archives
· Oral History Collection

BOOKS AND ARTICLES

Acheson, Dean. *Present at the Creation: My Years in the State Department.* New York: W. W. Norton, 1969.

Acklund, Len, and Steve McGuire, eds. *Assessing the Nuclear Age: Selections from the "Bulletin of the Atomic Scientists."* Chicago: Educational Foundation for Nuclear Science, 1986.

Albright, Joseph. *Bombshell: The Secret Story of America's Unknown Atomic Spy Conspiracy.* New York: Times Books, 1997.

Alperovitz, Gar. *Atomic Diplomacy: Hiroshima and Potsdam; The Use of the Atomic Bomb and the American Confrontation with Soviet Power.* New York: Penguin Books, 1985. Originally published 1965.

———. *The Decision to Use the Atomic Bomb.* New York: Vintage, 1995.

Alsop, Joseph, and Stewart Alsop. *We Accuse! The Story of the Miscarriage of Justice in the Case of J. Robert Oppenheimer.* New York: Simon and Schuster, 1954.

Alvarez, Luis. *Alvarez: Adventures of a Physicist.* New York: Basic Books, 1987.

Arney, William Ray. *Experts in the Age of Systems.* Albuquerque: University of New Mexico Press, 1991.

Aron, Raymond. *Memoirs: Fifty Years of Political Reflection.* New York: Holmes and Meier, 1990.

Bacher, Robert F. *Robert Oppenheimer, 1904–1967.* Los Alamos, NM: Los Alamos Historical Society, 1999.

Badash, Lawrence. "Science and McCarthyism." *Minerva* 38 (2000): 53–80.

———. *Scientists and the Development of Nuclear Weapons: From Fission to the Limited Test Ban Treaty, 1939–1963.* Atlantic Highlands, NJ: Humanities Press International, 1995.

Badash, Lawrence, Joseph O. Hirschfelder, and Herbert P. Broida, eds. *Reminiscences of Los Alamos, 1943–1945.* Dordrecht, Holland: Reidel, 1980.

Balogh, Brian. *Chain Reaction: Expert Debate and Public Participation in American Commercial Nuclear Power, 1945–1975.* Cambridge: Cambridge University Press, 1991.

Barkai, Abraham. *Branching Out: German-Jewish Immigration to the United States, 1820–1914.* New York: Holmes and Meier, 1994.

Barley, Stephen, and Beth Bechky. "In the Backrooms of Science: The Work of Technicians in Science Labs." *Work and Occupations* 21 (1994): 85–126.

Barnard, Chester I., J. Robert Oppenheimer, Charles A. Thomas, Harry A. Winne, and David E. Lilienthal. *A Report on the International Control of Atomic Energy, Prepared for the Secretary of State's Committee on Atomic Energy by a Board of Consultants.* Washington, DC: U.S. Government Printing Office, 1946.

Bartlett, Frances P. "The Atomic Brain Trust." *American Mercury* 81 (July 1955): 75–78.

Baruch, Bernard M. *Baruch: The Public Years.* New York: Holt, Rinehart and Winston, 1960.

Barzun, Jacques. *The House of Intellect.* New York: Harper and Row, 1959.

Bauman, Zygmunt. *Modernity and Ambivalence.* Oxford: Polity Press, 1991.

———. *Modernity and the Holocaust.* Ithaca, NY: Cornell University Press, 1989.

Baxter, James Phinney. *Scientists against Time.* Boston: Little, Brown, 1946.

Becker, Howard S. "Notes on the Concept of Commitment." *American Journal of Sociology* 66 (1960): 32–40.

———. "Personal Change in Adult Life." *Sociometry* 27 (1964): 40–53.

Behind Tall Fences: Stories and Experiences about Los Alamos at Its Beginning. Los Alamos, NM: Los Alamos Historical Society, 1996.

Bell, Daniel. *The Coming of Post-industrial Society: A Venture in Social Forecasting.* New York: Basic Books, 1976.

Belth, Nathan C. *A Promise to Keep: A Narrative of the American Encounter with Anti-Semitism.* New York: Times Books, 1979.

Bendix, Reinhard. *From Berlin to Berkeley: German-Jewish Identities.* New Brunswick, NJ: Transaction Books, 1986.

Bernal, J. D. *The Social Function of Science.* London: Routledge, 1939.

Bernstein, Barton J., ed. *The Atomic Bomb: The Critical Issues.* Boston: Little, Brown, 1976.

———. "In the Matter of J. Robert Oppenheimer." *Historical Studies in the Physical Sciences* 12 (1982): 195–252.

———. "Oppenheimer and the Radioactive Poison Plan." *Technology Review* 88 (1985): 14–17.

———. "The Oppenheimer Loyalty-Security Case Reconsidered." *Stanford Law Review* 42 (July 1990): 1383–1484.

———. "A Post-war Myth: 500,000 U.S. Lives Saved." In *Hiroshima's Shadow,* edited by Kai Bird and Lawrence Lifschultz, 130–34. Stony Creek, CT: Pamphleteers Press, 1998.

Bernstein, Jeremy. *Experiencing Science.* New York: Basic Books, 1978.

———. *The Life It Brings: One Physicist's Beginnings.* New York: Ticknor and Fields, 1987.

Bethe, Hans A. "Oppenheimer: 'Where He Was There Was Always Life and Excitement.'" *Science* 155 (1967): 1080–84.

———. "Review of *Brighter Than a Thousand Suns.*" *Bulletin of the Atomic Scientists* 14, no. 10 (July 1958): 426–28.

———. "The Technological Imperative." In Acklund and McGuire, *Assessing the Nuclear Age,* 73–78.

Bird, Kai, and Martin J. Sherwin. *American Prometheus: The Triumph and Tragedy of J. Robert Oppenheimer.* New York: Knopf, 2005.

Birmingham, Stephen. *"Our Crowd": The Great Jewish Families of New York.* New York: Harper and Row, 1967.

Blackett, P. M. S. *Atomic Weapons and East-West Relations.* Cambridge: Cambridge University Press, 1956.

———. *Fear, War, and the Bomb: Military and Political Consequences of Atomic Energy.* New York: Whittlesey House, 1949.

Blumberg, Stanley A., and Gwinn Owens. *Energy and Conflict: The Life and Times of Edward Teller.* New York: G. P. Putnam's Sons, 1976.

Blumberg, Stanley A., and Lewis G. Panos. *Edward Teller: Giant of the Golden Age of Physics.* New York: Charles Scribner's Sons, 1990.

Bohr, Aage. "The War Years and the Prospects Raised by the Atomic Weapons." In *Niels Bohr: His Life and Work as Seen by His Friends and Colleagues,* edited by S. Rozenthal, 191–214. Amsterdam: North-Holland Publishing Co., 1967.

Bohr, Niels. "A Challenge to Civilization." *Science* 102 (October 12, 1945): 363–64.

Bontecou, Eleanor. "President Eisenhower's 'Security' Program." *Bulletin of the Atomic Scientists* 9, no. 6 (July 1953): 215–17, 220.

Born, Max. *My Life: Recollections of a Nobel Laureate.* London: Taylor and Francis, 1978.

Boyer, Paul. *By the Bomb's Early Light: American Thought and Culture at the Dawn of the Atomic Age.* New York: Pantheon, 1985.

———. *Fallout: A Historian Reflects on America's Half-Century Encounter with Nuclear Weapons.* Columbus: Ohio State University Press, 1998.

Bridgman, Percy. *Reflections of a Physicist.* New York: Philosophical Library, 1955.

Brixner, Berlyn. "A Scientific Photographer at Project Y." In *Behind Tall Fences,* 90–100.

Brode, Bernice. *Tales of Los Alamos: Life on the Mesa, 1943–1945.* Los Alamos, NM: Los Alamos Historical Society, 1997.

Brown, John Mason. *Through These Men: Some Aspects of Our Passing History.* New York: Harper, 1956.

Bruford, W. H. *The German Tradition of Self-Cultivation: 'Bildung' from Humboldt to Thomas Mann.* Cambridge: Cambridge University Press, 1975.

Bush, Vannevar. *Science Is Not Enough.* New York: William Morrow, 1967.

Cahill, Spencer. "Toward a Sociology of the Person." *Sociological Theory* 16 (1998): 131–48.

Canaday, John. *The Nuclear Muse: Literature, Physics and the First Atomic Bombs.* Madison: University of Wisconsin Press, 2000.

Cassidy, David C. *J. Robert Oppenheimer and the American Century.* New York: Pi Press, 2005.

Chevalier, Haakon. *The Man Who Would Be God.* Berlin: Seven Seas, 1963. Originally published 1959.

———. *Oppenheimer: The Story of a Friendship.* New York: George Braziller, 1965.

Childs, Herbert. *An American Genius: The Life of Ernest Orlando Lawrence.* New York: E. P. Dutton, 1968.

Christman, Albert B. *Target Hiroshima: Deak Parsons and the Creation of the Atomic Bomb.* Annapolis, MD: Naval Institute Press, 1998.

Church, Fermor S., and Peggy Pond Church. *When Los Alamos Was a Ranch School.* Los Alamos, NM: Los Alamos Historical Society, 1974.

Church, Peggy Pond. *The House at Otowi Bridge: The Story of Edith Warner and Los Alamos.* Albuquerque: University of New Mexico Press, 1960.

Clark, Ian. *Limited Nuclear War: Political Theory and War Conventions.* Oxford: Martin Robertson, 1982.

Cohen, Karl. "A Re-examination of the McMahon Act." *Bulletin of the Atomic Scientists* 4, no. 1 (January 1948): 7–10.

Cohen, Naomi W. *Encounter with Emancipation: The German Jews in the United States, 1830–1914.* Philadelphia: Jewish Publication Society, 1984.

Coleman, Peter. *The Liberal Conspiracy: The Congress for Cultural Freedom and the Struggle for the Mind of Postwar Europe.* New York: Free Press, 1989.

Collins, Harry. *Gravity's Shadow: The Search for Gravitational Waves.* Chicago: University of Chicago Press, 2004.

———. "LIGO Becomes Big Science." *Historical Studies in the Physical and Biological Sciences* 32, no. 2 (2003): 235–96.

Collins, Harry, and Jay A. Labinger, eds. *The One Culture? A Conversation about Science.* Chicago: University of Chicago Press, 2001.

Compton, Arthur Holly. *Atomic Quest: A Personal Narrative.* New York: Oxford University Press, 1956.

Conant, James B. *Modern Science and Modern Man.* Garden City, NY: Doubleday, 1953.

———. *On Understanding Science: An Historical Approach.* New Haven, CT: Yale University Press, 1947.

———. *Science and Common Sense.* New Haven, CT: Yale University Press, 1963. Originally published 1951.

Congress for Cultural Freedom. *Science and Freedom: The Proceedings of a Conference Convened by the Congress for Cultural Freedom and Held in Hamburg on July 23rd–26th, 1953.* Boston: Beacon Press, 1955.

Cooley, Charles Horton. "Sociability and Personal Ideas." In *Human Nature and the Social Order,* rev. ed., 81–135. New York: Charles Scribner's Sons, 1922.

Coser, Lewis A. *Men of Ideas: A Sociologist's View.* New York: Free Press, 1965.

Cotkin, George. "The Tragic Predicament: Post-war American Intellectuals, Acceptance, and Mass Culture." In *Intellectuals in Politics: From the Dreyfus*

Affair to Salman Rushdie, edited by Jeremy Jennings and Anthony Kemp-Welch, 248–70. London: Routledge, 1997.

Cousins, Norman. "The Standardization of Catastrophe." *Saturday Review of Literature,* August 10, 1946, 16–18.

Cousins, Norman, and Thomas K. Finletter. "A Beginning for Sanity: A Review of the Acheson-Lilienthal Report." *Bulletin of the Atomic Scientists* 2 (1946): 11–14.

Curtis, Charles Pelham. *The Oppenheimer Case: The Trial of a Security System.* New York: Simon and Schuster, 1955.

Davis, Nuel Pharr. *Lawrence and Oppenheimer.* New York: Simon and Schuster, 1969.

Day, Michael A. "I.I. Rabi: The Two Cultures and the Universal Culture of Science." *Physics in Perspective* 6, no. 4 (2004): 428–76.

———. "Oppenheimer on the Nature of Science." *Centaurus* 43 (2001): 73–112.

de Santillana, Giorgio. *The Crime of Galileo.* London: Heinemann, 1958.

———. "Galileo and Oppenheimer." In *Reflections on Men and Ideas.* Cambridge, MA: MIT Press, 1968. First published in *The Reporter,* December 26, 1957.

———. "Notes sur la science en Amérique." *La Table Ronde* 105 (September 1956): 62–66.

Diamond, Sigmund. *Compromised Campus: The Collaboration of Universities with the Intelligence Community, 1945–1955.* New York: Oxford University Press, 1992.

Dinnerstein, Leonard. *Antisemitism in America.* Oxford: Oxford University Press, 1994.

———, ed. *Antisemitism in the United States.* New York: Holt, Rinehart and Winston, 1971.

Dower, John W. "The Bombed: Hiroshimas and Nagasakis in Japanese Memory." *Diplomatic History* 19, no. 2 (Spring 1995): 275–95.

———. *Embracing Defeat: Japan in the Wake of World War Two.* London: Allen Lane, 1999.

Dyson, Freeman. *Disturbing the Universe.* New York: Harper Colophon, 1979.

———. "The Scholar-Soldier." In *Weapons and Hope,* 135–48. New York: Harper and Row, 1984.

Easlea, Brian. *Fathering the Unthinkable: Masculinity, Scientists and the Nuclear Arms Race.* London: Pluto Press, 1983.

Einstein, Albert. *Ideas and Opinions.* New York: Crown, 1982. Originally published 1954.

Elias, Norbert. *The Germans.* New York: Columbia University Press, 1996.

———. *Mozart: Portrait of a Genius.* Cambridge: Polity Press, 1993.

———. *The Society of Individuals.* Edited by Michael Schröter, translated by Edmund Jephcott. Oxford: Basil Blackwell, 1991.

Else, Jon. *The Day after Trinity: J. Robert Oppenheimer and the Atomic Bomb.* Documentary film. Santa Monica, CA: Pyramid Films, 1980.

Ermenc, J. J. *Atomic Bomb Scientists: Memoirs, 1939–1945.* Westport, CT: Meckler, 1984.

Ezrahi, Yaron. *The Descent of Icarus: Science and the Transformation of Contemporary Democracy.* Cambridge, MA: Harvard University Press, 1990.

———. "Einstein and the Light of Reason." In Holton and Elkana, *Albert Einstein,* 253–78.

Federal Bureau of Investigation. *J. Robert Oppenheimer, FBI Security File.* Wilmington, DE: Scholarly Resources, 1978. Microfilm.

Feenberg, Andrew. *Questioning Technology.* London: Routledge, 1999.

Fermi, Laura. *Atoms in the Family: My Life with Enrico Fermi.* Albuquerque: University of New Mexico Press, 1982. Originally published 1954.

Fermi, Rachel, and Esther Samra. *Picturing the Bomb: Photographs from the Secret World of the Manhattan Project.* New York: Harry N. Abrams, 1995.

Feshbach, Herman, Tetsuo Matsui, and Alexandra Oleson, eds. *Niels Bohr: Physics and the World; Proceedings of the Niels Bohr Centennial Symposium, Boston, MA, USA, November 12–14, 1985.* London: Harwood, 1988.

Feuer, Lewis S. *Einstein and the Generations of Science.* New York: Basic Books, 1974.

———. *The Scientific Intellectual: The Psychological and Sociological Origins of Modern Science.* New York: Basic Books, 1963.

Feynman, Richard. 1980. "Los Alamos from Below." In Badash, Hirschfelder, and Broida, *Reminiscences of Los Alamos,* 105–32.

———. "The Pleasure of Finding Things Out." *The Listener,* November 26, 1981, 635–36.

Fine, Lenore, and Jesse A. Remington. *The Corps of Engineers: Construction in the United States.* Washington, DC: Office of the Chief of Military History, U.S. Army, 1972.

Fisher, Phyllis. *Los Alamos Experience.* New York: Japan Publications, 1985.

Fleming, Donald, and Bernard Bailyn, eds. *The Intellectual Migration.* Cambridge, MA: Harvard University Press, 1969.

Forman, Paul. "Behind Quantum Electronics: National Security as Basis for Physical Research in the United States, 1940–1960." *Historical Studies in the Physical and Biological Sciences* 18 (1987): 149–229.

———. "Física, modernidad y nuestra evasión de la responsibilidad." *Arbor: Ciencia, Pensamiento y Cultura* 147 (1994): 51–74.

———. "Social Niche and Self-Image of the American Physicist." In *The Restructuring of Physical Sciences in Europe and the United States, 1945–1960,* edited by Michelangelo de Maria, Mario Grilli, and Fabio Sebastiani, 96–104. Singapore: World Scientific, 1988.

Foucault, Michel. "Truth and Power." In Foucault, *Power/Knowledge: Selected Interviews and Other Writings, 1972–1977,* edited by Colin Gordon, 109–33. New York: Pantheon, 1980.

Frayn, Michael. *Copenhagen.* London: Methuen, 1998.

Fuller, Steve. *Thomas Kuhn: A Philosophical History for Our Times.* Chicago: University of Chicago Press, 2000.

Galison, Peter. "Physics between War and Peace." In Mendelsohn, Smith, and Weingart, *Science, Technology, and the Military,* 1:47–86.

Galison, Peter, and Barton Bernstein. " 'In Any Light': Scientists and the Decision to Build the Superbomb, 1952–1954." *Historical Studies in the Physical and Biological Sciences* 19 (1989): 267–347.

Gardner, Howard. *Leading Minds: An Anatomy of Leadership.* New York: Basic Books, 1995.

Garfinkel, Harold. "Conditions of Successful Degradation Ceremonies." *American Journal of Sociology* 61, no. 5 (March 1956): 420–24.

Gay, Peter. "Encounter with Modernism: German Jews in German Culture, 1888–1914." *Midstream* 21 (1975): 23–65.

Gerber, David A., ed. *Anti-Semitism in American History.* Urbana: University of Illinois Press, 1986.

Giddens, Anthony. *The Giddens Reader.* Edited by Philip Cassell. London: Macmillan, 1993.

Gillmor, Dan. *Fear, the Accuser.* New York: Abelard-Schuman, 1954.

Gilman, Sander L. *Jewish Self-Hate: Anti-Semitism and the Hidden Language of Race.* Baltimore: Johns Hopkins University Press, 1986.

———. *Smart Jews: The Construction of the Image of Jewish Superior Intelligence.* Lincoln: University of Nebraska Press, 1996.

Gilpin, Robert. *American Scientists and Nuclear Weapons Policy.* Princeton, NJ: Princeton University Press, 1962.

Giovanitti, Len, and Fred Freed. *The Decision to Drop the Bomb.* New York: Coward McCann, 1965.

Glazer, Nathan. *The Social Basis of American Communism.* New York: Harcourt, Brace and World, 1961.

Goffman, Erving. *Asylums: Essays on the Social Situation of Mental Patients and Other Inmates.* New York: Doubleday, 1961.

———. *The Presentation of Self in Everyday Life.* Garden City, NY: Doubleday Anchor, 1959.

Goldberg, Stanley. "General Groves and the Atomic West: The Making and Meaning of Hanford." In Hevly and Findlay, *The Atomic West,* 39–89.

———. "Groves and the Scientists: Compartmentalization and the Building of the Bomb." *Physics Today* 48, no. 8 (August 1995): 38–43.

Goldman, Harvey. *Max Weber and Thomas Mann: Calling and the Shaping of Self.* Berkeley and Los Angeles: University of California Press, 1988.

———. *Politics, Death, and the Devil: Self and Power in Max Weber and Thomas Mann.* Berkeley and Los Angeles: University of California Press, 1992.

Goodchild, Peter. *Edward Teller: The Real Dr. Strangelove.* Cambridge, MA: Harvard University Press, 2004.

———. *J. Robert Oppenheimer: Shatterer of Worlds.* Boston: Houghton Mifflin, 1981.

Goudsmit, Samuel A. *Alsos: The Failure in German Science.* London: Sigma Books, 1947.

Gowing, Margaret. *Britain and Atomic Energy, 1939–1945.* New York: St. Martin's Press, 1964.

———. "Niels Bohr and Nuclear Weapons." In *Niels Bohr: A Centenary Volume,* edited by A. P. French and P. J. Kennedy, 266–77. Cambridge, MA: Harvard University Press, 1985.

Green, Harold P. "The Oppenheimer Case: A Study in the Abuse of Law." *Bulletin of the Atomic Scientists* 33 (1977): 12–16, 56–61.

Greenberg, Clement. "Avant Garde and Kitsch." *Partisan Review* 6 (1939): 34–49.

Greenberg, Daniel S. *The Politics of Pure Science.* Chicago: University of Chicago Press, 1999. Originally published 1967.

———. *Science, Money, and Politics: Political Triumph and Ethical Erosion.* Chicago: University of Chicago Press, 2001.

Groves, Leslie R. *Now It Can Be Told: The Story of the Manhattan Project.* New York: Harper and Brothers, 1962.

Gusterson, Hugh. *Nuclear Rites: A Weapons Laboratory at the End of the Cold War.* Berkeley and Los Angeles: University of California Press, 1996.

Guttchen, Robert S. *Felix Adler.* New York: Twayne Publishers, 1974.

Hager, Thomas. *Force of Nature: The Life of Linus Pauling.* New York: Simon and Schuster, 1995.

Halberstam, David. *The Fifties.* New York: Villard Books, 1993.

Hales, Peter Bacon. *Atomic Spaces: Living on the Manhattan Project.* Urbana: University of Illinois Press, 1997.

———. "Topographies of Power: The Forced Spaces of the Manhattan Project." In *Mapping American Culture,* edited by Wayne Franklin and Michael Steiner, 251–90. Iowa City: University of Iowa Press, 1992.

Harré, Rom. *Personal Being.* Cambridge, MA: Harvard University Press, 1984.

Hart, David M. *Forged Consensus: Science, Technology, and Economic Policy in the United States, 1921–1953.* Princeton, NJ: Princeton University Press, 1998.

Hawkins, David. "Toward Trinity." Part 1 of *Project Y: The Los Alamos Story.* Los Angeles: Tomash Publishers, 1983.

Haynes, Roslynn D. *From Faust to Strangelove: Representations of the Scientist in Western Literature.* Baltimore: Johns Hopkins University Press, 1994.

Heilbron, John L., and Robert W. Seidel. *Lawrence and His Laboratory: A History of the Lawrence Berkeley Laboratory.* Vol. 1. Berkeley and Los Angeles: University of California Press, 1989.

Hein, Laura. "Learning about Patriotism, Decency, and the Bomb." In *Living with the Bomb: American and Japanese Cultural Conflicts in the Nuclear Age,* edited by Laura Hein and Mark Shelden, 279–86. Armonk, NY: M. E. Sharpe, 1997.

Herken, Gregg. *Brotherhood of the Bomb: The Tangled Lives and Loyalties of Robert Oppenheimer, Ernest Lawrence, and Edward Teller.* New York: Henry Holt and Co., 2002.

———. "The Oppenheimer Case: An Exchange." With a reply by Daniel J. Kevles. *New York Review of Books,* March 25, 2004, 46–47.

———. "The University of California, the Federal Weapons Labs, and the Founding of the Atomic West." In Hevly and Findlay, *The Atomic West,* 119–35.

———. *The Winning Weapon: The Atomic Bomb in the Cold War, 1945–1950.* Princeton, NJ: Princeton University Press, 1988.

Herman, Ellen. *The Romance of American Psychology: Political Culture in the Age of Experts.* Berkeley and Los Angeles: University of California Press, 1995.

Hershberg, James G. *James B. Conant: Harvard to Hiroshima and the Making of the Nuclear Age.* Stanford, CA: Stanford University Press, 1993.

———. "'Over My Dead Body': James Bryant Conant and the Hydrogen Bomb." In Mendelsohn, Smith, and Weingart, *Science, Technology, and the Military,* 2:379–430.

Hevly, Bruce, and John M. Findlay, eds. *The Atomic West.* Seattle: University of Washington Press, 1998.

Hewlett, Richard G., and Oscar E. Anderson Jr. *The New World, 1939/1946.* Vol. 1 of *A History of the United States Atomic Energy Commission.* University Park: Pennsylvania State University Press, 1962.

Hewlett, Richard G., and Francis Duncan. *Atomic Shield, 1947/1950.* Vol. 2 of *A History of the United States Atomic Energy Commission.* Berkeley and Los Angeles: University of California Press, 1991.

Hijiya, James A. "The *Gita* of J. Robert Oppenheimer." *Proceedings of the American Philosophical Society* 144, no. 2 (June 2000): 123–67.

Hoch, Paul K. "The Crystallization of a Strategic Alliance: The American Physics Elite and the Military in the 1940s." In Mendelsohn, Smith, and Weingart, *Science, Technology, and the Military,* 87–116.

Hoddeson, Lillian. "Mission Change in the Large Laboratory: The Los Alamos Implosion Program, 1943–1945." In *Big Science: The Growth of Large-Scale Research,* edited by Peter Galison and Bruce Hevly, 265–89. Stanford, CA: Stanford University Press, 1992.

Hoddeson, Lillian, Paul W. Henriksen, Roger A. Meade, and Catherine Westfall. *Critical Assembly: A Technical History of Los Alamos during the Oppenheimer Years, 1943–1945.* Cambridge: Cambridge University Press, 1993.

Hofstadter, Richard. *Anti-intellectualism in American Life.* London: Cape, 1964.

Hollinger, David A. *Science, Jews, and Secular Culture: Studies in Mid-Twentieth-Century American Intellectual History.* Princeton, NJ: Princeton University Press, 1996.

Holloway, Rachael L. *In the Matter of J. Robert Oppenheimer: Politics, Rhetoric and Self-Defense.* Westport, CT: Praeger, 1993.

Holton, Gerald. "Heisenberg, Oppenheimer, and the Transition to Modern Physics." In *Transformation and Tradition in the Sciences,* edited by Everett Mendelsohn, 155–73. Cambridge: Cambridge University Press, 1984.

———. "The Roots of Complementarity." In *Thematic Origins of Scientific Thought,* 115–61. Cambridge, MA: Harvard University Press, 1973.

———. "Young Man Oppenheimer." *Partisan Review* 48 (July 1981): 380–88.

Holton, Gerald, and Yehuda Elkana, eds. *Albert Einstein: Historical and Cultural Perspectives.* Mineola, NY: Dover Publications, 1982.

Hoopes, Roy. *Americans Remember the Home Front: An Oral Narrative of the World War II Years.* New York: Berkley Press, 2002.

Howe, Irving. *World of Our Fathers: The Journey of the East European Jews to America and the Life They Found and Made There.* London: Phoenix Press, 2000. Originally published 1976.

Howes, Ruth H., and Caroline L. Herzenberg. *Their Day in the Sun: Women of the Manhattan Project.* Philadelphia: Temple University Press, 1999.

Hughes, Thomas P. *American Genesis: A Century of Innovation and Technological Enthusiasm, 1870–1970.* New York: Viking, 1989.

———. "Technological Momentum in History: Hydrogenation in Germany, 1898–1933." *Past and Present* 44 (1969): 106–32.

Hunner, Jon. "Family Secrets: The Growth of Community at Los Alamos, New Mexico, 1943–1957." PhD diss., University of New Mexico, 1996.

———. *Inventing Los Alamos: The Growth of an Atomic Community.* Norman: University of Oklahoma Press, 2004.

Janowitz, Morris. "Military Elites and the Study of War." In *War: Studies from Psychology, Sociology, Anthropology,* edited by Leon Bramson and George W. Goethals, 345–57. New York: Basic Books, 1968.

Jasanoff, Sheila. *The Fifth Branch: Science Advisers as Policymakers.* Cambridge, MA: Harvard University Press, 1990.

Jaspers, Karl. *The Atom Bomb and the Future of Man.* Chicago: University of Chicago Press, 1961.

Jerome, Fred. *The Einstein File: J. Edgar Hoover's Secret War against the World's Most Famous Scientist.* New York: St. Martin's Press, 2002.

Jette, Eleanor. *Inside Box 1663.* Los Alamos, NM: Los Alamos Historical Society, 1977.

Johnson, George. *Strange Beauty: Murray Gell-Mann and the Revolution in Twentieth-Century Physics.* New York: Alfred A. Knopf, 1999.

Jones, Vincent C. *Manhattan: The Army and the Atomic Bomb.* Washington, DC: United States Army Center of Military History, 1985.

Jungk, Robert. *Brighter Than a Thousand Suns: A Personal History of the Atomic Scientists.* Translated by James Cleugh. Harmondsworth, UK: Penguin, 1965. Originally published 1958.

————. *The Nuclear State.* Translated by Eric Mosbacher. London: J. Calder, 1979.

Kaiser, David. "More Roots of Complementarity: Kantian Aspects and Influences." *Studies in History and Philosophy of Science and Technology* 23, no. 2 (1992): 213–39.

————. "The Postwar Suburbanization of American Physics." *American Quarterly* 56 (2004): 851–88.

Kamen, Martin D. *Radiant Science, Dark Politics: A Memoir of the Nuclear Age.* Berkeley and Los Angeles: University of California Press, 1985.

Kanon, Joseph. *Los Alamos.* New York: Broadway Books, 1997.

[Kennan, George F.] "The Sources of Soviet Conduct." *Foreign Affairs* 25, no. 4 (July 1947): 566–82.

Kerr, Clark. *The Uses of the University.* Cambridge, MA: Harvard University Press, 1972. Originally published 1963.

Kevles, Daniel J. "Cold War and Hot Physics: Science, Security, and the American State, 1945–1956." *Historical Studies in the Physical and Biological Sciences* 20 (1990): 239–64.

————. "The National Science Foundation and the Debate over Postwar Research Policy, 1942–1945: A Political Interpretation of *Science—the Endless Frontier.*" *Isis* 68 (1977): 5–26.

————. *The Physicists: The History of a Scientific Community in Modern America.* New York: Vintage Books, 1979.

————. "The Strange Case of Robert Oppenheimer." Review of *Brotherhood of the Bomb,* by Gregg Herken. *New York Review of Books,* December 4, 2003, 37–40.

Kipphardt, Heinar. *In der Sache J. Robert Oppenheimer: Ein Stück und seine Geschichte.* Reinbek bei Hamburg: Rowohlt, 1987.

————. *In the Matter of J. Robert Oppenheimer: A Play Freely Adapted on the Basis of Documents.* New York: Hill and Wang, 1968. Originally published 1964.

Kirstein, Peter N. "False Dissenters: Manhattan Project Scientists and the Use of the Atomic Bomb." *American Diplomacy,* March 2001, http://www.unc.edu/depts/diplomat/archives_roll/2001_03-06/kirstein_manhattan/kirstein_manhattan.html.

Knust, Herbert. "From Faust to Oppenheimer: The Scientist's Pact with the Devil." *Journal of European Studies* 13, nos. 1–2 (1983): 122–41.

Kraut, Benny. *From Reform Judaism to Ethical Culture: The Religious Evolution of Felix Adler.* Cincinnati, OH: Hebrew Union College Press, 1979.

Kugelmass, J. Alvin. *J. Robert Oppenheimer and the Atomic Story*. New York: Julian Messner, 1953.

Kunetka, James W. *City of Fire: Los Alamos and the Atomic Age, 1943–1945*. Englewood Cliffs, NJ: Prentice-Hall, 1978.

———. *Oppenheimer: The Years of Risk*. Englewood Cliffs, NJ: Prentice-Hall, 1982.

Kuznick, Peter J. *Beyond the Laboratory: Scientists as Political Activists in 1930s America*. Chicago: University of Chicago Press, 1987.

Lakoff, Sanford A. "Ethical Responsibility and the Scientific Vocation." In *Science and Ethical Responsibility: Proceedings of the U.S. Student Pugwash Conference, University of California, San Diego, June 19–26, 1979*, edited by Sanford A. Lakoff, 19–31. Reading, MA: Addison-Wesley, 1980.

———. "The Trial of Dr. Oppenheimer." In *Knowledge and Power: Essays on Science and Government*, edited by Sanford A. Lakoff, 65–86. New York: Free Press, 1966.

Landau, Susan. "Joseph Rotblat: The Road Less Traveled." *Bulletin of the Atomic Scientists* 52, no. 1 (January–February 1996): 47–54.

Lanouette, William. *Genius in the Shadows: A Biography of Leo Szilard; The Man behind the Bomb*. With Bela Silard. Chicago: University of Chicago Press, 1992.

Lapp, Ralph E. *The New Priesthood: The Scientific Elite and the Uses of Power*. New York: Harper and Row, 1965.

Larsen, Rebecca. *Oppenheimer and the Atomic Bomb*. New York: F. Watts, 1988.

Larson, Magali Sarfatti. "The Production of Expertise and the Constitution of Expert Power." In *The Authority of Experts: Studies in History and Theory*, edited by Thomas Haskell, 28–80. Bloomington: Indiana University Press, 1984.

Laurence, William L. 1946. "The Bikini Tests and Public Opinion." *Bulletin of the Atomic Scientists* 2 (1946): 2, 17.

———. *Dawn over Zero: The Story of the Atomic Bomb*. London: Museum Press, 1947.

———. *Men and Atoms: The Discovery, the Uses and the Future of Atomic Energy*. New York: Simon and Schuster, 1959.

———. *The Story of the Atomic Bomb*. Washington, DC: War Department, 1945.

Lawren, William. *The General and the Bomb: A Biography of General Leslie Groves, Director of the Manhattan Project*. New York: Dodd, Mead, 1988.

Leslie, Stuart. *The Cold War and American Science: The Military-Industrial-Academic Complex at MIT and Stanford*. New York: Columbia University Press, 1993.

Lewis, Richard S., and Jane Wilson, eds. *Alamogordo plus Twenty-five Years*. With Eugene Rabinowitch. New York: Viking Press, 1971.

Libby, Leona Marshall. *The Uranium People*. New York: Crane Russak, 1979.

Lieberman, Joseph I. *The Scorpion and the Tarantula: The Struggle to Control Atomic Weapons, 1945–1949*. Boston: Houghton Mifflin, 1970.

Light, Jennifer. "When Computers Were Women." *Technology and Culture* 40 (1999): 455–83.

Lilienthal, David E. *The Atomic Energy Years, 1945–1950.* Vol. 2 of *The Journals of David E. Lilienthal.* New York: Harper and Row, 1964.

———. *Change, Hope, and the Bomb.* Princeton, NJ: Princeton University Press, 1963.

———. "How Can Atomic Energy Be Controlled?" *Bulletin of the Atomic Scientists* 2 (October 1947): 14–15, 18.

———. *This I Do Believe.* New York: Harper and Brothers, 1949.

———. *TVA: Democracy on the March.* New York: Pocket Books, 1945.

———. *Unfinished Business, 1968–1981.* Edited by Helen M. Lilienthal. Vol. 7 of *The Journals of David E. Lilienthal.* New York: Harper and Row, 1983.

———. *Venturesome Years, 1950–1955.* Vol. 3 of *The Journals of David E. Lilienthal.* New York: Harper and Row, 1966.

Lovell, Bernard. "Patrick Maynard Stuart Blackett, Baron Blackett of Chelsea, 18 November 1897–13 July 1974." *Biographical Memoirs of Fellows of the Royal Society* 21 (1975): 1–115.

Low, Ian. "Science for Peace." *New Scientist,* July 24, 1975, 208–10.

Lynch, Michael, and David Bogen. *The Spectacle of History: Speech, Text, and Memory at the Iran-Contra Hearings.* Durham, NC: Duke University Press, 1996.

MacKenzie, Donald, and Boelie Elzen. "The Charismatic Engineer." 1991. In *Knowing Machines: Essays on Technical Change,* edited by Donald A. MacKenzie, 131–57. Cambridge, MA: MIT Press, 1996.

Major, John. *The Oppenheimer Hearing.* New York: Scarborough Books, 1983. Originally published 1971.

Makhijani, Arjun. "'Always' the Target?" *Bulletin of the Atomic Scientists* 31, no. 3 (May–June 1995): 23–27.

Manley, John. "Assembling the Wartime Labs." *Bulletin of the Atomic Scientists* 30 (May 1974): 43–48.

Marcuse, Herbert. "Industrialization and Capitalism." *New Left Review* 30 (March–April 1965): 3–17.

Margulis, Lynn. "Sunday with J. Robert Oppenheimer." In Lynn Margulis and Dorion Sagan, *Slanted Truths: Essays on Gaia, Symbiosis, and Evolution,* 5–28. New York: Springer-Verlag, 1997.

Marshall, Barbara. "Politics in Academe: Göttingen University and the Growing Impact of Political Issues, 1918–33." *European History Quarterly* 18, no. 3 (July 1988): 291–320.

Masco, Joseph P. "Nuclear Borderlands: The Legacy of the Manhattan Project in Post–Cold War New Mexico." PhD diss., University of California, San Diego, 1999.

Mason, Katrina R. *Children of Los Alamos: An Oral History of the Town Where the Atomic Age Began.* New York: Twayne, 1995.

Masters, Dexter, and Katharine Way, eds. *One World or None.* London: Purnell and Sons, 1946.

McCray, W. Patrick. "Project Vista, Caltech and the Dilemmas of Lee DuBridge." *Historical Studies in the Physical and Biological Sciences* 34, no. 2 (2000): 339–70.

McHugh, Peter. *Defining the Situation: The Organization of Meaning in Social Interaction.* New York: Bobbs-Merrill Co., 1968.

McLellan, David S. *Dean Acheson: The State Department Years.* New York: Dodd, Mead, 1976.

McMillan, Elsie Blumer. *The Atom and Eve.* New York: Vantage Press, 1995.

McMillan, Priscilla. *The Ruin of J. Robert Oppenheimer and the Birth of the Modern Arms Race.* New York: Viking Press, 2005.

Mendelsohn, Everett, Merritt Roe Smith, and Peter Weingart, eds. *Science, Technology, and the Military.* Vols. 1 and 2. Dordrecht: Kluwer Academic Publishers, 1988.

Merton, Robert K. "The Normative Structure of Science." 1942. In Merton, *The Sociology of Science: Theoretical and Empirical Investigations,* edited by Norman W. Storer, 267–78. Chicago: University of Chicago Press, 1973.

Meyerhoff, Hans. "Through the Liberal Looking Glass—Darkly." *Partisan Review* 22, no. 2 (Spring 1955): 238–48.

Michelmore, Peter. *The Swift Years: The Robert Oppenheimer Story.* New York: Dodd, Mead, 1969.

Mills, C. Wright. *The Sociological Imagination.* Oxford: Oxford University Press, 2000. Originally published 1959.

———. *White Collar: The American Middle Classes.* London: Oxford University Press, 1956. Originally published 1951.

Moore, Ruth E. *Niels Bohr: The Man, his Science, and the World They Changed.* New York: Alfred A. Knopf, 1966.

Moss, Norman. *Klaus Fuchs: The Man Who Stole the Atom Bomb.* London: Grafton Books, 1987.

Mosse, George L. *German Jews beyond Judaism.* Bloomington: Indiana University Press, 1985.

Mukerji, Chandra. *A Fragile Power: Scientists and the State.* Princeton, NJ: Princeton University Press, 1989.

Mullet, Shawn. "Strong Rope, Leaky Buckets, and Atomic Espionage." Manuscript, 2002.

Mumford, Lewis. *The Myth of Machine.* Vol. 2, *The Pentagon of Power.* New York: Harcourt, Brace, 1970.

Murdoch, Dugald. *Niels Bohr's Philosophy of Physics*. Cambridge: Cambridge University Press, 1987.

Nathanson, Charles E. "The Social Construction of the Soviet Threat: A Study in the Politics of Representation." *Alternatives* 8 (1988): 443–83.

Nelson, Steve, James R. Barrett, and Rob Ruck. *Steve Nelson, American Radical*. Pittsburgh: University of Pittsburgh Press, 1981.

Neuse, Steven M. *David E. Lilienthal: The Journey of an American Liberal*. Knoxville: University of Tennessee Press, 1996.

Newhouse, John. *War and Peace in the Nuclear Age*. New York: Alfred A. Knopf, 1989.

Newman, Steven Leonard. "The Oppenheimer Case: A Reconsideration of the Role of the Defense Department and National Security." PhD diss., New York University, 1977.

Nichols, Kenneth D. *The Road to Trinity*. New York: William Morrow, 1987.

Nieburg, H. L. *In the Name of Science*. Chicago: Quadrangle Books, 1966.

Norris, Robert S. *Racing for the Bomb: General Leslie R. Groves, the Manhattan Project's Indispensable Man*. South Royalton, VT: Steerforth Press, 2002.

Nye, Mary Jo. *Blackett: Physics, War, and Politics in the Twentieth Century*. Cambridge, MA: Harvard University Press, 2004.

———. "What Price Politics? Scientists and Political Controversy." *Endeavour* 23 (1999): 148–54.

Olesko, Kathryn Mary. *Physics as a Calling: Discipline and Practice in the Königsberg Seminar for Physics*. Ithaca, NY: Cornell University Press, 1991.

Olwell, Russell. "Physical Isolation and Marginalization in Physics: David Bohm's Cold War Exile." *Isis* 90, no. 4 (1999): 738–56.

O'Neill, Kevin. "Building the Bomb." In *Atomic Audit: The Costs and Consequences of U.S. Nuclear Weapons since 1940*, edited by Stephen I. Schwartz, 33–103. Washington, DC: Brookings Institution Press, 1998.

Ophir, Adi, and Steven Shapin. "The Place of Knowledge: A Methodological Survey." *Science in Context* 4 (1991): 3–21.

Oppenheimer, J. Robert. 1946. "The Atom Bomb as a Great Force for Peace." *New York Times Magazine*, June 9, 1946, 7, 59–60.

———. "Atomic Weapons and American Policy." *Foreign Affairs* 31, no. 4 (July 1953): 525–35.

———. "Atomic Weapons and the Crisis in Science." *Saturday Review of Literature*, November 24, 1945, 9–11.

———. "Comments on the Military Value of the Atom." *Bulletin of the Atomic Scientists* 7, no. 2 (February 1951): 43–45.

———. *The Flying Trapeze: Three Crises for Physicists*. London: Oxford University Press, 1964.

———. "The Forbearance of Nations." *New York Herald Tribune–Washington Post*, Paris edition, December 6, 1966, 16.

————. "The Future of Civilization in the Scientific Age." *France-Asie,* March–April 1961, 1807–15.

————. "International Control of Atomic Energy." In *The Atomic Age: Scientists in National and World Affairs,* edited by Morton Grodzins and Eugene Rabinowitch, 53–63. New York: Basic Books, 1963. Originally published in *Bulletin of the Atomic Scientists* 1, no. 12 (June 1, 1946): 1–5.

————. "In the Keeping of Unreason." *Bulletin of the Atomic Scientists* 16, no. 1 (January 1960): 18–22.

————. "Knowledge and the Structure of Culture." Helen Kenyon Lecture. Vassar College, Poughkeepsie, NY, October 29, 1958. Manuscript copy.

————. "The Need for New Knowledge." In *Symposium on Basic Research,* edited by Dael Wolfle, 1–15. Washington, DC: American Association for the Advancement of Science, 1959.

————. "The New Weapon: The Turn of the Screw." In Masters and Way, *One World or None,* 53–60.

————. "Niels Bohr and Atomic Weapons." *New York Review of Books,* December 17, 1964, 6–8.

————. "On Albert Einstein." *New York Review of Books,* March 17, 1966, 4–5.

————. *The Open Mind.* New York: Simon and Schuster, 1955.

————. *Science and the Common Understanding.* London: Oxford University Press, 1954.

————. "Science and the Human Community." In *Issues in University Education: Essays by Ten American Scholars,* edited by Charles Frankel, 48–62. New York: Harper and Brothers, 1959.

————. "Some Reflections on Science and Culture." Lecture, 1959. Chapel Hill: University of North Carolina, 1960.

Oreskes, Naomi, and Paul Rainger. "Science and Security before the Atomic Bomb: The Loyalty Case of Harold U. Sverdrup." *Studies in History and Philosophy of Modern Physics* 31 (2000): 309–69.

Orr, James J. *The Victim as Hero: Ideologies of Peace and National Identity in Postwar Japan.* Honolulu: University of Hawaii Press, 2001.

Ortega y Gasset, José. *The Revolt of the Masses.* New York: W. W. Norton, 1932.

Packard, George R., III. *Protest in Tokyo: The Security Treaty Crisis of 1960.* Princeton, NJ: Princeton University Press, 1966.

Pais, Abraham. *Einstein Lived Here.* Oxford: Clarendon Press, 1994.

————. *Niels Bohr's Times, in Physics, Philosophy, and Polity.* Oxford: Clarendon Press, 1991.

————. *'Subtle is the Lord . . .': The Science and Life of Albert Einstein.* Oxford: Clarendon Press, 1982.

————. *A Tale of Two Continents: A Physicist's Life in a Turbulent World.* Princeton, NJ: Princeton University Press, 1997.

Palevsky, Mary. *Atomic Fragments: A Daughter's Questions.* Berkeley and Los Angeles: University of California Press, 2000.

Pascal, Roy. *Culture and the Division of Labour.* Occasional Papers in German Studies, no. 5. Warwick: University of Warwick, 1974.

Peierls, Rudolf E. *Atomic Histories.* Woodbury, NY: AIP Press, 1997.

———. *Bird of Passage: Recollections of a Physicist.* Princeton, NJ: Princeton University Press, 1985.

Pells, Richard H. *The Liberal Mind in a Conservative Age: American Intellectuals in the 1940s and 1950s.* New York: Harper and Row, 1985.

———. *Radical Visions and American Dreams: Culture and Social Thought in the Depression Years.* Middletown, CT: Wesleyan University Press, 1973.

Piccard, Paul J. "Scientists and Public Policy: Los Alamos, August–November, 1945." *Western Political Quarterly* 18, no. 2 (June 1965): 251–62.

Pleat, F. David. *Infinite Potential: The Life and Times of David Bohm.* Reading, MA: Helix Books, 1997.

Polanyi, Michael. *Science, Faith and Society.* Chicago: University of Chicago Press, 1946.

Polenberg, Richard, ed. *In the Matter of J. Robert Oppenheimer: The Security Clearance Hearing.* Ithaca, NY: Cornell University Press, 2002.

Popper, Karl, *The Open Society and Its Enemies.* London: Routledge, 1945.

Porter, Theodore M. *Trust in Numbers: The Pursuit of Objectivity in Science and Public Life.* Princeton, NJ: Princeton University Press, 1995.

Powers, Thomas. *Heisenberg's War: The Secret History of the German Bomb.* London: Cape, 1993.

Price, Matt. "Roots of Dissent: The Chicago Met Lab and the Origins of the Franck Report." *Isis* 86 (1995): 222–44.

Rabi, Isidor I. "Playing Down the Bomb: Blackett versus the Atom." *Atlantic Monthly* 183 (April 1949): 21–24.

Rabi, Isidor I., Robert Serber, Victor F. Weisskopf, Abraham Pais, and Glenn T. Seaborg. *Oppenheimer.* New York: Charles Scribner, 1969.

Rabinow, Paul. *French DNA: Trouble in Purgatory.* Chicago: University of Chicago Press, 1999.

———. *French Modern: Norms and Forms of the Social Environment.* Cambridge, MA: MIT Press, 1999.

———. *Making PCR: A Story of Biotechnology.* Chicago: University of Chicago Press, 1996.

Radest, Howard B. *Toward Common Ground: The Story of the Ethical Culture Societies in the United States.* New York: Frederick Ungar Publishing Co., 1969.

Ravetz, Jerome. "Tragedy in the History of Science." In *Changing Perspectives in the History of Science: Essays in Honour of Joseph Needham,* edited by Mikuláš Teich and Robert Young, 204–22. London: Heinemann, 1973.

Reeves, Thomas C. *The Life and Times of Joe McCarthy: A Biography*. London: Madison Books, 1997.

Regis, Ed. *Who Got Einstein's Office? Eccentricity and Genius at the Institute for Advanced Study*. Reading, MA: Addison-Wesley, 1987.

Reingold, Nathan. "Vannevar Bush's New Deal for Research: Or the Triumph of the Old Order." *Historical Studies in the Physical and Biological Sciences* 17 (1987): 299–344.

Rhodes, Richard. *Dark Sun: The Making of the Hydrogen Bomb*. New York: Simon and Schuster, 1995.

———. "'I Am Become Death': The Agony of J. Robert Oppenheimer." *American Heritage* 20, no. 6 (October 1977): 72–82.

———. *The Making of the Atomic Bomb*. New York: Simon and Schuster, 1986.

Rieff, Philip, ed. *On Intellectuals: Theoretical Studies, Case Studies*. New York: Doubleday/Anchor, 1970.

Riesman, David, and Nathan Glazer. "The Intellectuals and the Discontented Classes." *Partisan Review* 22, no. 1 (Winter 1955): 47–72.

Riesman, David, Reuel Denney, and Nathan Glazer. *The Lonely Crowd: A Study of the Changing American Character*. New Haven, CT: Yale University Press, 1950.

Rigden, John S. 1995. "J. Robert Oppenheimer—before the War." *Scientific American* 273 (July 1995): 68–73.

———. *Rabi: Scientist and Citizen*. New York: Basic Books, 1987.

Ringer, Fritz, K. *The Decline of the German Mandarins: The German Academic Community, 1890–1933*. Hanover, NH: University Press of New England, 1990. Originally published 1969.

Rischin, Moses. *The Promised City: New York's Jews, 1870–1914*. Cambridge, MA: Harvard University Press, 1962.

Roensch, Eleanor (Jerry) Stone. *Life within Limits*. Los Alamos, NM: Los Alamos Historical Society, 1993.

Rogin, Michael Paul. *The Intellectuals and McCarthy: The Radical Specter*. Cambridge, MA: MIT Press, 1967.

Rose, John P. *The Evolution of U.S. Army Nuclear Doctrine, 1945–1980*. Boulder, CO: Westview Press, 1980.

Rose, Paul Lawrence. *Heisenberg and the Nazi Atomic Bomb Project: A Study in German Culture*. Berkeley and Los Angeles: University of California Press, 1998.

Rosenthal, Debra. *At the Heart of the Bomb: The Dangerous Allure of Weapons Work*. Reading, MA: Addison-Wesley, 1990.

Rossi, Bruno. *Moments in the Life of a Scientist*. Cambridge: Cambridge University Press, 1990.

Rotblat, Joseph. "Leaving the Bomb Project." *Bulletin of the Atomic Scientists* 41 (August 1985): 15–19.

————. *Pugwash: The First Ten Years; History of the Conferences on Science and World Affairs.* London: Heinemann, 1967.

————. *Scientists in the Quest for Peace: A History of the Pugwash Conferences.* Cambridge, MA: MIT Press, 1972.

Rothman, Hal K. *On Rims and Ridges: The Los Alamos Area since 1880.* Lincoln: University of Nebraska Press, 1992.

Royal, Denise. *The Story of J. Robert Oppenheimer.* New York: St. Martin's Press, 1969.

Sachar, Howard M. *A History of the Jews in America.* New York: Alfred A. Knopf, 1992.

Sanders, Jane A. *Cold War on Campus: Academic Freedom at the University of Washington, 1946–1964.* Seattle: University of Washington Press, 1979.

————. "The University of Washington and the Controversy over J. Robert Oppenheimer." *Pacific Northwest Quarterly* 70, no. 1 (January 1979): 8–19.

Sanger, S. L. *Working on the Bomb: An Oral History of WWII Hanford.* Portland, OR: Continuing Education Press, 1995.

Sarantakes, Nicholas Evan. "Alliance in Doubt: American Reaction to the 1960 US-Japanese Security Treaty Crisis." *American Diplomacy* 4, no. 3 (Autumn 1999), http://www.unc.edu/depts/diplomat/AD_Issues/amdipl-13/sarantakes1.html.

Sarton, George. "The History of Science." In *The Life of Science: Essays in the History of Civilization,* 29–58. New York: Henry Schuman, 1948.

Saunders, Frances Stonor. *Who Paid the Piper? The CIA and the Cultural Cold War.* London: Granta Books, 1999.

Schrecker, Ellen. *Many Are the Crimes: McCarthyism in America.* Boston: Little, Brown, 1998.

————. *No Ivory Tower: McCarthyism and the Universities.* New York: Oxford University Press, 1986.

Schwarz, Jordan A. *The Speculator: Bernard M. Baruch in Washington, 1917–1965.* Chapel Hill: University of North Carolina Press, 1981.

Schweber, Silvan S. "The Empiricist Temper Regnant: Theoretical Physics in the United States, 1920–1950." *Historical Studies in the Physical and Biological Sciences* 17 (1988): 17–98.

————. *In the Shadow of the Bomb: Bethe, Oppenheimer, and the Moral Responsibility of the Scientist.* Princeton, NJ: Princeton University Press, 2000.

————. "The Mutual Embrace of Science and the Military: ONR and the Growth of Physics in the United States after World War Two." In Mendelsohn, Smith, and Weingart, *Science, Technology, and the Military,* 1:3–45.

————. *QED and the Men Who Made It: Dyson, Feynman, Schwinger and Tomonoga.* Princeton, NJ: Princeton University Press, 1994.

————. "Reflections on the Sokal Affair: What Is at Stake?" *Physics Today* 50, no. 3 (March 1997): 73–74.

Scott-Smith, Giles. *The Politics of Apolitical Culture: The Congress for Cultural Freedom, the CIA and Post-war American Hegemony.* London: Routledge, 2002.

Segrè, Emilio. *Enrico Fermi, Physicist.* Berkeley and Los Angeles: University of California Press, 1970.

———. *A Mind Always in Motion: The Autobiography of Emilio Segrè.* Berkeley and Los Angeles: University of California Press, 1993.

Seidel, Robert. "The Postwar Political Economy of High-Energy Physics." In *Pions to Quarks: Particle Physics in the 1950s,* edited by Laurie Brown, Max Dresden, and Lillian Hoddeson, 497–507. Cambridge: Cambridge University Press, 1989.

Seitz, Frederick. *On the Frontier: My Life in Science.* New York: American Institute of Physics, 1994.

Serber, Robert. *The Los Alamos Primer: The First Lectures on How to Build an Atomic Bomb* Edited by Richard Rhodes. Berkeley and Los Angeles: University of California Press, 1992.

———. *Peace and War: Reminiscences of a Life on the Frontiers of Science.* With Robert P. Crease. New York: Columbia University Press, 1998.

Shapin, Steven. "Cordelia's Love: Credibility and the Social Studies of Science." *Perspectives on Science* 3 (1995): 255–75.

———. "Don't Let That Crybaby in Here Again." *London Review of Books,* September 7, 2000, http://www.lrb.co.uk/v22/n17/shap01_.html.

———. "The House of Experiment in Seventeenth-Century England." *Isis* 79 (1988): 373–404.

———. "How to Be Antiscientific." In *The One Culture? A Conversation about Science,* edited by Harry Collins and Jay A. Labinger. Chicago: University of Chicago Press, 2001.

———. "'The Mind Is Its Own Place': Science and Solitude in Seventeenth-Century England." *Science in Context* 4 (1991): 191–218.

———. "Personal Development and Intellectual Biography: The Case of Robert Boyle." *British Journal of the History of Science* 26 (1993): 335–45.

———. "The Philosopher and the Chicken: On the Dietetics of Disembodied Knowledge." In *Science Incarnate: Historical Embodiments of Natural Knowledge,* edited by Christopher Lawrence and Steven Shapin, 21–50. Chicago: University of Chicago Press, 1998.

———. "'A Scholar and a Gentleman': The Problematic Identity of the Scientific Practitioner in Early Modern England." *History of Science* 29 (1991): 279–327.

———. *A Social History of Truth: Civility and Science in Seventeenth-Century England.* Chicago: University of Chicago Press, 1994.

———. "Who Is the Industrial Scientist? Commentary from Academic Sociology and from the Shop-Floor in the United States, ca. 1900–ca. 1960." Paper presented to the 123rd Nobel Symposium, "Science and Industry in the 20th Century," Royal Swedish Academy of Sciences, November 21–23, 2002.

Shapin, Steven, and Barry Barnes. "Head and Hand: Rhetorical Resources in British Pedagogical Writing, 1770–1850." *Oxford Review of Education* 2, no. 3 (1976): 231–54.

Sheehan, Helena. *Marxism and the Philosophy of Science: A Critical History.* Atlantic Highlands, NJ: Humanities Press International, 1993.

Shepley, James R., and Clay Blair Jr. *The Hydrogen Bomb: The Men, the Menace, the Mechanism.* New York: David McKay, 1954.

Sherry, Michael S. *In the Shadow of War: The United States since the 1930s.* New Haven, CT: Yale University Press, 1995.

Sherwin, Martin. "Niels Bohr and the First Principles of Arms Control." In Feshbach, Matsui, and Oleson, *Niels Bohr,* 319–29.

———. *A World Destroyed: The Atomic Bomb and the Grand Alliance.* New York: Vintage Books, 1977.

Shils, Edward. "Freedom and Influence: Observations on the Scientists' Movement in the United States." *Bulletin of the Atomic Scientists* 8 (January 1957): 13–18.

———. "Science and Scientists in the Public Arena." *American Scholar* 56 (Spring 1987): 185–202.

Shortland, Michael, and Richard Yeo, eds. *Telling Lives in Science: Essays on Scientific Biography.* Cambridge: Cambridge University Press, 1996.

Shurcliff, W. A. *Bombs at Bikini: The Official Report of Operation Crossroads.* New York: William H. Wise, 1947.

Simpson, Alan. "The Re-trial of the Oppenheimer Case." *Bulletin of the Atomic Scientists* 10, no. 10 (December 1954): 387–88.

Sims, Benjamin. 1999. "Concrete Practices: Testing in an Earthquake-Engineering Laboratory." *Social Studies of Science* 29, no. 4 (August 1999): 483–518.

Smith, Alice Kimball *A Peril and a Hope: The Scientists' Movement in America, 1945–1947.* Cambridge, MA: MIT Press, 1970. Originally published 1965.

———. "Scientists and the Public Interest, 1945–46." *Newsletter on Science, Technology, and Human Values* 24 (June 1978): 24–32.

Smith, Alice Kimball, and Charles Weiner, eds. *Robert Oppenheimer: Letters and Recollections.* Stanford, CA: Stanford University Press, 1980.

Smith, Martin Cruz. *Stallion Gate.* New York: Ballantine Books, 1986.

Snow, C. P. *Science and Government.* Cambridge, MA: Harvard University Press, 1961.

———. *The Two Cultures.* Cambridge: Cambridge University Press, 1993. Originally published 1959.

Sorkin, David. "Wilhelm von Humboldt: The Theory and Practice of Self-Formation (*Bildung*), 1791–1810." *Journal of the History of Ideas* 44 (1983): 55–73.

Spanier, John W., and Joseph L. Nogee. *The Politics of Disarmament: A Study in Soviet-American Gamesmanship.* New York: Praeger, 1962.

Stern, Philip M. *The Oppenheimer Case: Security on Trial.* With Harold P. Green. New York: Harper and Row, 1969.

Stouffer, Samuel A., Edward A. Suchman, Leland C. DeVinney, Shirley A. Star, and Robin M. Williams Jr. *The American Soldier.* Vol. 1, *Adjustment during Army Life.* New York: Science Editions, 1965. Originally published 1949.

Strickland, Donald. *Scientists in Politics: The Atomic Scientists Movement, 1945-46.* Lafayette, IN: Purdue University Studies, 1968.

Strout, Cushing. *Conscience, Science, and Security: The Case of Dr. J. Robert Oppenheimer.* Chicago: Rand McNally, 1963.

Stuewer, Roger H., ed. *Nuclear Physics in Retrospect: Proceedings of a Symposium on the 1930s.* Minneapolis: University of Minnesota Press, 1979.

Sudoplatov, Pavel, and Anatoli Sudoplatov. *Special Tasks: The Memoirs of an Unwanted Witness—a Soviet Spymaster.* With Jerrold L. and Leona P. Schecter. New York: Little, Brown, 1994.

Swann, E. D. "Planning of Science in War." In Association of Scientific Workers, *Planning of Science: Report of Proceedings of the Open Conference Held at Caxton Hall, January 30th-31st, 1943,* 12-16. London: Association of Scientific Workers, 1943.

Sylves, Richard T. *The Nuclear Oracles: A Political History of the General Advisory Committee of the Atomic Energy Commission, 1947 1977.* Ames: Iowa State University Press, 1987.

Szasz, Ferenc Morton. "Great Britain and the Saga of J. Robert Oppenheimer." *War in History* 2, no. 3 (1995): 320-33.

Szilard, Leo. *Leo Szilard: His Version of the Facts; Selected Recollections and Correspondence.* Edited by Spencer R. Weart and Gertrud Weiss Szilard. Cambridge, MA: MIT Press, 1978.

Taylor, Bryan C. "The Politics of the Nuclear Text: Reading Robert Oppenheimer's *Letters and Recollections.*" *Quarterly Journal of Speech* 78 (1992): 429-49.

Teller, Edward. *The Legacy of Hiroshima.* With Allen Brown. Westport, CT: Greenwood Press, 1975. Originally published 1962.

———. *Memoirs: A Twentieth-Century Journey in Science and Politics.* With Judith L. Shoolery. Oxford: Perseus Press, 2001.

———. "The Role of the Scientist." In *Better a Shield than a Sword: Perspectives on Defense and Technology,* 229-35. New York: Free Press, 1987.

———. "Seven Hours of Reminiscences." *Los Alamos Science* 4, no. 7 (Winter/Spring 1983): 190-96.

———. "The Work of Many People." *Science,* February 25, 1955, 267-75.

Thayer, Harry. *Management of the Hanford Engineer Works in World War Two: How the Corps, DuPont and the Metallurgical Laboratory Fast Tracked the Original Plutonium Works.* New York: ASCE Press, 1996.

Thomas, W. I. "Situational Analysis: The Behavior Pattern and the Situation." 1927. In Thomas, *On Social Organization and Social Personality,* edited by Morris Janowitz, 154–67. Chicago: University of Chicago Press, 1966.

Thompson, E. P. "Time, Work-Discipline and Industrial Capitalism." *Past and Present* 38 (1967): 56–97.

Thorpe, Charles. "Against Time: Scheduling, Momentum, and Moral Order at Wartime Los Alamos." *Journal of Historical Sociology* 17, no. 1 (March 2004): 31–55.

———. "Disciplining Experts: Scientific Authority and Liberal Democracy in the Oppenheimer Case." *Social Studies of Science* 32, no. 4 (August 2002): 525–62.

———. "Violence and the Scientific Vocation." *Theory, Culture and Society* 21, no. 3 (June 2004): 59–84.

Thorpe, Charles, and Steven Shapin. "Who Was J. Robert Oppenheimer? Charisma and Complex Organization." *Social Studies of Science* 30, no. 4 (2000): 545–90.

Trilling, Diana. "The Oppenheimer Case: A Reading of the Testimony." *Partisan Review* 21, no. 6 (November–December 1954): 604–35.

Trilling, Lionel. "The Leavis-Snow Controversy." In *Beyond Culture,* 133–58. Harmondsworth, UK: Penguin, 1966.

Truslow, Edith C., and Ralph Carlisle Smith. "Beyond Trinity." Part 2 of *Project Y: The Los Alamos Story.* Los Angeles: Tomash Publishers, 1983.

Turner, Stephen. *Liberal Democracy 3.0: Civil Society in an Age of Experts.* London: Sage, 2003.

Ulam, Stanislaw. *Adventures of a Mathematician.* New York: Scribner, 1976.

Ulam, Stanislaw, H. W. Kuhn, A. W. Tucker, and Claude E. Shannon. "John von Neumann, 1903–1957." In Fleming and Bailyn, *The Intellectual Migration,* 235–69.

Ungar, Sheldon. "The Great Collapse, Democratic Paralysis and the Reception of the Bomb." *Journal of Historical Sociology* 5, no. 1 (March 1992): 84–103.

United States Advisory Committee on Human Radiation Experiments. *Final Report of the Advisory Committee on Human Radiation Experiments.* Oxford: Oxford University Press, 1996.

United States Atomic Energy Commission. *In the Matter of J. Robert Oppenheimer: Transcript of Hearing before Personnel Security Board and Texts of Principal Documents and Letters.* Cambridge, MA: MIT Press, 1971. Originally published 1954.

Vaughan, Diane. *The Challenger Launch Decision: Risky Technology, Culture and Deviance at NASA.* Chicago: University of Chicago Press, 1996.

———. "The Role of the Organization in the Production of Techno-scientific Knowledge." *Social Studies of Science* 29 (1999): 913–43.

Vilar, Jean. *Le Dossier Oppenheimer.* Geneva: Éditions Gonthier, 1965.

Visvanathan, Shiv. "Atomic Physics: The Career of an Imagination." In *Science, Hegemony and Violence: A Requiem for Modernity,* edited by Ashis Nandy, 113–66. Oxford: Oxford University Press, 1996.

Walker, J. S. "The Decision to Use the Atomic Bomb: A Historiographical Update." *Diplomatic History* 14 (1990): 97–114.

Walker, Mark. *German National Socialism and the Quest for Nuclear Power, 1939–1949.* Cambridge: Cambridge University Press, 1989.

Walter, Maila L. *Science and Cultural Crisis: An Intellectual Biography of Percy Williams Bridgman (1882–1961).* Stanford, CA: Stanford University Press, 1990.

Wang, Jessica. *American Science in an Age of Anxiety: Scientists, Anticommunism, and the Cold War.* Chapel Hill: University of North Carolina Press, 1999.

———. "Liberals, the Progressive Left, and the Political Economy of Post-war American Science: The National Science Foundation Debate Revisited." *Historical Studies in the Physical and Biological Sciences* 26 (1995): 139–66.

———. "Science, Security, and the Cold War: The Case of E.U. Condon." *Isis* 83 (1992): 238–69.

———. "Scientists and the Problem of the Public in Cold War, 1945–1960." In "Science and Civil Society," edited by Lynn K. Nyhart and Thomas H. Broman, *Osiris* 17 (2002): 323–47.

Weber, Max. *From Max Weber: Essays in Sociology.* Edited by H. H. Gerth and C. Wright Mills. New York: Oxford University Press, 1958.

Weiner, Charles. "A New Site for the Seminar: The Refugees and American Physics in the Thirties." In Fleming and Bailyn, *The Intellectual Migration,* 152–89.

Weisgall, Johnathan M. *Operation Crossroads: The Atomic Tests at Bikini Atoll.* Annapolis, MD: Naval Institute Press, 1994.

Weisskopf, Victor. *The Joy of Insight: Passions of a Physicist.* New York: Basic Books, 1991.

Welsh, Ian. *Mobilising Modernity: The Nuclear Moment.* London: Routledge, 2000.

Wersky, Gary. *The Visible College: A Collective Biography of British Scientists and Socialists of the 1930s.* London: Free Association Books, 1988.

White, W. L. *Bernard Baruch: Portrait of a Citizen.* New York: Harcourt, Brace, 1950.

Whitley, Richard. *The Intellectual and Social Organization of the Sciences.* Oxford: Clarendon Press, 1984.

Wigner, Eugene. "Memoir of the Uranium Project." In *The Collected Works of Eugene Paul Wigner,* edited by Alvin M. Weinberg with Alfred M. Perry, vol. 5, *Nuclear Energy, Part A: The Scientific Papers,* 23–130. Berlin: Springer-Verlag, 1992.

———. *The Recollections of Eugene P. Wigner as Told to Andrew Szanton.* New York: Plenum Press, 1992.

Willhelm, Sidney M. "Scientific Unaccountability and Moral Accountability." In
     The New Sociology: Essays in Social Science and Social Theory in Honor of C.
     Wright Mills, edited by Irving Louis Horowitz, 181–87. New York: Oxford
     University Press, 1964.
Williams, Robert C., and Philip L. Cantelon, eds. *The American Atom: A*
     Documentary History of Nuclear Policies from the Discovery of Fission to the
     Present, 1939–1984. Philadelphia: University of Pennsylvania Press, 1984.
Williamson, Samuel R., Jr., and Steven L. Rearden. *The Origins of U.S. Nuclear*
     Strategy, 1945–1953. New York: St. Martin's Press, 1993.
Wilson, Jane S., and Charlotte Serber, eds. *Standing By and Making Do: Women of*
     Wartime Los Alamos. Los Alamos, NM: Los Alamos Historical Society, 1988.
Wilson, Robert R. "The Conscience of a Physicist." In Lewis and Wilson,
     Alamogordo plus Twenty-five Years, 67–76. New York: Viking Press, 1971.
———. "Hiroshima: The Scientists' Social and Political Reaction." *Proceedings of*
     the American Philosophical Society 140, no. 3 (September 1996): 350–57.
———. "Niels Bohr and the Young Scientists." In Acklund and McGuire,
     Assessing the Nuclear Age, 35–41.
———. "A Recruit for Los Alamos." *Bulletin of the Atomic Scientists* 31 (March
     1975): 41–47.
Wilson, Thomas Williams. *The Great Weapons Heresy.* Boston: Houghton Mifflin,
     1970.
Winston, Andrew S. "'As His Name Indicates': R. S. Woodworth's Letters of
     Reference and Employment for Jewish Psychologists in the 1930s." *Journal of*
     the History of the Behavioral Sciences 32, no. 1 (January 1996): 30–43.
Wrong, Dennis. "The Oversocialized Conception of Man in Modern Sociology."
     American Sociological Review 26 (1961): 183–93.
Yoneyama, Lise. *Hiroshima Traces: Time, Space, and the Dialectics of Memory.*
     Berkeley and Los Angeles: University of California Press, 1999.
York, Herbert F. *The Advisors: Oppenheimer, Teller, and the Superbomb.* San
     Francisco: W. H. Freeman, 1976.
———. *Making Weapons, Talking Peace: A Physicist's Odyssey from Hiroshima to*
     Geneva. New York: Basic Books, 1987.
Zachary, G. Pascal. *Endless Frontier: Vannevar Bush, Engineer of the American*
     Century. New York: Free Press, 1997.
Zerubavel, Eviatar. *Hidden Rhythms: Schedules and Calendars in Social Life.*
     Chicago: University of Chicago Press, 1981.
Zimmerman, Bill, Len Radinsky, Mel Rothenberg, and Bart Myers. *Towards a*
     Science for the People. Brookline, MA: A People's Press, 1972.